Risk Assessment

Risk Assessment

Procedures and Protocols

Edward A. McBean, PhD, PEng, PE, FCAE, FEC, FCSCE, DWRE

University of Guelph

WILEY

This edition first published 2019

Registered Office
John Wiley & Sons, Inc., 111 River Street, Hoboken, NJ 07030, USA

Editorial Office
111 River Street, Hoboken, NJ 07030, USA

For details of our global editorial offices, customer services, and more information about Wiley products visit us at www.wiley.com.

Wiley also publishes its books in a variety of electronic formats and by print-on-demand. Some content that appears in standard print versions of this book may not be available in other formats.

Library of Congress Cataloging-in-Publication Data

Names: McBean, Edward A., author.
Title: Risk assessment : procedures and protocols / by Edward A. McBean.
Description: Hoboken, NJ : John Wiley & Sons, 2019. | Includes bibliographical references and index. |
Identifiers: LCCN 2017059958 (print) | LCCN 2018004221 (ebook) | ISBN 9781119289081 (pdf) |
ISBN 9781119289098 (epub) | ISBN 9781119289067 (cloth)
Subjects: LCSH: Environmental risk assessment.
Classification: LCC GE145 (ebook) | LCC GE145 .M48 2018 (print) | DDC 519–dc23
LC record available at https://lccn.loc.gov/2017059958

Cover design by Wiley
Cover image: ©Shawn Knol/Getty Images

Set in 10/12pt Warnock by SPi Global, Pondicherry, India

Printed in the United States of America

V068017_061418

Dedicated to Matthew, Derek, and Melissa as they are the future.

Contents

Preface

The intent in this book is to carefully describe the methodologies, typical mathematical notation, and assumptions typically used in risk assessment calculations. Subsequently, the book's chapters describe various statistical analysis procedures that are used for estimating the parameters used in risk assessment methodologies. Numerous examples and descriptions of the bases of the methodologies are provided.

Unlike much of the professional literature in statistics, this text makes concerted efforts to describe statistical techniques in terms comprehensible to the nonstatistician. This is accomplished by downplaying mathematical notation, comprehensively explaining the development of equations, and emphasizing example applications. Thus, as example problems of interpretations of environmental monitoring results are described, the text demonstrates through use of simple examples, how the procedures are utilized. References are provided, with particular emphasis on works describing applications reported in the technical literature. Problems included at the end of the chapters stress fundamentals and increase the usefulness of this book as a classroom text, intended for senior undergraduate and graduate students in environmental engineering and environmental sciences.

The collection and laboratory analyses of samples needed to characterize environmental quality are expensive. Further, as society expresses increasing concern for environmental protection and as instrumentation technology evolves to allow detection of contaminants at ever-lower concentrations, expenditures for monitoring environmental quality will continue to increase with time. As a direct consequence of the substantial costs of environmental monitoring, it is essential to use available environmental quality data as effectively as possible. Effective utilization involves answering questions such as, "Is the environmental quality acceptable?," "Is the environmental quality improving or deteriorating?," and "Is the risk acceptable and/or need to be managed?" Responding to these types of questions requires interpretation of data, and this stage of assessment is beset with difficulties. Some difficulties with interpreting environmental-monitoring results include:

i) Since environmental data are frequently expensive to accumulate, the data sets being interpreted are typically modest in size.
ii) The data may involve a vector of chemical and biological constituent measurements because consideration must typically be given to a range of constituents. Correlation between the constituents may help the infilling of missing data or the identification of outlier data.
iii) Early detection of any deterioration in environmental quality is highly desirable because early detection may provide the opportunity for controlling the problem, at a lower cost, before the problem magnifies. Any procedure for identifying early warning signals must not, however, falsely identify a problem of apparent environmental deterioration when one

does not actually exist, a so-called false positive, nor should it fail to identify a problem when one does exist, a so-called false negative.

iv) The vagaries of nature typically introduce significant noise, and sources of variability such as seasonal effects, make the identification of trends more difficult.

v) The derivation of quantitative risk assessments is, in many ways, data dependent. However, a key question is whether the information returned by risk estimates warrant additional data collection efforts?

The net result of difficulties such as the five examples mentioned above is that making sense of the data relevant to risk assessment methodologies necessarily involves statistical interpretation. Statistical interpretation procedures must be sensitive to small changes in environmental quality and yet recognize the potentially substantial costs of any additional data collection requirement.

The need for the statistical interpretation of data is widespread. The range of concerns for each environmental media – air quality, surface water quality, groundwater quality, and soil contamination – is similar in many respects. And yet, there is no single statistical analysis procedure universally applicable to the variety of problems associated with environmental quality data. Instead, the practitioner needs to have an array of statistical procedures available. A multitude of statistical analysis tests are available, but each of the tests possesses assumptions that may, or may not, be appropriate for specific circumstances and hence, having a number of techniques is frequently required. Computer programs now becoming widely available facilitate the use of various procedures (e.g. ProUCL, EPA, 2013). The difficulty remains for the student and practitioner to learn which conditions dictate a particular procedure and which conditions render it highly inappropriate.

Following the introduction (Chapter 1), the book is organized into three parts as follows:

Part I – Chapters 2 through 4 develop the fundamental calculation procedures and methodologies for risk characterization.

Part II – Chapters 5 through 7 describe the characteristics of common distributions as utilized to describe data.

Part III – Chapters 8 through 12 describe the bases used in hypothesis testing to determine when there are differences in environmental quality at various locations. Problems of censored data are considered as they influence the utilization of alternative tests. Chapter 12 focuses on nonparametric procedures, an alternative to the parametric procedures utilized in previous chapters.

Author of the Book

Edward A. McBean (B.A.Sc. from the University of British Columbia and S.M., C.E., and Ph.D. from the Massachusetts Institute of Technology) is a Professor of Water Resources, Canada Research Chair (CRC) in Water Supply Security in the School of Engineering and now a University Research Leadership Chair, at the University of Guelph. Dr. McBean has also been the Assistant Dean of the College of Physical and Engineering Science at the University of Guelph for an 8-year period. Dr. McBean started his career in risk assessment as a faculty member for two decades at the University of Waterloo and the University of California, followed by a decade as a Vice-President of Conestoga-Rovers and Associates (CRA) and President of CRA Engineering Inc., both environmental engineering firms. For the most recent decade, Dr. McBean has been a Canada Research Chair of Water Supply Security at the University of Guelph. He is an author of 2 previous books, editor of 17 books, and author of more than 350 papers in the refereed technical literature. Dr. McBean has been the recipient of many awards including Fellow of the Canadian Academy of Engineering, Fellow of the Canadian Society of Civil Engineering, Fellow of Engineering Institute of Canada, the K.Y. Lo Award for his work in international engineering, the Research and Development Award of the Professional Engineers of Ontario, and the Camille Dagenais Award for outstanding contributions to engineering.

Acknowledgments

The work reported in this book is the assembly of work completed by many. A much earlier version of this book was published under the title of Statistical Procedures for Analyses of Environmental Monitoring Data and Risk Assessment by Prentice-Hall in 1998, but this book is no longer in print. The copyright for this early book was requested, and received, from the publisher. Extensive rewriting and examples have been introduced into this book. Most importantly, the emphasis of this current book is on risk assessment and the information provided on statistical data in support of risk assessment methodologies.

In addition to the literature, the experience and efforts of many individuals has been important in developing this book, and for this assistance I am truly grateful. In particular, the talent and expertise of Frank Rovers, a co-author of the previous book, is gratefully acknowledged. Frank has been an inspiration as well as a colleague and friend throughout my professional career.

As well, I gratefully acknowledge the work and effort of many colleagues and students, who have been very important to the learning expressed in this book. These individuals are too many to name, but specific important examples include:

- The employees of CRA (now GHD) who so generously provided examples. The advice and assistance of many is acknowledged, with special mention of Wes Dyck, John Donald, and Klaus Schmidtke.
- All the people who examined drafts of the book and whose comments for improvement were valuable. In particular, the comments from many students are gratefully acknowledged.
- The efforts of Melissa McBean and Munir Bhatti who prepared many of the figures in this book are also gratefully acknowledged.

To all of the above, I owe sincere thanks for their assistance.

In an undertaking of the magnitude of this text, it is not possible to avoid errors, and for this I apologize; but this book is supplied in good faith. While I have taken reasonable care to avoid errors, I accept no liability for any damage, consequential or otherwise, that may be caused by the use of this book. Any suggestions for corrections, criticisms, or suggestions for improvements will be greatly appreciated and I will work towards improvements in any future versions of this book. As well, many of the coefficients continue to be updated, and hence, the reader must refer to the Internet and other sources of the information for these updates; the emphasis in this book must necessarily be on the methodological processes and explanations thereof. I would also welcome any additional information and data that would make future editions of the book more complete.

About the Companion Website

This book is accompanied by a companion website:

www.wiley.com/go/McBean/RiskAssessProcedure_1e

Scan this QR code to visit the companion website

The BCS is a compilation of answers to various questions in the book itself. The intent in providing these is to expand the opportunity for the students to "learn through using the information provided in the book." Many of the questions are from old examination questions generated over the years while using the book in "draft" form.

1

Background to Risk Assessment and Management

1.1 The Case for Risk Assessment, Leading to Risk Management

Many different definitions of risk exist, using such terms as hazard, danger, and exposure. Therefore, to put oneself "at risk" means to participate voluntarily, or involuntarily, in an activity or event that could lead to injury, damage, or loss. Hence, "risk" refers to the possibility of experiencing harm from a hazard, and thus risk assessment involves evaluating the likelihood or frequency of experiencing a risk.

Since risks are about events that cause problems, risk identification typically starts with attention to the source of the problem or risk. According to US EPA, a risk assessment involves the evaluation of scientific information, including such dimensions as a dose–response relationship, and the extent of human exposure to a chemical. "Someone or something that creates or suggests a hazard" would be an example.

There is a distinction between hazard and risk; we need to understand this so that attention of resources can be directed to actions based primarily on the level of risk rather than just the existence of a hazard.

It is also important to be mindful of the distinction between risk and uncertainty. Risk can be characterized by consequences or impacts multiplied by probability. On the other hand, uncertainty is a term that applies to, for example, the predictions of possible events (e.g. the chemical concentration to which someone is exposed). Perhaps some of the events are only partially observable. Uncertainty might exist due to limited knowledge – perhaps we just do not know enough about how to describe a circumstance or a future outcome. The result is that both risk and uncertainty are highly relevant to the subject of risk assessment; both dimensions must be incorporated. Given this differentiation between "risk" and "uncertainty," the first set of chapters in this book will focus on "risk" where some possible outcomes are undesired or may result in a significant loss. Alternatively, attention to uncertainty is discussed primarily in Chapters 4 through 12.

Another important element of risk assessment is that society is moving toward more involvement in many decisions (e.g. where is a landfill to be sited?). This involves working toward rejecting having risks imposed on the public, and hence, an appreciation of the attitudes of the public is essential. There is an association of risk with chance or probability. One of the important dimensions of this is that it requires the quantification of the probability. It also means that the procedures by which risk assessment are developed must be logical and transparent to allow open dialogue with the public about how the risks are calculated.

As apparent from the above, there is a multistage process in which the elements of risk are calculated: first, the dimensions of risk are assessed, which is then followed by a decision whether risk management is required (is the risk sufficiently severe that efforts to control or

Risk Assessment: Procedures and Protocols, First Edition. Edward A. McBean.

Companion website: www.wiley.com/go/McBean/RiskAssessProcedure_1e

Table 1.1 Steps related to risk management.

1) Establish the context of the risk – Criteria appropriate to judge the risks need to be carefully laid out. This will include consultations and communications with the needs of pertinent stakeholders to be initiated early in the process
2) Identify the risks – The hazards (what can go wrong) and the causes of such hazards need to be carefully delineated
3) Analyze the risks – The likelihood (probability) and severity (magnitude) of any and all consequences need to be characterized
4) Evaluate the risks – The risk estimates need to be considered in relation to the criteria set out in the beginning of the procedure
5) Treat risks – Practical options feasible to manage the high priority risks and monitor the low priority risks need to be determined
6) Monitor and review – Because circumstances may change, the basis must be monitored and periodically reviewed
7) Periodically communicate and consult – Extensive experience has consistently demonstrated that leaving communication and consultation steps to the end of the risk management process will usually create stakeholder outrage and avoidable conflicts

manage the risk are required?). This is accomplished by the first step in this process being to identify the risks, followed by the determination whether management of the risks is necessary and feasible. This will include consideration of the probability of recurrence (Table 1.1).

The development of risk assessment and management processing is definitely "a work in progress" as the profession continues to learn about the issues. There is substantial uncertainty that still remains. Regardless, it is important to understand the underlying principles; the assembly of data will continue to evolve. An example of a challenge is an individual who decides to undertake a potentially dangerous activity such as skydiving. The scientific approach to risk analysis does not necessarily allow the person to decide on how to deal or manage this type of risk although there may be strong evidence that certain activities are dangerous. Instead, most of the attention herein will be on risk in its most fundamental form. The approach in this book will involve the following: (i) identify a hazard, (ii) analyze the risk associated with that hazard, and (iii) determine if the elimination, or control, of the risk is warranted.

Risk enters the purview of the public in a wide spectrum of dimensions. Examples of risks that we may encounter on a daily basis are indicated in Table 1.2. Each day we weigh risks of different types of activities. We weigh the risk of injury when driving a car or riding a bicycle. We assess the number of people engaged in the activities and our perceived rate of injury and we decide whether or not to participate.

In words, innumerable aspects of everyday society involve a risk. It should be apparent that we cannot make all of the risks become zero but we can, and often do, avoid some risks. Different approaches are needed for different circumstances and different degrees of data availability. One of our concerns will be in establishing how to assemble and interpret data as a basis from which risk assessment can be developed. The strategies must consider both humans and the environment. In this context, the book develops the concepts and then builds the concepts and techniques into a strategy to undertake risk assessments.

This book primarily focuses on environmental risk. As will be shown, risks are pervasive. We can use parallels to other risks we face in society to develop reasonable trade-offs to allow decisions on risk management.

Ultimately, what is needed is to describe exposure assessments to human health and the environment, but the complications of risk assessment will be apparent when confronted with the spectrum of dimensions such as the degrees to which a chemical bioaccumulates and

Table 1.2 Examples of risks encountered on a daily basis.

Examples of risks	Basis for risk
Turning on the light	Possible electrocution
Soap in the shower	There may be a chemical that fluoresces in the soap and/or the potential for increased slippage as a result of the soap
Falling on stairs	Concern with falling
Coffee/tea as a result of the caffeine	Due to elevation of blood pressure
Sweetener in coffee/tea	Sugar/fat/heart disease exacerbated
Peanut butter	The aflatoxin (a mold) and liver cancer where, particularly in the developing world, the storage of peanuts is an issue
Riding a bicycle to work	Increased potential for involvement in an accident
Drinking of water	As a result of the disinfection byproducts arising from chlorination of water
Brick and cinder blocks	Radon and hence cancer potential
X-rays for disease identification	May cause cancer
Air travel	Through a combination of the radiation hazard as well as the potential for a plane crash

biomagnifies, the importance of ecological modeling, and dose–response methodologies. The preceding should be sufficient to demonstrate that the field of risk assessment and management is very data intensive and requires careful appreciation of data variability.

The emphasis herein will be on risk assessment. Risk management procedures go far beyond risk assessment and must deal with multifaceted aspects of a problem and must be responsive to the severity of conditions as well as to how much money is available and the degree to which management is feasible.

1.2 The Need for Risk Quantification

Key words in the preceding definition of risk in Section 1.1 are "voluntary" and "involuntary." Voluntary risks are hazards associated with activities that we undertake voluntarily such as walking down the stairs, riding a motorcycle, smoking cigarettes, or skydiving. Involuntary risks are negative impacts that may occur without a person's prior consent or knowledge. Involuntary risks include, as examples, lightning strikes, tornadoes, floods, and exposure to environmental contaminants.

Risks may also be characterized as "statistically verifiable" or "statistically nonverifiable." As the first name implies, statistically verifiable risks are risks for voluntary or involuntary activities that have been determined from direct observation (e.g. statistics related to motor vehicle accidents). These types of risks can be compared to one another since data exists. Statistically nonverifiable risks are risks from involuntary activities that are based on very limited datasets and mathematical equations. For example, we know that the risk of a nuclear energy generation incident killing a person is low but because there are very few events that have occurred, it is difficult to establish probabilities. Hence, while aspects of statistically verifiable and nonverifiable risks are similar, they are also very different. This means that although we must consider both, we cannot necessarily compare them.

Another complication is that the magnitudes of risks associated with different activities and phenomena vary widely. For example, one's chances of being struck by lightning are low compared with fire-related deaths. Both of these are largely involuntary risks, as they represent risks over which we have limited control.

On the other hand, voluntary risks are associated with activities that are largely controllable. There are risks associated with living in our society and individuals are (or should be) constantly evaluating various actions with varying degrees of focus and caution. Essentially, these types of assessments are often just part of one's experience. People consider risks as they relate to voluntary activities, and generally, people try to stay out of harm's way. For example, you could travel to work by bicycle, but there is a risk of being struck by a vehicle while en route. This risk could be avoided by taking a bus to work, but there are other risks associated with bus travel. As well, if we were to drive a vehicle to work, we could take risk precautions by reducing speed and/or wearing a seat belt.

One of the most challenging stumbling blocks regarding risk is that the vast majority of people do not understand and quantify the risks they face on a day-to-day basis. Most people behave as if life is largely free of risk (or, at a minimum, the thrill of participation makes the risk acceptable). Undertaking activities that are considered to be "risky" is frequently taken as irrational and should be avoided. Risks imposed on us by others are generally considered to be unacceptable. There are dangers associated with travel; people are aware of the potential for airplane accidents, as an example. However, there are dangers associated with staying home – 25% of all fatal accidents occur in people's homes. Everything we do involves risks. It enters all aspects of our daily lives, contributing to the way we live. Generally, people do not use formal procedures to estimate risks but instead rely on personal attitudes and experiences, media reporting, and input from friends and colleagues.

The logical procedure for minimizing risks is to quantify all risks and then choose those that are smaller in preference to those that are larger. However, this does not translate to mean that we should not try to minimize our risk exposures, but it is important to recognize that attempts to minimize anything is best accomplished using a quantitative procedure. We cannot minimize our risks by simply avoiding those we are aware of, because those we are aware of are likely those publicized by the media. Many more risks may exist as unknowns.

The result is that decision-making processes regarding risk are highly complex, with multiple features that influence the degree of hazard and danger. These features should also serve to affect the actions we take to minimize risks. The means that will be developed in this book to approach risk assessment, and therefore influence how we assess risks, will be primarily quantitative.

The intent in this book is to provide the framework for the process of quantifying specific types of risks and to allow assessment of the trade-offs. In most respects, the emphasis herein is related to environmental risks although not exclusively. To compare risks, we must be capable of calculating or quantifying the risks.

Risks will usually be expressed as a probability of effects associated with a particular activity (e.g. drinking water with low levels of arsenic). In this book, we will be evaluating the probabilities of various types of events or exposure scenarios. Scientific notation will be used to present quantitative methodologies and information. Hence, there is a component of risk management that involves data compilation and evaluation, and, in many respects, this is a question of uncertainty, a concept that was discussed in Section 1.1. As well, there will be the development of exposure assessments, toxicity assessments, and, ultimately, a risk characterization.

A comment made by many is that the world seems like a hazardous place. However, as Wilson (1979) indicated, if we look back at the world of a century ago, we realize that life expectancies have increased substantially from 50 to 70 years. Therefore, the sum of all risks must be less than it was historically. However, now we are more aware of risks due to increased societal focus on such dimensions as media, publicity, and education.

1.3 Environmental Risk

Concern with the quality of the environment is pervasive. Members of the public, politicians, the media, lawyers, scientists, engineers, and so on are all watchful of current environmental quality levels and the perceived trends in these levels. Much of the concern with environmental quality is real and appropriate, arising from a legacy of inadequate environmental protection, but some is only perception. As a result, attention to and concern about environmental quality is a growing phenomenon, one that must be addressed by professionals throughout the environmental industry (in the broadest context, to food, to water, to the air we breathe, etc., so not just a question for environmental engineers).

During the past several decades, the public has been increasingly demanding that the risks associated with the exposure to environmental contaminants be reduced. The demands have been directed in part toward government agencies responsible for soil, water, and air quality for the protection of human health and ecosystems. The result has been the need for dramatic increases in the requirements for professionals to analyze and interpret environmental quality data. Suggestions on tools and approaches that can be utilized to assess risk management are described in the chapters that follow.

Directions and opportunities will be used to demonstrate, by example, how to access various types of data that are available and valuable (e.g. sources of epidemiology and toxicology). The emphasis here is on using the data – scientists and engineers are not always involved in the collection of the data.

1.4 A Measure of Quantifying Risk: Loss of Life Expectancy

There are many ways of expressing or quantifying risks. One of these is loss of life expectancy (LLE). LLE is the average time period by which a person's life is shortened by the risk under consideration. Thus, LLE is the product of the probability that a risk will cause death and the consequences in terms of lost life expectancy if it does cause death.

The LLE is a highly useful concept but there are limitations as to how to interpret it. As way of explanation, if we assume the life expectancy of a man is 80 years, and he is currently 50 years old, and if that person takes a risk that has a 1% chance of being immediately fatal, this risk causes an LLE of $0.01 \times (80 - 50)$, or 0.3 years. It is important to understand that this does not mean that the man will die 0.3 years sooner as a result of accepting the risk but if 100 people of this man's age took this risk, $(0.01) \times 100 = 1$ person might die immediately, having his life shortened by 30 years, while the other 99 would not have their lives shortened as a result of that risk. The result is that the average lost lifetime for the 100 people would be 0.3 years. This is the LLE associated with that particular risk.

While it is important to consider this concept, the real world is much more complicated. Society is made up of many people of many ages, and the effects must be summed over a lifetime, making the calculations much more complex than the simple LLE example described above. The intent in this book is to provide guidance on how the various methodologies may be developed and to demonstrate how the calculations may be done. In the initial chapters, the equations that can be used to make the calculations for risk will be developed. The second part of the book involves the development of statistical data tools to allow the calculations to be completed. Risk assessment generally focuses on extremes, and available data are typically rather minimal. For reasons that will be explained later, the increased risk of death is typically on a scale such that the risk of one-in-a-million is considered acceptable and hence rare. Risk assessment is a step in the risk management process. It is generally more appropriate to characterize risks by providing an upper limit on a risk reflecting uncertainty and then assessing which risks are acceptable.

The assessment of procedures for managing risk in some management responses will be obvious. For example, if the concentration of a chemical in drinking water causes an unacceptable health risk, then either water treatment must occur or an alternative water source must be identified. Hence, ways of managing risk are, in many ways, a natural outgrowth of risk assessment. It is important to understand the underlying principles because they are not subject to change and apply in a broad sense to risk assessment. Some are quantitative and some remain qualitative.

Statistical interpretation of environmental quality data has a major role to play in such areas as qualifying effects, assessing consequences, measuring risks, and interpreting evidence. The intent is that this book will assist the student and the practitioner in all of these areas.

The book was prepared for use for instruction to senior undergrad and grad students in technical disciplines. Guidance is provided to the technical literature for those wishing a greater understanding of the technical details, but the mathematical basis and structure of the arguments are hopefully at the level of detail that can be followed from the material presented herein.

1.5 Reliance on Environmental Data

1.5.1 Characteristics of Data

The basis of environmental risk is upon environmental data. Data interpretation is necessary to allow the development of risk assessment. For environmental risk, the need is to assess the potential severity and the probability of occurrence, hence fundamental in terms of understanding environmental data. Risk assessment must be based on defensible data and interpretation of that data.

Collection of samples from the field, the need for laboratory analyses of these samples, and the time requirements for interpreting the resulting information can represent enormous expenditures for government agencies, corporations, and citizens. Consequently, when assessing the resulting data, diligence must be utilized to ensure the interpretation of the data is comprehensive and defensible.

There are many reasons for the increased focus on environmental quality. These reasons include the increases in human population and urban densification that have resulted in more locally concentrated pollutant releases and deteriorated environmental quality. The reasons also include the result of enhanced laboratory technologies (e.g. LC-MS-MS), which can now detect concentrations of substances such as pharmaceuticals, at levels that previously were not quantifiable. With the intensive monitoring efforts now being carried out, instances of deteriorated environmental quality are now identified, which previously might have remained undetected.

The focus on environmental quality has resulted in greatly enhanced needs to understand the evolution of environmental quality conditions. Part of the understanding of the evolution is gained from the enhanced statistical interpretation capabilities now available to interpret environmental quality data. However, by no means is statistical interpretation by itself sufficient. Other needs may include mathematical modeling of air quality and surface and subsurface water and soil quality. Thus, in addressing the statistical analyses of environmental quality data, the statistical analyses described in this book must be tempered with many other considerations. The use of statistics is only one of the tools we employ when interpreting environmental quality data. Statistical analyses are not an interpretation of the facts, but, when properly used, these analyses make the facts easier to see and allow other evidence to enter into judgments about environmental phenomena (Unwin et al., 1985). This is relevant because we rely on environmental data that is notoriously challenging, for reasons that will be discussed in many of the following chapters.

As will be apparent, an extensive theoretical basis exists for statistical methodologies. However, much of the statistics literature is premised on the availability of very lengthy data records, a circumstance that seldom exists for environmental phenomena. In interpreting the data, we must therefore understand the limitations and assumptions of the statistical procedures and how these affect data interpretation. In addition to brief datasets, the difficulties of dealing with environmental data may include the presence of numerous parameters, high degrees of variability in some constituents, and "censored" data where censored indicates that the magnitude of a constituent is known only as being less than some specified magnitude. As well, analyses must reflect the frequent occurrence of incompatible data due to different sampling methods, different laboratory analytical methods, and/or different times or spatial intensities of monitoring. Frequently, some of the information available is anecdotal. For example, some data are the result of nonstatistical sampling, grab samples, and vagueness in terms of where, how, and when the samples were collected.

Conceptually, environmental processes are characterized as complex, multifaceted, chaotic, and dynamic. The result is that we are limited in our ability to use many of the procedures described by statisticians in the available literature (or, at a minimum, we have to be inventive in the application of these procedures). One of the reasons for preparation of this book is to respond to some of these concerns. This is accomplished using examples and discussion of the advantages and disadvantages of the various statistical analysis procedures and how statistics can be utilized to assist in these endeavors. Clear, complete, statistically accurate, and understandable information is essential in making informed decisions. Broadly, statistics attempts to characterize order in "disorder," such as the quantification of equation error or measurement error. One must always temper statistics with a clear understanding of the problem, so that spurious information is not introduced, nor valid information omitted. Statistical analyses of data are not an interpretation of the facts; statistical analyses are just another way of making the facts easier to see and therefore interpret (McBean and Rovers, 1990).

1.5.2 Indications of the Sources of Variability in Environmental Data

The nature of the variability of environmental quality data greatly influences statistical analyses of the data. The specifics of statistical analyses will depend upon the way the phenomena of interest is defined and sampled. In general, the ability of a sample of environmental quality data to characterize the population from which it is drawn is related to such aspects as (i) the size of the sample, (ii) the degree to which it was selected at random, (iii) the degree of independence between the observations comprising the sample, and (iv) how the data can be employed.

If extensive monitoring requirements are specified in governmental legislation, this will translate into sizable expenditures in terms of time and dollars to collect the data. Nevertheless, samples of environmental quality are just that, namely, subsets of the populations that are of interest. The result is that in many assessments of environmental phenomena, estimates of groundwater quality, for example, must be developed from only very brief data records, brief being with respect to both temporal and spatial dimensions. The result is that there may be substantial quantities of data, but there is only a finite amount that is usable for specific applications.

These features stress the difficulty with statistical interpretation of environmental data in relation to risk assessments. There is a need for the analysis procedure to be sensitive to small changes (e.g. early detection of contamination is desirable), and yet a point of diminishing marginal returns also occurs. There may be more data collected than is necessary to make a decision, and thus, some of the money spent in data collection may be unnecessary.

The result is that on many occasions we must be inventive in terms of how statistical analyses for risk assessment should proceed. Many standard statistical analysis techniques that are valid for long records have little utility in situations possessing only a brief record or, at a minimum, the techniques must be modified to correct for biases created because of the brevity of the record. A further complicating factor is that many of the data records are highly variable or "noisy" due to, for example, seasonal phenomena. An additional consideration arising in part with increasing instrument technology is that we can measure features that previously were reported only as censored (less than) data. Further, a number of chemicals have maximum concentration levels (MCLs) to which humans and the environment can be exposed to while not incurring injury, which are very close to the technological instrumentation capability in being detected. The result is that problems associated with statistical analyses of censored datasets are increasing.

For the variety of reasons indicated above, there is not a single approach to statistical analysis. Instead, what is frequently needed is a series of approaches, each of which possess useful attributes that may be appropriate, in addressing a particular question. Statistical interpretation of environmental quality data therefore has a major role to play in such areas as qualifying effects, assessing consequences, measuring risks, and interpreting the ramifications.

1.5.3 Independence of Successive Data Values

Time-series analyses are pertinent to the problem of estimating trends and cycles (e.g. seasonal variations). For example, consider the situation where there is a tendency for a groundwater monitoring well that is yielding high chloride concentrations today to also yield similar values tomorrow and for nearby monitoring wells to also yield sampling aliquots with elevated chloride concentrations. Hence, these types of sampling results are not necessarily independent from each other. Similarly, replicate sampling (e.g. the splitting of field or laboratory samples into several samples) does not create independent samples. As a result, there are differing degrees of independence of monitoring results, and these aspects must be assessed during the statistical analyses of the resulting data.

Many statistical analysis procedures assume independence of data. Dependent samples exhibit less variability, and statistics determined from dependent data will therefore have underestimated sample variances. Dependence can severely influence the results of testing any hypothesis. See Example 1.1 for an indication of hypothesis testing.

Concerns with hypothesis testing include those dealing with the independence/dependence of successive samples. One approach to minimize dependence in samples is to allow sufficient time between sampling times to allow the "real-time memory" of the system to be exceeded. For example, the statistical analysis of annual peak flows (the highest flow in an individual year) involves statistical analysis of independent events since the peak flow in 1 year is unlikely to be related to the peak flow in a later year.

For some situations we can avoid the problem of dependent data. We might "deseasonalize" the data by removing the periodic characteristic(s) associated with the seasons. However, the ability to remove seasonality in environmental quality data records may be constrained by the brief length of environmental data available, which limits our ability to isolate the seasonal variability from the other sources of variability, which are present in the dataset.

Example 1.1 Example of Statistical Hypothesis Testing
Consider the question of whether a landfill is leaking leachate, which will contaminate the underlying groundwater, as depicted in Figure 1.1.

One way to consider this question is to monitor the groundwater quality both upgradient (Point A) and downgradient (Point B) from the landfill. We might then compare the quality at A and B to determine if there is a difference. Thus, a hypothesis might be the following:

Hypothesis – There is no statistically significant difference between the quality of the groundwater at the two locations.

Outcome 1: If we accept the hypothesis, then we conclude that there is insufficient information to indicate the landfill is leaking.
Outcome 2: If we reject the hypothesis, evidence exists to indicate that the landfill is leaking.

The question of hypothesis testing in environmental phenomena is a recurring one. The details of hypothesis testing will be a recurring question throughout this book, and the quantitative aspects of hypothesis testing will be left to later chapters.

1.5.4 Uncertainties and Errors in Environmental Quality Data

There are different levels of "observational" data. For example, "proxy" data are observations of one variable that have a high probability of being indicative of levels of another variable. Such data are indirect "observations." Another example is remotely sensed data by which many "indices" are derived from satellite data including temperature and vegetation. Image classification techniques may allow patterns in "data" to be recognized as signatures of certain environmental features.

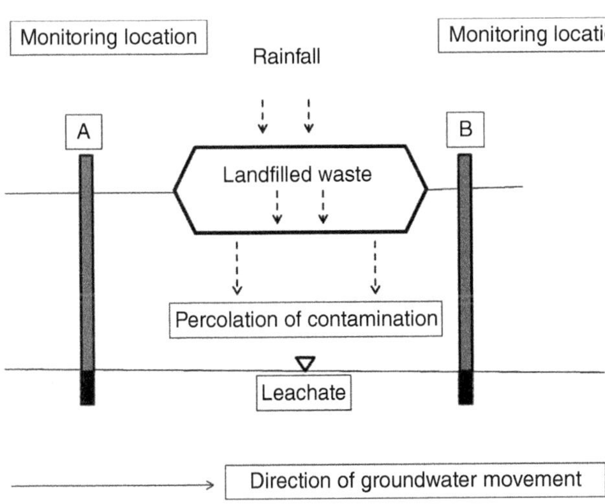

Figure 1.1 Schematic depiction of monitoring in the vicinity of a landfill.

In the strictest sense, there are very few types of data that are direct "observations" of a variable. Data on animal (or human) demography are collected by observing various signs, such as spores, tracks, nests, houses, income tax reports, and so on. Even something as obviously "observable" as the digital elevation is subject to interpretation, according to the methods used to produce the measurements. The validity of these "data" depends on such features as one's confidence in a particular measurement method and the reliability of calibration techniques of the instruments, all of which must then be considered when interpreting the data.

Quantifying observations involves the employment of sensory techniques, indirect measurements of related variables, and various levels of processing. It can then be argued that data are observed only in the context of the experimental design in which they are produced. It becomes a matter of interpretation as to what degree of processing is appropriate to produce a quantified observation that will then be called "data."

Errors in sampling procedures, inadequate sample storage and preservation techniques, and laboratory analytical errors are examples of errors in environmental datasets. As a demonstration of the multifaceted initiation points for such errors to exist in a dataset, further examples of the sources of error in the collection and analysis of groundwater quality data are listed in Table 1.3 and for air quality listed in Table 1.4.

There is always a degree of uncertainty associated with each discrete measurement of environmental quality. In interpreting data, each discrete measurement is really a range of statistically probable values instead of a single value. There are two subdivisions of reproducibility criteria, namely, replication and repeatability. Replication is when two or more results are obtained by the same operator in a given laboratory using the same apparatus of successive determinations on identical test material, within a short period of time. Often this is done

Table 1.3 Examples of sources of error in the sampling and analysis of groundwater quality data.

Sampling of a nonhomogeneous region in which wells and springs intersect more than one chemical type in water
Piezometers and wells that are inadequately flushed out prior to sampling of groundwater may result in a subsequent sample of the groundwater not being representative of conditions in the adjacent soil environment
Monitoring wells that are subject to temporal variations in water levels resulting in variations in chemical concentrations may exhibit significant sampling error
Cross contamination of a sample may occur at the time of sampling as a result of an unclean container into which the sample is placed
An error in the laboratory protocol of the experiment during the laboratory samples
Improper preservation techniques. For example, groundwater samples are often particularly susceptible to changes in pressure and improper sample storage. Improper preservation techniques can result in a chemical alteration of the sample as it adjusts to new equilibrium conditions. For example, the pH levels in groundwater samples have been noted as increasing up to 1.0 pH units due to CO_2 escape to the atmosphere during storage

Table 1.4 Examples of possible sources of error in the collection of air quality data.

Instrumentation error may arise due to poor calibration or "drift" of the calibration of the instrument with time
Channel error incurred during transmission from the monitoring locations to a central data processing unit
Fluctuations in meteorological conditions or quantities being released by the emission source, resulting in nonrepresentative air quality conditions

during quality assurance and quality control testing of a laboratory to assure that the lab results are trustworthy. Alternatively, repeatability is a quantitative expression of the random error associated in the long run with a single operator in a given laboratory obtaining successive results with the same apparatus under constant operating conditions on identical test material. Obviously, the requirements for quality assurance and quality control can be substantial.

Many of the statistical analyses described in this book are concerned with sampling errors and the estimation of population characteristics from samples of data. The fact that sampling errors are inherent in random data does not mean, however, that statistical manipulations and sophistication can in any way overcome faulty data. The quality of any statistical analysis is no better than the quality of the data utilized. Furthermore, statistical considerations should not be used to replace judgment and careful thought in analyzing data. Statistics must be regarded as a tool or an aid to understanding, but never as a replacement for careful thought.

1.6 Some Summary Indications of Approaches for Statistical Analyses

The concern with the statistical interpretation of data is widespread. However, the variability encountered in one circumstance is quite possibly very different from that encountered in another circumstance. There is no convenient "recipe" that can be utilized to stipulate an approach that can be universally applied. Instead, statistical analyses of environmental data have become the science of collecting, analyzing, and interpreting data with the findings at each decision point assisting in identifying the next stage of analysis. Statistics is concerned with the scientific methods used in collecting, organizing, analyzing, summarizing, and presenting data as well as with drawing valid conclusions and making reasonable decisions. Statistical analyses do not consist of a standard set of rules. Instead, analyses involve successive tests and refinements, with each test gaining an improvement in understanding of the data (and its information content). The findings of the tests may well be that additional statistical analyses are needed.

Since there is no convenient recipe, the practitioner needs to have available at his or her disposal a set of approaches, with each approach having utility in application to a class of problems. In selecting the procedure for use in a particular application, there are no absolute rules, only guidelines. To a large extent, the selection of the best procedure involves careful scrutiny of the characteristics of the problem at hand and the assumptions implicit in the particular statistical interpretation.

An analyst must still understand the basics in terms of both how to characterize a problem and the method by which the results may be interpreted. With the automated calculation procedures available in today's software, it is all too easy to employ a computer package without understanding the basis for the statistical procedures. The intent in this book is to explain the features of the various techniques in terms understandable to the nonstatistician and to provide some order to the available procedures by presenting the advantages and disadvantages of the limitations of the procedures. The discussion included in this book describes some of the sources of error and how these error sources should be reflected in the statistical analyses and in their interpretation. Proper application of statistical methods by someone who understands the utility and limitations of these methods can be most helpful in revealing the information that the data hold.

The emphasis in the chapters dealing with the statistical analyses of data is on examples to develop the equations and relationships to demonstrate how the various procedures may be used. References are provided for those wishing to pursue specific features in greater depth. Much of the theoretical background is omitted, with the focus being on the engineering applications to environmental quality data.

The various statistical methods are tools for data analysis that, like any tools, have proper and improper applications. The person using a statistical method is responsible to see that it is applied properly since the value of the results obtained depends on it. Even in the best circumstances, however, statistical data analysis can provide only evidence, never proof. Inferential statistics are aimed at distinguishing between random noise in the data and the real effects that are of interest. Only through careful consideration and interpretation of all the evidence can one hope to even begin finding answers to questions about causes for, and effects on, environmental quality.

1.7 Overview of Book Content

Morrison and Henkel indicated that "Alas statistical interference is not scientific inference. To have the latter we will have much more than the facade that claims of (statistical) significance provide. There are, of course, no computational formulas for scientific inference: the questions are much more difficult and the answers much less definite than those of statistical inference."

In this context the contents of the book are organized in the following manner:

1) Part II consists of Chapters 2 and 3. Chapter 2 describes ways of calculating risks, while Chapter 3 deals with public perceptions of risks and some of the more advanced challenges in risk assessment. Public perceptions of risks are a very important area of concern as many of the issues of environmental risk are directly related to the public's willingness to accept risk.
2) Part III of the book deals with the statistical analyses of data. For questions focused on risk perception, we have minimal data unlike, for example, situations of rolling of a dice. Chapter 4 focuses strictly on the fundamentals of statistical characterization of data since that is the underpinning of much of environmental risk assessment. Chapters 5 through 7 examine the attributes of commonly used probability distributions, as appropriate to environmental quality data. Chapter 8 uses these distributions to develop alternative types of control charts and to identify data outliers. Chapter 9 examines the capabilities of correlation and regression to better summarize data behavior. Chapters 10 through 12 examine different procedures for hypothesis testing. Chapter 10 considers relatively standard tests, whereas Chapter 11 examines procedures for multiple comparisons and Chapters 12 describe procedures appropriate when the data include numerous censored (less than) data.

1.8 References

McBean, E. and Rovers, F. (1990). Flexible selection of statistical discrimination tests for field-monitored data. In: *Groundwater and Vadose Zone Monitoring* (ed. D.M. Nielsen and A.I. Johnson), 256–265. Philadelphia, PA: American Society for Testing and Materials.

Unwin, J., Miner, R.A., Srevers, G., and McBean, E. (1985). Groundwater Quality Data Analysis. *National Council for Air and Stream Improvement Technical Bulletin, No. 462.*

Wilson, R. (1979). Analyzing the daily risks of life. *Technology Review* 81 (February 1979): 41–46.

1.9 Problems

1.1 The probability of an earthquake of Richter scale 8 (i.e. a very severe earthquake) is estimated as 10^{-12} occurring some time over the next 70 years at location A. The probability of a similar earthquake is 10^{-9} at location B and 10^{-5} at location C.

i) Would you consider the likelihood at location C as being de minimis risk?
ii) How much greater is the risk for C than for A?

1.2 Explain what the term means when you indicate "a 1 in 25-year storm" in your approach to managing the risk of flooding.

Part I

Methodologies for Risk Characterization

2

Introduction to Risk Assessment

A risk is a potential for a loss. The loss is the realization of that negative potential. A risk is running across a busy street blindfolded. A loss is getting hit by a car while doing that.
—Riskviews

2.1 Challenges in Risk Assessment

As discussed in Chapter 1, the public are increasingly involved in decision-making processes on issues of risk, and consequently, there is increased pressure to decrease the risks to which people are exposed, especially in the environmental realm. Success in environmental risk assessment often means carrying out a process that considers public interests and concerns and allows for public consultation and feedback. The public may exert a substantial influence on identification and selection of environmental risk management options. It is not only important to acknowledge and appreciate public sentiment but also to understand how the public perceives risk.

In society, we face very small risks every day; some examples of risks that increase the chance of death by one-in-a-million are listed in Table 2.1. It is interesting to use these types of activities, which are relatively familiar to the public, to characterize various types of risk. Further discussion of the bases for these numbers will be provided in Section 2.3, but for current purposes, the values in Table 2.1 are activities that have been reported (Burgman, 2005) as increasing the risk of death by one-in-a-million.

Given the values in Table 2.1, and assuming there are 240 working days per year, which of the following activities would you rank as most risky?

A. Riding a bicycle to work (20 km each way) every working day over a 4-year period.
B. Working 1 year in a coal mine (40 h/week).
C. Living 4 km from a nuclear reactor for 5 years.
D. Working for 1 year as a firefighter.
E. Working for 2 months in a nuclear power plant.

Consider the following to generate the risks:

A. (biking to work for 4 years): 20 km × two directions × 240 days/year × 1/15 × 4 years = 2560. In words, this means there is a 2560×10^{-6} chance of death (or a probability of death of 2.5 of 1000) due to riding the bicycle.
B. (working in a coal mine for 1 year): 40 h/week × 52 weeks/year × 1 year = 2080.
C. (live 4 km from a nuclear reactor for 5 years): 1/50 × 0.02 × 5 year = 0.1

Risk Assessment: Procedures and Protocols, First Edition. Edward A. McBean.

Companion website: www.wiley.com/go/McBean/RiskAssessProcedure_1e

Table 2.1 Activities that increase the risk of death by one-in-a-million.

Activity	Cause of death
Spending 1 h in a coal mine	Black lung disease
Spending 1 h as an agricultural worker	Accident
Spending 2.5 days as a firefighter	Accident
Traveling 15 km by bicycle	Accident
Traveling 500 km by car	Accident
Flying 1500 km by jet	Accident
Living 2 months in a brick building	Cancer from radiation
Working for 1 year in a nuclear power plant	Cancer from radiation
Living 50 years within 8 km of a nuclear reactor	Radiation (accident)
Living 20 years near a PVC plant	Cancer from vinyl chloride

Source: From Burgman (2005).

(Note that one could argue that living within 4 km is riskier but information for this degree of characterization is not available).

D. (firefighter for a year): 240 working days/year/2.5 = 96.

E. (work 2 months in a nuclear power plant): 1/6 × 1 = 0.17.

Based on statistical evidence, the activities ordered from most risky to least risky are A, B, D, E, and C. Laypeople rarely order these activities correctly because the public seldom has, or understands, the statistical evidence needed to accurately evaluate risk and instead relies on inferences based on what they remember hearing or seeing. This effect is called judgmental bias.

Selective perception also plays a role in how people evaluate risk; facts are distorted when they are viewed through a personal lens of social influence (interpretation of facts are based on social beliefs), and new evidence is often deemed reliable only if it is consistent with the beliefs initially held.

Other challenges to accurate risk perception among the general public include:

- *Fear of the unknown*: The general public often lack the knowledge required to understand the complex environmental processes and technologies that are typically the subjects of risk assessment. In the perception of risks related to these processes and/or chemicals or other hazards that are unfamiliar with the result, they tend to be exaggerated.
- *Visibility*: The news media have a strong influence on public perception of risk and tend to direct significant attention to rare circumstances that may involve sensational elements. For this reason, the public is often disproportionately aware of these concerns and perceives associated risks to be greater than they are in reality.
- *Voluntariness*: People tend to tolerate higher risks when they participate in a voluntary activity (e.g. smoking or riding a bike). Environmental risks are generally involuntarily assumed, and people are less willing to tolerate risk in this case.
- *Control*: People tend to tolerate higher risks when they believe they are in control of the outcome, compared with instances when they believe they have little control (e.g. driving a car versus traveling as a passenger on an airplane). In addition, people often have an inflated view of their ability to control their fate when accepting voluntary risks.
- *Cost*: Opinions on the necessity of risk management measures may vary depending on who will pay for implementation of the management measures. The public less frequently supports measures that will be paid for with public funds than those that will be funded

privately; it is sometimes the view that no expense is too great to reduce or eliminate risk if someone else will bear the cost.

- *Dread*: Risk is often judged based on the degree to which people dread the outcome instead of the likelihood that it will occur. For example, a member of the public will typically underestimate the risk of death from a more common cause, such as cancer, and overestimate the risk of death from rare causes, such as disease epidemics.
- *Understanding*: The public is largely ignorant of even the simplest statistical concepts, and so they lack the background necessary to understand risk estimates. The public view of risk may be highly simplistic: An activity can be rated as 100% safe, or 100% dangerous, and everything in between suggests a 50/50 change of occurring. In addition, the lay public may have trouble interpreting scientific and mathematical symbols and notations (for example, recognizing that 10^{-9} is greater than 10^{-12}).

When dealing with environmental risk in the public purview, it is oftentimes beneficial to frame the concept of risk in a familiar context. People encounter risk on a day-to-day basis. Many times a day, we unconsciously assess activities such as crossing the street or driving a car, considering factors such as our perceived risk of injury, when deciding whether or not to participate. Comparing values of environmental risk to those for more familiar risks can help to normalize the public's perception of unfamiliar environmental risks (see Section 2.3).

2.2 Categories of Risk

Risks can be categorized based on the likelihood that the outcome of interest will occur, as well as on the impact or severity of the outcome. Table 2.2 lists "likelihood" categories and Table 2.3 lists categories that might be defined to describe the impact of an outcome on the success of an engineering project, for example.

Table 2.2 Possible categorization of likelihood of risk.

Category	Description
1	Very high, almost certain to occur
2	High, likely to occur
3	Medium, as likely as not to occur
4	Low, may occur occasionally
5	Very low, unlikely to occur

Table 2.3 Possible categorization of risk impact (on project success).

Category	Description
Very high	Threatens the success
High	Substantial impact on time, costs, or quality
Medium	Notable impact on time, cost, or quality
Low	Minor impact on time, cost, or quality
Very low	Negligible impact

A similar categorization scheme divides risks into four broad categories:

1) Familiar high risks that exact a large toll and are described by significant statistical evidence (e.g. vehicular accidents).
2) Risks of low probability whose consequences are so large that they must be taken seriously (e.g. a large earthquake).
3) An extension of the second category, events whose probabilities are so very low that they have rarely, if ever, happened, yet whose consequences are so severe that they must be taken seriously (e.g. catastrophic failure of a nuclear power plant).
4) A collection of risks that are hard to evaluate because they show up as increases in naturally occurring phenomena (e.g. cancer risks).

This text focuses primarily on category 4 as listed above. Chapter 3 provides some insights into methodologies that can be used to address some of the aspects of categories 2 and 3.

The negative effects of environmental hazards risks (category 4) almost always appear as an increase in naturally occurring phenomena. There is significant pressure to reduce environmental risks to very low levels, but questions arise when the costs of risk management outweigh the benefits. This question will be explored.

2.3 De Minimis Risk

There is a limit to the degree to which environmental risk should be reduced, in the interest of time, money, and effort. The compromise is "de minimis," or negligible risk, usually interpreted as about one-in-a-hundred thousand, or one-in-a-million. This is based on de minimis risk, examples of which were described in Section 2.1. Additional examples of de minimis risk are listed in Table 2.4. Members of the public wish to minimize their exposure

Table 2.4 Examples of increasing risks of death by one-in-a-million.

Smoking 1.4 cigarettes
Drinking 0.5 l of wine
Living 2 days in New York or Boston
Traveling 6 min by canoe
Traveling 10 miles by bicycle
Traveling 300 miles by car
Flying 1000 miles by jet (accident)
Flying 6000 miles by jet (cancer/cosmic radiation)
Living 2 months in Denver (cosmic radiation)
Living 2 months in average stone or brick building due to radon
One chest X-ray in a good hospital
Living 2 months with a cigarette smoker
Eating 40 tablespoons of peanut butter due to aflatoxin
Drinking City of Miami drinking water for 1 year
Drinking thirty 12 oz cans of diet soda
Living 20 years near a PVC plant due to vinyl chloride
Eating 100 charcoal-broiled steaks by benzopyrene

Source: From Slovic et al. (1979).

to risks and generally wish to drive the risks to zero. However, when we live in society, we are continuously exposed to risks. When we ride a bicycle, we might be struck by a vehicle. When we take an airplane trip, there is the potential for a plane crash, and so on. As a result, the public generally has an understanding of the risks they encounter on a day-to-day basis. In response, a possible approach to explain risks is to rely upon communicating the risks of environmental impacts, by putting them into context, comparing the environmental risks with other risks about which members of the public are relatively familiar.

In environmental risk management, reducing risk to one-in-a-million reflects an attempt to meet or exceed the safety standards of activities the public undertakes voluntarily on a regular basis. Comparing environmental risks with more familiar risks may help to bring the magnitude of environmental risks into perspective and increase public awareness and acceptance of these risks.

It is important to appreciate that the above estimates are approximate. The following examples demonstrate the types of procedures sometimes used to identify activities that increase one's risk of death by one in one million, demonstrating just how approximate the values in Table 2.4 are.

Example 2.1 Consider airplane travel. How many kilometers must be traveled to increase one's risk of death by one-in-a-million?

Assume the following information is developed from a review of available data:

- 150 billion km traveled per year by airplane.
- 100 people on average are killed annually during air travel.

The risk of death per kilometer traveled is 100/150 billion km = 6.67×10^{-10}/km.

A total risk of death equal to 1×10^{-6} is associated with X km of air travel, where

$$X = 1 \times 10^{-6} / 6.67 \times 10^{-10} /\text{km} = 1500\,\text{km}$$

In words, traveling 1500 km in an airplane increases one's chance of dying by one-in-a-million.

Example 2.2 Consider cigarette smoking. How many cigarettes must be smoked to increase one's risk of death by one-in-a-million?

Assume the following information is developed from a review of available data:

- Smokers consume an average of 3000 cigarettes/year.
- 15% of the population will die from smoking.
- Average life expectancy is 70 years.

The average annual risk of death from smoking is 0.15/70 years = 2×10^{-3}/year.

The risk of death per cigarette smoked is 2×10^{-3}/year/3000 cigarettes/year = 6.67×10^{-7}/cigarette.

A total risk of death equal to 1×10^{-6} is associated with smoking X cigarettes, where

$$X = 1 \times 10^{-6} / 6.67 \times 10^{-7} /\text{cigarette} = 1.5\ \text{cigarettes}$$

The two components of risk are the "exposure aspects" and the "toxicity" of the chemical. These elements are depicted in schematic form in Figure 2.1. Examples of the pathways are illustrated in Figure 2.2.

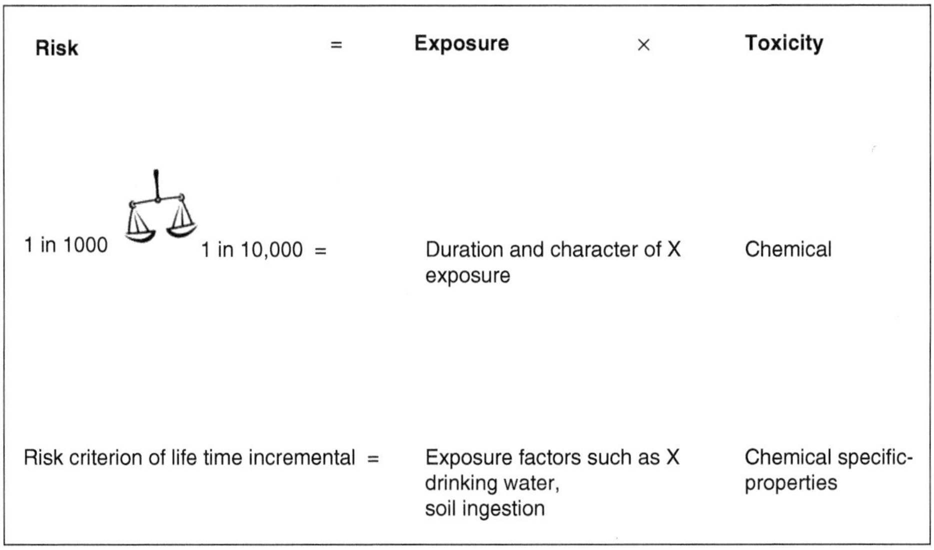

Figure 2.1 The two components of risks.

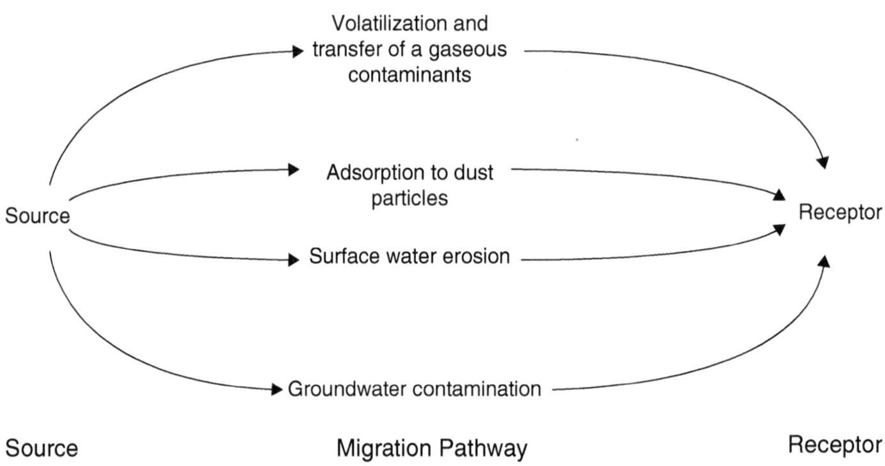

Figure 2.2 Schematic depiction of migration pathways.

2.4 Toxicological Versus Epidemiological Data

Hazard assessment is the process of determining whether exposure to a stressor may cause an increase in the incidence of specific adverse health effects (e.g. cancer, birth defects) and whether the adverse health effect is likely to occur in humans.

A wide variety of studies and analyses are used to support a hazard identification analysis. In dealing with the public, it is important to have a methodology that is traceable and transparent and developed from first principles so that, for example, the risk of developing cancer from consumption of arsenic-impacted groundwater can be estimated. The estimates of risk are frequently derived from data collected by toxicologists and/or epidemiologists.

Most toxicological studies follow the classic paradigm of experimental studies. In such an experiment, all environmental conditions are fixed to the extent possible, and only one or at most a very few are varied. The notion is that differences in outcome will result only from planned and controlled changes in exposure, usually with animal subjects. Our confidence in the results is strengthened to the extent that we are aware of all causal factors and they are controlled in the experiment. However, economic restrictions limit, for example, the number of animals that can be included in any toxicological study. Most environmental regulations that apply to carcinogens are set at levels that are intended to ensure a lifetime cancer risk of no greater than 10^{-4}–10^{-6}. If the experimental animal subjects in a lifetime bioassay feeding study were exposed to a carcinogenic chemical at a level comparable to that of even relatively polluted environments, the probability of a tumor being observed would normally be less than 10^{-4}, requiring a study size of 10 000 or more animals to detect a statistically significant result. Since this is infeasible, experiments are instead typically designed using relatively high concentrations of a chemical and a much smaller group of animal subjects to obtain a meaningful increase in adverse outcomes, and results are then extrapolated to the range of interest. The results of some animal experiments are also extrapolated to estimate impacts on humans, although there is some uncertainty in this approach. It must be established, for example, whether the metabolism of sheep is sufficiently similar to that of humans to allow extrapolation to humans from studies using sheep. Thus, even when best estimates of health impacts arising from environmental exposure are obtained from toxicological studies, considerable uncertainty often exists.

In contrast to toxicologists, epidemiologists observe the results of natural experiments, most of which are observational. There is no manipulation of experimental variables; the study design takes advantage of "natural experiments" in which groups of subjects, usually humans, have been exposed (inadvertently) to different levels of potentially hazardous materials. Clearly, there are limits to what can be learned from these natural experiments. For example, through observation it may be found that the workers in an industrial complex manufacturing chemical Y demonstrate a specific type of illness. Where there is a very profound effect on disease rates or large amounts of data exist, it may not be difficult to identify chemical Y as the cause. However, when the incremental effects are more modest and data are limited, conclusive findings can be challenging. The extrapolated uncertainty at low contaminant doses may vary as much as 40-fold, depending on the extrapolation model (Stallones, 1988). Epidemiological studies are often obtained from occupational groups. Here, the cancer risk is evaluated, for example, in working populations that are exposed to higher levels of contaminants than normally found in the general environment. Extrapolation of findings from high to much lower exposure levels is required and hence introduces uncertainty into the estimates (Asante-Duah, 1993).

2.5 Basics of Environmental Risk Assessment

The classical chemical and environmental risk assessment paradigm remains the most popular. Figure 2.3 illustrates the general framework.

Problem formulation is undertaken to establish the purpose of the risk assessment and the fundamental question(s) to be addressed. Outcomes of this phase are identification of the individual(s) or population(s) at risk and the endpoint of concern (e.g. illness), a conceptual model of the exposure scenario from source to receptor, and a plan for analysis of the risk (US EPA, 2012).

Development of an exposure scenario requires identification of the source, exposure pathway, and receptor of the contaminant of concern:

Source – transport pathways – receptor

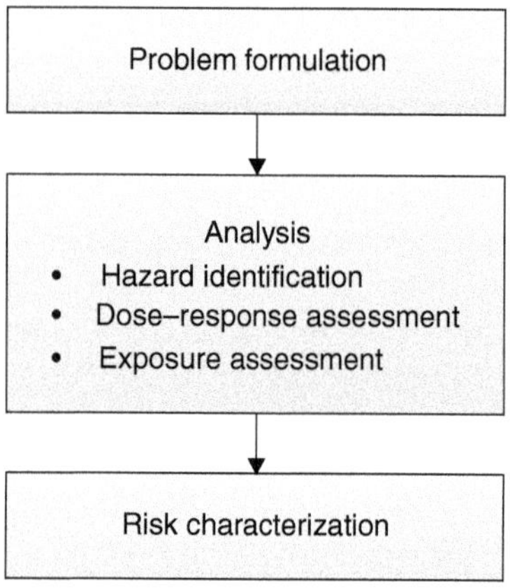

Figure 2.3 General framework of environmental risk management.

When any one of these three elements is absent, no risk exists (e.g. if no pathway exists to create exposure, no risk exists). In complex assessments, there may be primary and secondary sources, a variety of transport mechanisms (through surface water, groundwater, air, soil) and exposure routes (ingestion, inhalation, absorbance, injection), and a number of receptors at the end of each pathway that might affect their fate, transport, and persistence.

Risk analysis, following development of the exposure scenario, incorporates three steps: hazard identification, dose–response assessment, and exposure assessment. Generally, hazard identification involves identification of the cause of the endpoint of concern and the range of human health effect associated with the exposure. The dose–response assessment investigates the relationship between chemical dose and the probability of a response in a human. The primary purpose of the exposure assessment is to determine the chemical dose to which an individual or population is exposed by the defined route of exposure. Note that exposure differs from dose – while exposure describes the amount of chemical that is encountered and available for uptake, dose is the amount of the chemical that enters the body and is retained, and thereby available for the target organs, to cause, or potentially cause, illness or death.

The risk characterization step integrates the findings of the dose–response assessment and the exposure assessment to estimate the probability that the endpoint of concern will occur. In this phase, the risk may also be described according to its severity and consequences, and limitations of the analysis based on the quality of input data may be identified (ILSI, 2000).

2.6 Estimating Intake (Dose)

The intake rate, or dose, of a chemical is typically expressed in terms of the mass of a chemical that enters the body per unit of body weight (BW) per unit of time (e.g. mg/kg_{BW}/day). Average daily intake is calculated according to Equation (2.1). In a complex risk assessment, the intake calculation is performed for each chemical of concern and each exposure pathway:

$$\text{Intake}\left(\text{mg/kg}_{\text{BW}}/\text{day}\right) = \frac{(C \times IR \times EF \times ED)}{(BW \times AT)} \tag{2.1}$$

where

C = concentration in air, water, soil, etc. (mg/m^3, mg/l or mg/kg)
IR = intake rate of contaminated media (m^3/day, l/day, or kg/day)
EF = exposure frequency (days/year)
ED = exposure duration (years)

Table 2.5 Typical values used in estimating intake levels.

Parameter	Description	Units	Value
BW_A	Body weight, adult	kg	70
BW_C	Body weight, child	kg	15
IR_W	Drinking water ingestion (intentional)	l/day	2
IR_{Wa}	Drinking water ingestion (accidental)	l/day	0.1
IR_A	Air inhalation (adult/child)	m^3/day	20/15
IR_{Sr}	Soil ingestion (residential) (adult/child)	mg/day	100/200
IR_{So}	Soil ingestion (occupational)	mg/day	50
EF_R	Exposure frequency (residential)	days/year	365
EF_O	Exposure frequency (occupational)	days/year	250

BW = body weight (kg)
AT = averaging time (days)

The time over which intake is averaged (AT) varies depending on whether the chemical of concern is a carcinogen or a noncarcinogen (this will be explained in greater detail in Sections 2.7 and 2.8). Based on Equation (2.1), examples of intake will also vary depending on characteristics of the receptor(s) and their behavior. Potential receptors include residents (adult, child) and workers (commercial, industrial, construction, etc.). Table 2.5 lists additional information that may be useful in calculations of average daily intake.

Example 2.3 Assume soil is contaminated with chemical MNO at a concentration of 2.4 mg/kg. Receptors are exposed via soil ingestion over a period of 5 years (ED). Calculate the average daily intake for the receptors listed below, assuming AT = ED:

- A child who lives in a home built on a contaminated site.
- A construction worker who works on a contaminated site.

Child intake scenario:

$$\text{Intake}(\text{mg/kg}_{\text{BW}}/\text{day}) = \frac{(\text{C} \times \text{IR} \times \text{EF} \times \text{ED})}{(\text{BW} \times \text{AT})}$$
$$= \frac{(2.4\ \text{mg/kg} \times 10^{-6}\ \text{kg/mg} \times 200\ \text{mg/day} \times 365\ \text{day/year})}{(15\ \text{kg}_{\text{BW}} \times 5\ \text{year} \times 365\ \text{day/year})}$$
$$= 6.4 \times 10^{-5}\ \text{mg/kg}_{\text{BW}}/\text{day}$$

(averaged over 5 years)

Construction worker intake scenario:

$$\text{Intake}(\text{mg/ kg}_{\text{BW}}/\text{ day}) = \frac{(\text{C} \times \text{IR} \times \text{EF} \times \text{ED})}{(\text{BW} \times \text{AT})}$$
$$= \frac{(2.4\ \text{mg/ kg} \times 10^{-6}\ \text{kg/mg} \times 50\ \text{mg/ day} \times 250\ \text{day/ year})}{(70\ \text{kg}_{\text{BW}} \times 5\ \text{year} \times 365\ \text{day/ year})}$$
$$= 2.34 \times 10^{-6}\ \text{mg/kg}_{\text{BW}}/\text{day}$$

(averaged over 5 years and assuming the worker is on the job for 250 day/year)

2.7 Calculating the Risk for Noncarcinogens

The approach to environmental risk calculation differs depending on whether a substance has carcinogenic or noncarcinogenic effects. Carcinogens are substances that lead to the development of malignant (cancerous) cells, while noncarcinogens do not lead to the development of malignant cancerous cells but have other adverse health effects, including impacts on the development, size, or functioning of the body and its organs. Challenges in these procedures include assessment of:

- *Acute effects*: These are caused by a single exposure and result in an immediate health problem.
- *Chronic effects*: These are long lasting and can be the result of a single large dose or repeated smaller doses.

This section outlines the methodology used to quantify the risk associated with exposure to noncarcinogens.

In theory, noncarcinogens have a toxicity threshold below which adverse human health effects are not expected to occur and immediately above which responses begin to manifest. The threshold concept is the basis for establishing the "no observable adverse effect level" (NOAEL) for a chemical. The reference dose (RfD) and the "lowest observable adverse effect level" (LOAEL) are also important. Figure 2.4 illustrates the relationship among the RfD, NOAEL, and LOAEL. These terms are defined below:

- *NOAEL*: The highest exposure at which there are no statistically or biologically significant increases in frequency or severity of adverse effects in an exposed population over a control population.

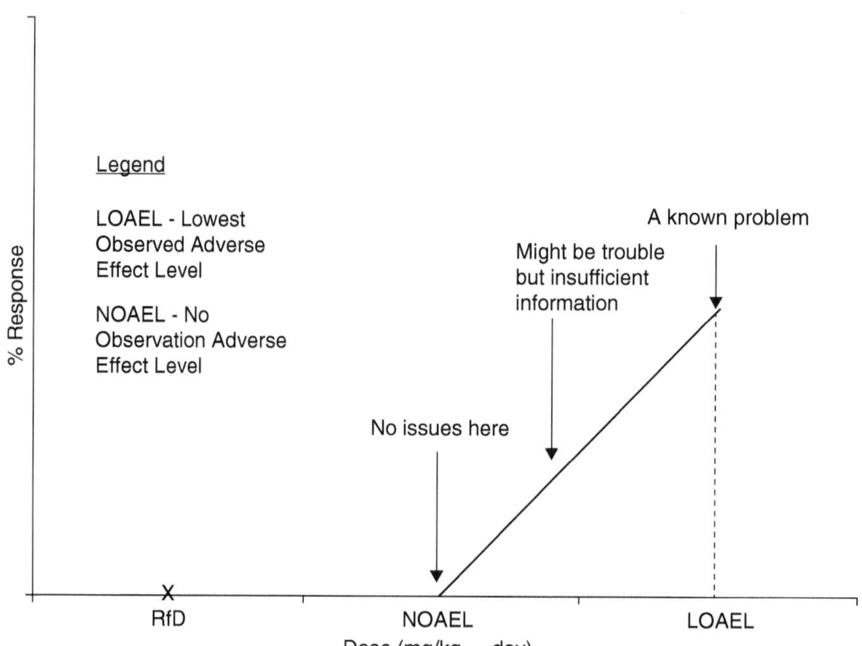

Figure 2.4 Schematic depiction of dose–response curve for noncarcinogenic effects.

- *RfD*: The dose likely to be without an appreciable risk of adverse effects over a lifetime of exposure. The RfD is determined by applying a number of uncertainty factors (UF) and a modifying factor to the NOAEL, such that

$$\mathrm{RfD} = \frac{\mathrm{NOAEL}}{(\mathrm{UF1} \times \mathrm{UF2} \times \cdots \times \mathrm{MF})}.$$

  UFs are typically equal to 10 and account for such aspects as extrapolation between species, extrapolation from high to low exposure levels, protection of sensitive subpopulations (e.g. asthmatics), incomplete data, use of the LOAEL instead of the NOAEL, etc.
  The MF accounts for additional uncertainties not explicitly accounted for in the UFs and has a default value of 1, but may be as high as 10, based on a professional assessment of uncertainties. This "safety factor" approach is used because data are often limited and sampling errors may be high, especially at very low concentrations.
- *LOAEL*: The lowest dose at which an increase in frequency/severity of adverse health effects is observed in an exposed population over a control population.

The Environmental Protection Agency (EPA) uses the RfD to define acceptable concentrations of chemicals in the environment with respect to oral and dermal exposure (ingestion and absorption). The reference concentration (RfC) is an analogous concept that applies to chemical exposure via inhalation. Table 2.6 lists the RfD and/or RfC for selected chemicals of concern.

Table 2.6 Reference dose and concentration for chemicals of concern.

Chemical	Hazard or effect	RFD (mg/kg/day)	RfC (ug/m^3)
Anthracene	Skin irritant	0.3	
Benzo(g,h,i)perylene	Limited information available Cocarcinogenic effects	0.000 57	
Fluoranthene	Defatting agent that affects skin Cocarcinogenic effect	0.041	
Naphthalene (co-carcinogen)	Inhalation – eye irrigation, headache, confusion Ingestion – abdominal pain, nausea, vomiting Dermal/eye – system effects – anemia, decreased hemoglobin	0.004	71
Phenanthrene	Inhaled or ingested irritant	0.000 57	
Pyrene	Cocarcinogenic	0.03	
Cis-1,2-dichloroethylene (*cis*-1,2-DCE)	Eye and respiratory tract irrigation, nervous system depression	0.01	1.1×10^2
Trans-1,2-dichloroethylene (*trans*-1,2-DCE)	Nervous system depression	0.01	
TPH (gas/diesel)	Neurological symptoms, lung damage (in rats)	0.06	
TPH (heavy oils)	Low vapor pressure, solubility, and toxicity CNS depression, kidney damage	6.0	

Sources: Ontario Ministry of Environment and Energy (1996); US EPA (1997); California Department of Toxic Substances Control (CDTSC) (1991).

The RfD is compared to the calculated chronic daily intake (CDI) of a chemical to assess the potential for adverse health effects. The CDI is calculated according to Equation (2.2) (which is comparable with Equation [2.1]):

$$\mathrm{CDI}\left(\mathrm{mg/kg_{BW}/day}\right)=\frac{\left(\mathrm{C}\times\mathrm{IR}\times\mathrm{EF}\times\mathrm{ED}\right)}{\left(\mathrm{BW}\times\mathrm{AT}\right)} \tag{2.2}$$

where
C = concentration in air, water, soil, etc. (mg/m^3, mg/l, or mg/kg)
IR = intake rate (m^3/day, l/day, or kg/day)
EF = exposure frequency (days/year)
ED = exposure duration (years)
BW = body weight (kg)
AT = averaging time (period over which the exposure is averaged, usually 365 days for noncarcinogens) (days) (For noncarcinogens, the AT is typically 365 days.)

The likelihood of adverse effects resulting from exposure to a noncarcinogen is represented by the hazard quotient (HQ), the ratio of the CDI to the RfD (Equation [2.3]):

$$\mathrm{HQ}=\frac{\mathrm{CDI}}{\mathrm{RfD}} \tag{2.3}$$

An HQ greater than 1 (i.e. the estimated CDI is greater than the RfD) indicates a concern for potential for adverse health effects among some groups, and generally, the greater the value, the greater the level of concern. If HQ < 1, adverse health effects are unlikely, even in sensitive subpopulations.

Example 2.4 Naphthalene is used in mothballs. It is a cocarcinogen (a suspected carcinogen), but is known to also cause other adverse health effects. Calculate the naphthalene intake via inhalation for an employee in a factory where the concentration of naphthalene in the air is 0.1 mg/m^3. The average employee works at the factory for 11.5 years. Is the employee likely to suffer an adverse effect from naphthalene exposure while working at the factory?

Assume the RfD for naphthalene via inhalation is 0.004 mg/kg/day and the employee works 250 days/year:

$$\begin{aligned}\mathrm{CDI}&=\frac{\left(\mathrm{C}\times\mathrm{IR}\times\mathrm{EF}\times\mathrm{ED}\right)}{\left(\mathrm{BW}\times\mathrm{AT}\right)}\\&=\frac{\left(0.1\ \mathrm{mg/m^3}\times 20\ \mathrm{m^3/day}\times 250\,\mathrm{day/year}\times 11.5\ \mathrm{year}\right)}{\left(70\ \mathrm{kg_{BW}}\times 11.5\ \mathrm{year}\times 365\ \mathrm{day/year}\right)}\\&=0.020\ \mathrm{mg/kg_{BW}/day}\end{aligned}$$

To estimate risk for noncarcinogens, we divide the CDI by the RfD to obtain the hazard quotient (HQ):

$$\begin{aligned}\mathrm{HQ}&=\frac{\mathrm{CDI}}{\mathrm{RfD}}\\&=\frac{0.020\,\mathrm{mg/kg_{BW}/day}}{0.004\ \mathrm{mg/kg_{BW}/day}}\\&=4.9\end{aligned}$$

The HQ is greater than 1, which indicates some concern that the employee will suffer adverse health effects from naphthalene exposure while working at the factory.

Example 2.5 Toluene is a solvent that, among other things, is used to dissolve paint, glue, etc. It is a noncarcinogen, but is known to cause other adverse health effects. Calculate the toluene intake via ingestion for a child who plays in a river (downstream of an industrial discharge) where the concentration of toluene in the water is 2 mg/l. The child plays in the river every day for 4 years and ingests water accidentally while playing. Is the child likely to suffer an adverse effect from oral exposure to toluene during the period she plays in the river? Assume the RfD for toluene via ingestion is 0.2 mg/kg/day:

$$\begin{aligned}\mathrm{CDI} &= \frac{(\mathrm{C} \times \mathrm{IR} \times \mathrm{EF} \times \mathrm{ED})}{(\mathrm{BW} \times \mathrm{AT})} \\ &= \frac{(2\ \mathrm{mg/l} \times 0.1\ \mathrm{l/day} \times 365\ \mathrm{day/year} \times 4\ \mathrm{year})}{(15\ \mathrm{kg} \times 4\ \mathrm{year} \times 365\ \mathrm{day/year})} \\ &= 0.013\ \mathrm{mg/kg/day}\end{aligned}$$

To estimate risk for noncarcinogens, we divide the CDI by the RfD to obtain the hazard quotient (HQ):

$$\begin{aligned}\mathrm{HQ} &= \frac{\mathrm{CDI}}{\mathrm{RfD}} \\ &= \frac{0.013\ \mathrm{mg/kg/day}}{0.2\ \mathrm{mg/kg/day}} \\ &= 0.06\end{aligned}$$

The HQ is less than 1, which indicates that the child is unlikely to suffer adverse health effects from oral exposure to toluene during the period she plays in the river.

Noncarcinogenic effects of different chemicals on the same organ system, even if by different mechanisms or exposure pathways, are considered to be additive. In this case, the HQ (each describing the risk associated with a single chemical by a single exposure pathway) are summed, resulting in a measure called the hazard index (HI), which characterizes the risk associated with multiple chemicals by multiple exposure pathways:

$$\mathrm{HI} = \mathrm{HQ}_1 + \mathrm{HQ}_2 + \cdots \tag{2.4}$$

The same criteria are applied to the HI as were outlined for the HQ: If $\mathrm{HI} > 1$, there is concern for potential adverse human health effects, and further evaluation or risk management may be required, but if $\mathrm{HI} < 1$, adverse human health effects are unlikely and no further action is required.

Example 2.6 The table below lists the RfD and calculated CDI for a group of chemicals found in contaminated soil. None of these chemicals are carcinogens, but all have the potential to cause other adverse human health effects; the primary target organ for each chemical is also given. Is an adult exposed to this group of chemicals at the rates listed below likely to experience adverse effects on the kidney? The liver?

Chemical	RfD (mg/kg$_{BW}$/day)	CDI (mg/kg$_{BW}$/day)	Target organ
A	0.05	0.003	Kidney
Y	2.0	12.0	Liver
B	0.3	0.16	Kidney
Z	0.3	0.10	Liver

$$\begin{aligned} HQ_A &= \frac{CDI_A}{RfD_A} \\ &= \frac{0.003\ mg/kg_{BW}/day}{0.05\ mg/kg_{BW}/day} \\ &= 0.06 \end{aligned}$$

$$HQ_Y = 6.0$$

$$HQ_B = 0.53$$

$$HQ_Z = 0.33$$

$$HI_{kidney} = HQ_A + HQ_B = 0.59$$

$$HI_{liver} = HQ_Y + HQ_Z = 6.33$$

$HI_{kidney} < 1$, so therefore adverse effects on the kidney are considered unlikely. $HI_{liver} > 1$, so there is concern for potential adverse effects. Additional evaluation and/or risk management would be required.

Risk-based screening is a tool used to help risk assessors and others involved in the remediation of contaminated sites in the decision-making process to determine if remediation is required. The risk-based screening level (RBSL) for a chemical (the concentration above which additional evaluation and risk management may be required) can be calculated using the intake equation (Equation [2.3]). In this case, C = RBSL, and CDI = RfD × HQ, where HQ is the acceptable HQ (usually 1). All other variables are known or assumed, and the equation is solved to obtain RBSL.

Example 2.7 Estimate the allowable concentration (RBSL) of toluene in a drinking water well, assuming that water from the well is used for drinking only. The well, designed to supply water for a period of 30 years, is the only source of drinking water for the farmer who lives on the property:

$$HQ \times RfD = \frac{(RBSL \times IR \times EF \times ED)}{(BW \times AT)}$$

$$RBSL = \frac{(HQ \times RfD \times BW \times AT)}{(IR \times EF \times ED)}$$

$$\begin{aligned} RBSL &= \frac{\left(1 \times 0.2\ mg/kg/day \times 70\ kg_{BW} \times 30\ year \times 365\ day/year\right)}{\left(2\ l/day \times 365\ day \times 30\ year\right)} \\ &= 7.0\ mg/l \end{aligned}$$

The allowable concentration of toluene in the well is 7.0 mg/l. If the concentration is found to exceed 7.0 mg/l, additional evaluation and risk management may be required.

2.8 Calculating Risks for Carcinogens

2.8.1 Background to Classification System for Carcinogens

This section outlines the methodology used to quantify the risk associated with exposure to carcinogens.

Carcinogens, and possible carcinogens, are identified through epidemiological and toxicological research, and several agencies including the International Agency for Research on Cancer (IARC) and the US EPA (or USEPA) are responsible in the United States for collecting and making available information about these chemicals. Both agencies use a similar classification system for chemicals based on cancer-causing potential. The EPA's "weight-of-evidence" scale and examples of chemicals included in each category are shown in Table 2.7. It will perhaps surprise you to learn that there are relatively few group A chemicals known to cause cancer in humans. Out of about 900 that have been evaluated, only about 100 are known human carcinogens (American Cancer Society, 2013).

There is no exposure threshold for carcinogens; if dose is greater than zero, risk is greater than zero (i.e. there is no dose below which the risk of developing cancer is zero for a carcinogen).

2.8.2 Calculating Risk from Carcinogens

This section outlines the methodology used to quantify the risk associated with exposure to carcinogens.

From Section 2.7, intake for noncarcinogens is described by the CDI over the period of exposure. For carcinogens, however, intake is described by the lifetime average daily dose (LADD),

Table 2.7 EPA weight-of-evidence classification system for carcinogenicity.

Group	Description	Examples
A	Human carcinogen, with sufficient evidence from epidemiological studies to treat as a known carcinogen	Benzene
B1 or B2	A probable human carcinogen with: • Limited evidence from epidemiological studies (B1) • Limited evidence from animal studies and inadequate evidence or no data from epidemiological studies (B2)	Benzo(a)pyrene
C	Possible human carcinogen, with limited evidence from animal studies in absence of human data	Tetrachloroethylene
D	Not classifiable as to human carcinogenicity, owing to inadequate human and animal evidence	Ethylbenzene, xylene
E	Evidence of noncarcinogenicity for humans, with no evidence of carcinogenicity in at least two adequate animal tests in different species or in both adequate animal and epidemiological studies	Borax (sodium borate)

Source: From http://www.cancer.org/cancer/cancercauses/othercarcinogens/generalinformationaboutcarcinogens/known-and-probable-human-carcinogens.

the total intake averaged over a lifetime (see Equation [2.5]). This is because carcinogenic risks are cumulative – each additional exposure increases the likelihood of developing cancer, even if the exposures are separated by a period of years:

$$\text{LADD}(\text{mg/kg}_{\text{BW}}/\text{day}) = \frac{(\text{C} \times \text{IR} \times \text{EF} \times \text{ED})}{(\text{BW} \times \text{AT})} \tag{2.5}$$

where
C = concentration in air, water, soil, etc. (mg/m^3, mg/l, or mg/kg)
IR = intake rate (m^3/day, l/day, or kg/day)
EF = exposure frequency (days/year)
ED = exposure duration (years)
BW = body weight (kg)
AT = averaging time (period over which the exposure is averaged, life expectancy × 365 days for carcinogens) (days)

As used in relation to CDI, the information listed in Table 2.5 is often useful in calculations of LADD.

Example 2.8 Assume an individual is exposed to chemical X in drinking water at a concentration of 2.4 mg/l over a period of 10 years. As the first step in risk characterization, calculate the average daily intake of chemical X assuming that it is a noncarcinogen. If chemical X is a carcinogen, how does your estimate of average daily intake change?

Noncarcinogen

$$\begin{aligned}\text{CDI} &= \frac{(\text{C} \times \text{IR} \times \text{EF} \times \text{ED})}{(\text{BW} \times \text{AT})}\\ &= \frac{(2.4\ \text{mg/l} \times 2\ \text{l/day} \times 365\ \text{day/year} \times 10\ \text{year})}{(70\ \text{kg}_{\text{BW}} \times 10\ \text{year} \times 365\ \text{day/year})}\\ &= 0.069\ \text{mg/kg}_{\text{BW}}/\text{day}\end{aligned}$$

Carcinogen

$$\begin{aligned}\text{LADD} &= \frac{(\text{C} \times \text{IR} \times \text{EF} \times \text{ED})}{(\text{BW} \times \text{AT})}\\ &= \frac{(2.4\ \text{mg/l} \times 2\ \text{l/day} \times 365\ \text{day/year} \times 10\ \text{year})}{(70\ \text{kg}_{\text{BW}} \times 70\ \text{year} \times 365\ \text{day / year})}\\ &= 0.010\ \text{mg/kg}_{\text{BW}}/\text{day}\end{aligned}$$

To quantify the risk associated with exposure to carcinogens, the LADD is multiplied by the cancer slope factor (CSF) for the chemical of concern, resulting in an estimate of the incremental excess lifetime cancer risk (IELCR) (Equation [2.6]):

$$\text{IELCR} = \text{LADD} \times \text{CSF} \tag{2.6}$$

The CSF (often also referred to as the "potency value") is derived from an estimate of "unit risk," which describes the incremental risk per unit intake of a carcinogen based on epidemiological

Table 2.8 Cancer data for chemicals of concern.

Chemical	Cancer type	CSF $(mg/kg/day)^{-1}$
Benzo(a)pyrene	Inhalation data not available. May cause lung and skin cancer	0.61
Benzo(b)fluoranthene	Lung, skin	0.61
Benzo(k)fluoranthene		0.061
Benzo(a)pyrene	Forestomach and esophagus (rat)	6.1
Chrysene	Skin	0.0061
Dibenz(a,h)anthracene	Lung, skin (rats)	6.1
Indeno(1,2,3-cd)pyrene	Skin	0.61
1,1-Dichloroethylene (1,1-DCE)	Kidney adenocarcinoma	0.175
Trichloroethylene (TCE)	Liver, kidney	0.006
Vinyl chloride	Liver, immune system	0.29
Benzidine	Urinary bladder, liver	235
Arsenic		1.5

and toxicological data and modeling. Unit risk is typically expressed in terms of a proportion (of a population) affected per unit increase in concentration of a chemical. For example, a unit risk of 3×10^{-5} per mg/l indicates that three excess cancer cases are expected to develop per 100000 people exposed to the chemical of concern at a concentration of 1 mg/l in drinking water. The CSF assumes typical intake rates and average adult body weight (see Table 2.5) and is usually expressed in units of $(mg/kg_{BW}/day)^{-1}$. The values of unit risk and CSF are upper bounds (95th percentile upper confidence limit) on the probability of a response (i.e. development of cancer) over a lifetime – that is, there is only a 5% chance that the probability of developing cancer would exceed the estimated value.

CSFs are highly variable from one chemical to another; Table 2.8 lists CSF values for selected known and suspected human carcinogens.

Table 2.9 provides the CSF and RfD for commonly present disinfection by-product (DBP) chemicals formed when water is chlorinated by reactions with entrained organics.

Additional CSF values and other data on known and suspected carcinogens are available from the following sources:

i) Jurisdictional-specific toxicity values.
ii) Integrated Risk Information System (IRIS):
   a) Contains only CSF factors and RfDs reviewed by EPA.
   b) No external peer review of safe dose estimates (RfDs, SFs).
iii) EPA Criteria Documents.
iv) Health Effects Assessment Summary Tables (HEAST):
   a) These tables summarize all toxicity data.
   b) No internal or external peer review of quality of data is provided.
v) Agency for Toxic Substances and Disease Registry (ATSDR).
vi) Peer-reviewed technical literature.

The IELCR describes the risk of developing cancer from exposure to a chemical of concern, which exists in excess of background risk. This concept illustrates the fact that risk assessment is most often concerned with the extremes or "tails" of a distribution, wherein lies one of the

Table 2.9 Cancer slope factors and reference dose values for commonly occurring DBPs.

DBP name	Last update[a]	Cancer group	Cancer slope factor (oral), CSF ((mg/kg/day)$^{-1}$)	Reference dose RfD (mg/kg/day)
Chloroform (TCM)	19 October 2001	Probable human carcinogen (B2)	NA	1×10^{-2}
Dibromochloromethane (DBCM)	1 March 1991	Possible human carcinogen (C)	8.4×10^{-2}	2×10^{-2}
Bromodichloromethane (BDCM)	1 March 1991	Probable human carcinogen (B2)	6.2×10^{-2}	2×10^{-2}
Bromoform (TBM)	1 March 1991	Probable human carcinogen (B2)	7.9×10^{-3}	2×10^{-2}
Monochloroacetic acid (MCAA)		Not classified as to human carcinogenicity (D)	NA	4×10^{-3}
Dichloroacetic acid (DCAA)	11 September 2003	Likely to be carcinogenic to humans	5×10^{-2}	4×10^{-3}
Trichloroacetic acid (TCAA)	30 September 2011	Suggestive evidence of carcinogenic potential	7×10^{-2}	2×10^{-2}

[a] For oral exposure only; NA, not available.

main challenges of risk assessment. Consider that in North America, the risk of developing cancer is 2 in 5, or 40%. If the IELCR to an individual from exposure to a certain chemical is 1×10^{-5}, that individual's risk of developing cancer is 0.400 01; the incremental excess risk is very small relative to the total risk, which indicates why the development of inputs to carcinogenic risk is so challenging.

The USEPA defines acceptable carcinogenic risk as 1×10^{-4} to 1×10^{-6}; risks greater than 1×10^{-4} are unacceptable and require the implementation of risk management measures. Generally, action is not warranted for risks less than 1×10^{-4}, but based on site-specific concerns, remedial action is sometimes taken when risk is between 1×10^{-4} and 1×10^{-6}.

2.8.3 Generalization to Allow Quantification of Exposure Assessment for Other Scenarios

Estimation of the intake rate involves a measure of exposure expressed as the mass of a substance in contact with a biological exchange boundary (e.g. skin, lungs, gastrointestinal tract) per unit of body weight per unit of time. The intake is estimated for each chemical(s) of concern for each exposure pathway, and hence is different when ingested, versus inhalation pathways, for example, as the target organ is different. As well, differences in receptors influence the potential to have a high IELCR. For example, children do not weigh as much as adults, nor do they breathe at the same rate, influencing the calculations.

As a consequence, exposure assessment must be modified to make the concentrations of chemicals of concern at the point of exposure.

As with most exposure risk calculations, the exposure factor options are premised on most likely exposure (MLE) and reasonable maximum exposure (RME), each of which are as indicated in Figure 2.5.

Additional information is the intake rate, and the exposure assumptions (potentially exposed populations or receptors), meaning residents (adult, child), workers (commercial, industrial, construction, utility, visitors), and activity patterns for the identified receptors may be appropriate to assemble.

Figure 2.5 Hypothetical data response curve.

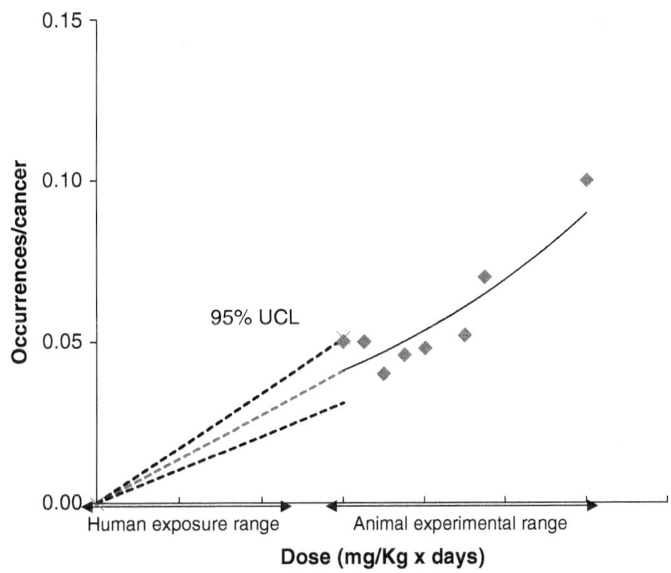

Table 2.10 Additional examples of exposure factors.

Parameter	Description	Units	
ATc	Averaging time (cancer)	Days	25 550
ATnc	Averaging time (noncancer)	Days	
EDr	Exposure duration – residential	Years	90 (5)
Edo	Exposure duration – occupational	Years	25
FI	Fraction of ingested soil coming from impacted area	Unitless	100%
Abs	Absorption	Unitless	50–100%

A – 365 days/year × 70 year lifetime.
Daily intake rate of indoor air.

Potentially useful exposure factors beyond those listed earlier in Table 2.5 are listed in Table 2.10. The listed factors demonstrate that the magnitudes are in response to different receptors, occupations, and so on.

The likelihood of illness is clearly a result of the exposures of the receptors to the contaminants. Two types of receptors are typically employed: an adult and a young child. In a risk assessment, the adult and the child are assigned behavioral patterns and living conditions such that they represent the maximum cumulative intakes for individuals at the point of exposure. The adult receptor is assumed to be in the 20–39-year age bracket, have a 70 kg weight, and a breathing rate of 23 m^3/day. The young child receptor is assumed to be a child about 2–3 years of age, have a 15 kg weight, and a breathing rate of 15 m^3/day.

The potentially exposed populations (receptors) include residents (adult, child), workers (commercial, industrial, construction, utility), and visitors. Different activity patterns for the identified receptors influence the exposure factors. Clearly, differences arise due to features including weight, dermal area, etc. As well, the exposure period differs for different receptors, e.g. construction workers would presumably be involved for a much smaller time (both days/week and the number of years); different activity patterns apply to the identified receptors.

The chemical intake depends on chemical concentration(s) in the media that have an exposure with media sufficient to reach the receptor (air, soil, water, food), the intake rate (e.g. soil ingested) or contact rate (e.g. dermal contact), frequency and duration of intake or contact, and absorption factor (applied dose versus absorbed dose).

2.8.3.1 Construction/Utility Worker

The construction/utility worker receptor is typically assumed to be an adult conducting ground intrusive activity (e.g. working in a trench during maintenance of a subsurface utility) to a depth that could, for example, expose groundwater. The base of the excavation could become flooded with impacted groundwater, and potential incidental ingestion (hand-to-mouth contact), dermal contact, and inhalation of groundwater vapors with a trench could occur. Other considerations might be the construction/utility work being exposed to soil through incidental ingestion and dermal contact with exposed skin on the head, hands, and forearms. In addition, inhalation of volatile chemicals in ambient air emitted from the soil could be possible, or windblown particulates may also occur.

Examples utilizing each of the intake equations for ingestion, inhalation, and soil are described in each of the examples described below. The intent is to demonstrate the calculation sequences as they might be applied in risk assessments.

Example 2.9 Benzidine is a chemical used to produce dyes for cloth, paper, leather, etc. It is a carcinogen known to cause bladder cancer in humans. Calculate the benzidine intake via inhalation for an employee in a factory where the concentration of benzidine in the air is 6×10^{-2} μg/m^3. The average employee works at the factory for 25 years, in 12-hour shifts, for 250 days/year. What is the risk that the employee will develop cancer from exposure to benzidine? The cancer slope factor for benzidine via inhalation is 235 (mg/kg$_{BW}$/day)$^{-1}$:

$$\text{LADD} = \frac{(C \times IR \times EF \times ED)}{(BW \times AT)}$$

$$= \frac{\left(6 \times 10^{-2}\ \mu\text{g/m}^3 \times 0.001\ \text{mg}/\mu\text{g} \times 20\ \text{m}^3/\text{day} \times 12/24 \times 250\ \text{day/year} \times 25\ \text{year}\right)}{\left(70\ \text{kg}_{BW} \times 70\ \text{year} \times 365\ \text{day/year}\right)}$$

$$= 2.10 \times 10^{-6}\ \text{mg/kg}_{BW}/\text{day}$$

To estimate risk for carcinogens, we apply the cancer slope factor to the LADD to obtain the IELCR:

$$\text{IELCR} = \text{Intake} \times \text{CSF}$$

$$= 2.10 \times 10^{-6}\ \text{mg/kg}_{BW}/\text{day} \times 235\left(\text{mg/kg}_{BW}/\text{day}\right)^{-1}$$

$$= 4.94 \times 10^{-4}$$

IELCR > 1×10^{-4}, indicating that risk management is required.

Example 2.10 Determine if a drinking water quality standard for arsenic (a carcinogen) of 0.050 mg/l, the water quality standard in Bangladesh, is sufficiently protective of human health over a lifespan of consumption for 70 years. Assume 2 l/day consumption. The CSF for arsenic is 1.5 (mg/kg/day)$^{-1}$:

$$\text{LADD} = \frac{(C \times IR \times EF \times ED)}{(BW \times AT)}$$

$$= \frac{\left(0.05\ \text{mg/l} \times 2\ \text{l/day} \times 365\ \text{day/year} \times 70\ \text{year}\right)}{\left(70\ \text{kg}_{BW} \times 70\ \text{year} \times 365\ \text{day/year}\right)}$$

$$= 1.43 \times 10^{-3}\ \text{mg/kg}_{BW}/\text{day}$$

$$\text{IELCR} = \text{LADD} \times \text{CSF}$$

$$= 1.43 \times 10^{-3}\ \text{mg/kg}_{BW}/\text{day} \times 1.5\left(\text{mg/kg}_{BW}/\text{day}\right)^{-1}$$

$$= 2.14 \times 10^{-3}$$

The risk of developing cancer from exposure to arsenic at a concentration of 0.05 mg/l is equal to 2.1×10^{-3}, which is far in excess of "de minimis" risk. While having a lower water quality standard would be highly desirable, technologies available for reaching a lower water quality are too costly for the rural populations in Bangladesh.

Carcinogenic effects of different chemicals on the same organ system, even if by different mechanisms or exposure pathways, or by the same chemical over multiple exposures are considered to be additive. In this case, the IELCR values (each describing the risk associated with a single chemical by a single exposure pathway) are summed, resulting in a measure called the cumulative target risk (CTR), which characterizes the risk of developing cancer of a certain organ system associated with exposure to multiple chemicals and/or by multiple pathways:

$$\mathrm{CTR} = \mathrm{IELCR}_1 + \mathrm{IELCR}_2 + \cdots \tag{2.7}$$

The same criteria are applied to the CTR as were outlined for the IELCR: If risk is less than 1×10^{-4}, risk management is generally not warranted, but remedial action is sometimes taken if the risk is between 1×10^{-4} and 1×10^{-6}. If $\mathrm{CTR} > 1 \times 10^{-4}$, risk management is required.

Example 2.11 The table below lists the CSF and calculated LADD for a group of chemicals found in contaminated soil. All of the chemicals are carcinogens. The primary target organ for each chemical is also given. Is the risk of developing cancer of the kidney in an adult exposed to this group of chemicals at the rates listed acceptable? Also comment on the risk of developing liver cancer.

Chemical	CSF $(\mathrm{mg/kg/day})^{-1}$	LADD (mg/kg/day)	Target organ
AB	0.029	4.7×10^{-5}	Kidney
YZ	7.3	9.5×10^{-6}	Liver
DE	0.73	8.6×10^{-5}	Kidney
WX	0.2	5.7×10^{-4}	Liver

$$\begin{aligned}\mathrm{IELCR}_{\mathrm{AB}} &= \mathrm{LADD}_{\mathrm{AB}} \times \mathrm{CSF}_{\mathrm{AB}}\\ &= 4.7 \times 10^{-5}\ \mathrm{mg/kg_{BW}/day} \times 0.029\left(\mathrm{mg/kg_{BW}/day}\right)^{-1}\\ &= 1.36 \times 10^{-6}\end{aligned}$$

$$\mathrm{IELCR}_{\mathrm{YZ}} = 6.94 \times 10^{-5}$$

$$\mathrm{IELCR}_{\mathrm{DE}} = 6.28 \times 10^{-5}$$

$$\mathrm{IELCR}_{\mathrm{WX}} = 1.14 \times 10^{-4}$$

$$\mathrm{CTR}_{\mathrm{kidney}} = \mathrm{IELCR}_{\mathrm{AB}} + \mathrm{IELCR}_{\mathrm{DE}} = 6.4 \times 10^{-5}$$

$$\mathrm{CTR}_{\mathrm{liver}} = \mathrm{IELCR}_{\mathrm{YZ}} + \mathrm{IELCR}_{\mathrm{WX}} = 1.83 \times 10^{-4}$$

$\mathrm{CTR}_{\mathrm{kidney}}$ is within the range of acceptable risk (1×10^{-4} to 1×10^{-6}), although additional investigation may be warranted. $\mathrm{CTR}_{\mathrm{liver}}$ is greater than 1×10^{-4} and risk management would be required.

Example 2.12

(I) *Ingestion of drinking water*

An adult drinks water daily containing 0.05 mg/l of benzene over a 30-year period. Estimate the intake dose of each compound using the appropriate intake equation and exposure parameters. Benzene is a carcinogen used as a building block for plastics, rubber, resins, nylon, polyester, dry cleaning, etc. and is a significant component in gasoline. Benzene enters water from discharge by factories and leaching from landfills.

Assume the adult is home for EF = 350 days/year:

$$\begin{aligned}\mathrm{LADD} &= \frac{\left(0.05\ \mathrm{mg/l} \times 2\ \mathrm{l/day} \times 30\ \mathrm{years}\right)}{\left(70\ \mathrm{kg_{BW}} \times 70\ \mathrm{years} \times 365\ \mathrm{day/year}\right)}\\ &= 5.87 \times 10^{-4}\ \mathrm{mg/(kg_{BW}day)}\end{aligned}$$

$$\text{Benzene CSF} = 0.029\,\text{mg}/(\text{kg}_{\text{BW}} - \text{day})^{-1}$$

An adult consumes water containing 0.05 mg/l of benzene for 30 years. Estimate the dose of benzene exposure and parameters:

$$\begin{aligned}\text{IELCR} &= 5.87 \times 10^{-4} \times 0.029 \\ &= 2.02 \times 10^{-2}\end{aligned}$$

This level of exposure of benzene indicates an unacceptably high human health exposure risk.

(II) *Inhalation of benzene in air*
Assume an individual lives his entire life beside a factory that employs benzene in its fabrication of chemicals. The individual breathes outdoor air containing 0.01 mg/m^3 of benzene. Estimate the dose of benzene using the appropriate intake equations and exposure parameters assuming residential conditions. Benzene is a carcinogen.
Estimate the dose of benzene and the exposure amounts:

$$\begin{aligned}\text{LADD} &= \frac{\left(0.01\,\text{mg/m}^3 \times 20\,\text{m}^3/\text{day} \times 70\,\text{years} \times 365\,\text{days/year}\right)}{\left(70\,\text{kg}_{\text{BW}} \times 70\,\text{years} \times 365\,\text{days/year}\right)} \\ &= 2.85 \times 10^{-3}\,\text{mg}/(\text{kg}_{\text{BW}}\text{day})\end{aligned}$$

$$\begin{aligned}\text{IELCR} &= 2.85 \times 10^{-3} \times 0.019 \\ &= 98.3 \times 10^{-3} \\ &= 0.098\end{aligned}$$

The magnitudes of the chemical intake depend on the chemical concentration in the media that comes in contact with the receptor (air, soil, water, food). The intake rate is calculated (as soil or water ingested or air inhaled) or contact rate (e.g. dermal contact) and reflects the frequency and duration of intake or contact.

The total risk is represented as

$$\text{Risk}_{\text{Total}} = \sum_{i=1}^{N} \text{risk}_i$$

And further, over the number of exposure pathway routes,

$$\text{Risk}_{\text{Total}} = \sum_{j=1}^{P}\sum_{i=1}^{N} \text{CDI}_j \times \text{SF}_i$$

to the same target organ, such that the total cancer risk across multiple pathways:

$$\text{Total exposure risk} = \text{Risk}_{\text{pathway 1}} + \text{Risk}_{\text{pathway 2}}$$

The preceding all relate to a linear (low-dose) cancer risk equation (i.e. relatively low intake rates) in comparison with those experienced by test animals.

Quantifying cancer risks is accomplished by calculating the incremental probability of the occurrence over a lifetime (or a portion thereof, should the situation warrant) as a result of exposure to the potential carcinogen. This is combined with incremental excess lifetime cancer.

As for noncarcinogens, a risk-based screening is a tool used to determine if remedial action is required with respect to contamination by carcinogenic chemicals. The RBSL is calculated using the intake equation. In this case, C = RBSL, and LADD = IELCR/CSF, where IELCR is the acceptable risk level (often 1×10^{-4} to 1×10^{-6}). All other variables are known or assumed, and the equation is solved to obtain RBSL.

Example 2.13 Estimate the allowable concentration (RBSL) of benzene in the drinking water well, assuming that water from the well is used for drinking only. The well, designed to supply water for a period of 30 years, is the only source of drinking water for the farmer who lives on the property. Assume a target risk of 1×10^{-5}. The CSF for benzene is 0.029 $(\text{mg/kg/day})^{-1}$:

$$\frac{\text{IELCR}}{\text{CSF}} = \frac{(\text{RBSL} \times \text{IR} \times \text{EF} \times \text{ED})}{(\text{BW} \times \text{AT})}$$

$$\text{RBSL} = \frac{((\text{IELCR}/\text{CSF}) \times \text{BW} \times \text{AT})}{(\text{IR} \times \text{EF} \times \text{ED})}$$

$$\text{RBSL} = \frac{\left(\left(1 \times 10^{-5} / 0.029 \left(\text{mg/kg}_{\text{BW}}/\text{day}\right)^{-1}\right) \times 70\ \text{kg}_{\text{BW}} \times 70\ \text{year} \times 365\ \text{day/year}\right)}{\left(2\ \text{l/day} \times 365\ \text{day} \times 30\ \text{year}\right)}$$

$$= 0.028\ \text{mg/l}$$

The allowable concentration of benzene in the well is 0.028 mg/l. If the concentration is found to exceed 0.028 mg/l, additional evaluation and risk management may be required.

EPA Guidance (OSWER Directive 9355.0 – 30) indicates that where cumulative carcinogenic site risk to an individual based on RME for both current and future land use is less than 10^{-4}, action is generally not warranted. However, action may be taken at less than 10^{-4} based on site-specific concerns, but all records of decisions for remedial actions taken at sites posing risks within 10^{-4}–10^{-6} must explain why remedial action is warranted.

Example 2.14 Human Health Risk Assessment from Arsenic Exposure

The metalloid arsenic is cited as ubiquitous in nature as it is present in all of the environmental compartments of the Earth. Out of the 98 naturally occurring elements, arsenic ranks as 20th and 12th for its presence in the Earth's crust and in the human body, respectively.

Human arsenic exposures, prevalent through dietary and nondietary sources in Bangladesh, present a major health risk to the Bangladeshi population. A quantitative human health risk assessment as a result of arsenic exposure through food and water intake, tea intake, accidental soil ingestion, and chewing of betel quid, while people meet their desirable dietary intake requirements throughout their lifetime, is described in Joseph et al. (2015a, b). The results show that food and water intake combined make up approximately 98% of the daily arsenic intake. Under an exposure scenario where arsenic concentration in water is at the WHO guideline (0.01 mg/l), food intake is the major arsenic intake pathway ranging from 67 to 80% of the average daily arsenic intake. However, the contribution from food drops to a range of 29–45% for an exposure scenario where water is at the Bangladeshi standard (0.05 mg/l).

Table 2.11 Balanced diet menu plans for Bangladeshi adults.

	Intake amount	
Food type	**Male adult (heavy activity level)**	**Female adult (moderate activity level)**
Rice	540 g	420 g
Lentils	220 g	45 g
Amaranth red leaf	250 g	250 g
Pointed gourd	125 g	NA
Onion	16 g	8 g
Other spices	24 g	12 g
Potato	300 g	150 g
Banana	200 g	100 g
Melon	100 g	NA
Fish	120 g	100 g
Milk	0.25 l	0.375 l
Chicken	50 g	NA
Cowpea	NA	125 g

Source: Adapted from Islam et al. (2013).

From the much longer Joseph et al. (2015a, b) articles, a subset of analyses is utilized herein as an example. The population of Bangladesh is 160 million people, most of whom live in villages. Assume for current purposes that the urban populations (where they extract deep groundwater for their water supply) is 20 million and the remaining 140 million utilize the shallow groundwater where the arsenic levels are much higher (for more information on the reasons for the arsenic and indications as to arsenic concentrations, interested readers are directed to Sorensen et al. (2014) and McBean (2013)).

Table 2.11 forms the basis for assessing As exposure from dietary intake. The estimated average daily As intake from food is summarized in Table 2.12.

The lifetime average daily dose (LADD) of As for the lifetime receptor can be predicted by summing up the annual As intake estimates over the lifespan and amortizing the total lifetime As intake across the total number of lifetime days. The incremental excess lifetime cancer risk (IELCR) is then obtained by multiplying the LADD with an oral slope factor (1.5 mg/kg_{BW}/day, as per US EPA).

Table 2.13 presents the estimated range of apportionment of the average daily As intake (across the various age groups in the population) among the top three As intake pathways for the rural Bangladeshi population, namely, food intake, total water intake, and soil ingestion. Evidently, when As concentration in consumed water is at 0.05 mg/l, water intake turns out to be the major As intake pathway (ranging from 54 to 70% across the various age groups). This is followed by food intake, ranging from 29 to 45% of the average daily As intake. However, under the condition that As concentration in consumed water is at 0.01 mg/l, consumption of food becomes the most important As intake pathway, with its contribution in average daily As intake ranging from 67 to 80% across the various age groups.

Table 2.14 indicates the change in average daily As intake through food between adults with a moderate activity level and adults with a heavy activity level, presenting the results from estimating the incremental risk of cancer cases in the Bangladeshi rural population.

Table 2.12 Estimated average daily As intake from food.

Food type	Average As concentration		Dry matter (%)	Intake amount				Average daily As intake (mg)	
				Male		Female		Male	Female
				Heavy		Moderate			
Rice	0.2	mg As/kg dw	90	540	g	420	g	0.097 20	0.075 60
Lentil	0.1		90	220		45		0.019 80	0.004 05
Cowpea	0.1		90	0		125		0.000 00	0.011 25
Amaranth red leaf	0.4		10	250		250		0.007 50	0.007 50
Pointed gourd	0.2		10	125		0		0.002 50	0.000 00
Onion	0.1		15	16		8		0.000 24	0.000 12
Other spices	0.2		100	24		12		0.004 80	0.002 40
Potato	0.4		20	300		150		0.012 00	0.006 00
Banana	0.01		30	200		100		0.006 00	0.003 00
Melon	0.01		15	100		0		0.001 50	0.000 00
Chicken	0.2		30	50		0		0.000 03	0.000 00
Fish	0.4		30	120		100		0.000 72	0.000 60
Milk	0.02	mg As/l	NA	0.25	l	0.375	l	0.008 75	0.013 13
						Total		*0.16*	*0.12*

Table 2.13 Estimated distribution of LADD/lifetime average daily As intake among the top three As intake pathways.

		At 0.01 mg/l As concentration (%)	At 0.05 mg/l As concentration (%)
Food intake	Male	68–77	32–40
	Female	67–80	29–45
Water intake	Male	23–31	59–70
	Female	19–33	54–70
Soil ingestion	Male	0.1–0.6	0.1–0.3
	Female	0.1–0.2	0.1–0.4

Table 2.14 Incremental number of cancer cases estimated for the Bangladeshi population due to As exposure.

		As concentration of 0.01 mg/l		As concentration of 0.05 mg/l	
		Moderate activity	Heavy activity	Moderate activity	Heavy activity
Composite LADD ($mg/kg_{BW}/day$)	Male	0.0033	0.0037	0.0068	0.0072
	Female	0.0029	0.0033	0.0062	0.0066
IELCR	Male	0.0049	0.0055	0.0102	0.0108
	Female	0.0043	0.0050	0.0092	0.0099
Relative to de minimis risk	Male	4870	5516	10 196	10 841
	Female	4320	5007	9234	9920

Excess cancer risks calculated as follows, for male with 0.01 mg/l concentration as example.

Table 2.14 lists various districts within the divisions of Bangladesh, the associate populations, and the average arsenic concentrations in the groundwater. Assuming 4 l/day of water consumption, the individual columns of Table 2.15 indicate the following:

a) Column V lists the chronic daily intake (CDI) of arsenic from water.
b) Column VI lists the CDI of arsenic from food.
c) Column VII lists the IECLR as a result of water and food ingestion.
d) Column VIII the number of cases of cancer (multiplication of values in column III by values in VII).

As a summary of the results,

a) The total population of the country is 135 million (as of 2011).
b) The total number of cancer cases is, from summing over column VIII, 1.2 million. (Note that since Dhaka draws its water from deep wells where there is no arsenic, there is no exposure to the residents of Dhaka from water, only food.)

As a result of the above, the arsenic-caused cancer cases are approximately 1% of the country population.

2.9 Ecological Risk Assessment

An ecological risk assessment is an explicit effort to estimate the probability and magnitude of the effects of ecological exposure to contamination. Ecological risk assessment accomplishes this by estimating the likelihood of undesirable ecological effects occurring as a result of human activities. The evaluation of the ecological risks may be appropriate to complete in many respects, including with regard to wildlife, aquatic organisms, and plants, including threatened and endangered species. The assessment must be based on peer-reviewed methods and models and employ the best available scientific and commercial data. As a result the process of ecological risk assessment is scientifically complex as it entails substantial data assembly on the feeding habitats and life cycle attributes of relevant ecological receptors, allowing development of informed environmental decisions as suggested in Figure 2.6.

Changes in ecology typically considered as undesirable are those that alter important structural or functional characteristics or components of ecosystems.

Stressor characterizations may be biological, physical, or chemical. Ecological effects include any impacts upon any level of the ecosystem derived from hazard assessment (acute/chronic toxicity) and consideration of biotransformations, biodegradation, reproductive effects, predator/prey interactions, production, and community biomass.

An indication of potential effort necessary to complete an ecological risk assessment is provided by considering the interactions implicit in Figure 2.6, which, while only a schematic, does indicate the attributes and interactions between species that are relevant to an ecological risk assessment.

Assessment endpoints for ecological risk assessment are set by ecological relevance and policy goals/society values (i.e. project ecosystem structure/function). Measurement endpoints are the measurable factors that respond to stressors and describe characteristics of ecosystem important to assessment endpoint. Hence, ecological risk assessments involve the integration of biological requirements and habits of species, geographic features that influence exposure, incident information (as available), and so on. As apparent, the process becomes very in depth and represents a substantial undertaking, very quickly. Guidance is provided in a number of technical

Table 2.15 Calculation sequence for cancer risks to populations of Bangladesh.

I	II	III	IV	V	VI	VII	VIII
District/Zila	**Average arsenic concentration (ug/l)**	**Population (2011 census)**	**Division**	**CDI (mg/kg/day) (water)**	**CDI cumulative (mg/kg/day)**	**IECLR**	**Number of cases**
Barguna	1	892 781	**Barisal**	6.67E−05	2.40E−03	3.60E−03	3210
Barisal	92	2 324 310		6.13E−03	8.46E−03	1.27E−02	29 507
Bhola	10	1 776 795		6.67E−04	3.00E−03	4.50E−03	7987
Jhalakati	23	682 669		1.53E−03	3.86E−03	5.80E−03	3956
Patuakhali	3	1 535 854		2.00E−04	2.53E−03	3.80E−03	5829
Pirojpur	30	1 113 257		2.00E−03	4.33E−03	6.50E−03	7231
Brahmanbaria	101	2 840 498	**Chittagong**	6.73E−03	9.06E−03	1.36E−02	38 617
Chandpur	366	2 416 018		2.44E−02	2.67E−02	4.01E−02	96 870
Chittagong	32	7 616 352		2.13E−03	4.46E−03	6.70E−03	50 991
Comilla	142	5 387 288		9.47E−03	1.18E−02	1.77E−02	95 328
Cox's Bazar	3	2 289 990		2.00E−04	2.53E−03	3.80E−03	8691
Feni	54	1 437 371		3.60E−03	5.93E−03	8.90E−03	12 785
Lakshmipur	179	1 729 188		1.19E−02	1.43E−02	2.14E−02	36 996
Noakhali	162	3 108 083		1.08E−02	1.31E−02	1.97E−02	61 214
Dhaka	41	12 043 997	**Dhaka**	0.00E + 00	2.33E−03	3.50E−03	42 094
Faridpur	140	1 912 969		9.33E−03	1.17E−02	1.75E−02	33 467
Gazipur	4	3 403 912		2.67E−04	2.60E−03	3.90E−03	13 258
Gopalganj	187	1 172 415		1.25E−02	1.48E−02	2.22E−02	26 022
Jamalpur	14	2 292 674		9.33E−04	3.26E−03	4.90E−03	11 223
Kishoreganj	52	2 911 907		3.47E−03	5.80E−03	8.70E−03	25 319
Madaripur	191	1 165 952		1.27E−02	1.51E−02	2.26E−02	26 345
Manikganj	24	1 392 867		1.60E−03	3.93E−03	5.90E−03	8211

Munshiganj	189	1 445 660		1.26E−02	1.49E−02	2.24E−02	32 376
Narayanganj	48	2 948 217		3.20E−03	5.53E−03	8.30E−03	24 455
Narsingdi	41	2 224 944		2.73E−03	5.06E−03	7.60E−03	16 898
Netrokona	40	2 229 642		2.67E−03	5.00E−03	7.50E−03	16 711
Rajbari	49	1 049 778		3.27E−03	5.60E−03	8.40E−03	8813
Shariatpur	151	1 115 824		1.01E−02	1.24E−02	1.86E−02	20 749
Sherpur	22	1 358 325		1.47E−03	3.80E−03	5.70E−03	7736
Tangail	20	3 605 083		1.33E−03	3.66E−03	5.50E−03	19 810
Bagerhat	156	1 476 090	**Khulna**	1.04E−02	1.27E−02	1.91E−02	28 186
Chuadanga	79	1 129 015		5.27E−03	7.60E−03	1.14E−02	12 865
Jessore	70	2 764 547		4.67E−03	7.00E−03	1.05E−02	29 014
Jhenaidah	46	1 771 304		3.07E−03	5.40E−03	8.10E−03	14 339
Khulna	35	2 318 527		2.33E−03	4.66E−03	7.00E−03	16 218
Kushtia	104	1 946 838		6.93E−03	9.26E−03	1.39E−02	27 051
Magura	28	918 419		1.87E−03	4.20E−03	6.30E−03	5781
Meherpur	116	655 392		7.73E−03	1.01E−02	1.51E−02	9893
Narail	88	721 668		5.87E−03	8.20E−03	1.23E−02	8873
Satkhira	133	1 985 959		8.87E−03	1.12E−02	1.68E−02	33 354
Bogra	18	3 400 874	**Rajshahi**	1.20E−03	3.53E−03	5.30E−03	18 008
Dinajpur	3	2 990 128		2.00E−04	2.53E−03	3.80E−03	11 348
Gaibandha	22	2 379 255		1.47E−03	3.80E−03	5.70E−03	13 550
Joypurhat	1	913 768		6.67E−05	2.40E−03	3.60E−03	3285
Kurigram	22	2 069 273		1.47E−03	3.80E−03	5.70E−03	11 785
Lalmonirhat	1	1 256 099		6.67E−05	2.40E−03	3.60E−03	4516
Naogaon	6	2 600 157		4.00E−04	2.73E−03	4.10E−03	10 648
Natore	1	1 706 673		6.67E−05	2.40E−03	3.60E−03	6135
Nilphamari	2	1 834 231		1.33E−04	2.46E−03	3.70E−03	6777

(Continued)

Table 2.15 (Continued)

I	II	III	IV	V	VI	VII	VIII
District/Zila	**Average arsenic concentration (ug/l)**	**Population (2011 census)**	**Division**	**CDI (mg/kg/day) (water)**	**CDI cumulative (mg/kg/day)**	**IECLR**	**Number of cases**
Pabna	32	2 523 179		2.13E−03	4.46E−03	6.70E−03	16 893
Panchagarh	3	987 644		2.00E−04	2.53E−03	3.80E−03	3748
Rajshahi	8	2 595 197		5.33E−04	2.86E−03	4.30E−03	11 146
Rangpur	8	2 881 086		5.33E−04	2.86E−03	4.30E−03	12 374
Sirajganj	31	3 097 489		2.07E−03	4.40E−03	6.60E−03	20 428
Thakurgaon	1	1 390 042		6.67E−05	2.40E−03	3.60E−03	4997
Habiganj	22	2 089 001	**Sylhet**	1.47E−03	3.80E−03	5.70E−03	11 897
Moulvibazar	20	1 919 062		1.33E−03	3.66E−03	5.50E−03	10 545
Sunamganj	48	2 467 968		3.20E−03	5.53E−03	8.30E−03	20 472
Sylhet	22	3 434 188		1.47E−03	3.80E−03	5.70E−03	19 558
		135 647 693					1 196 381

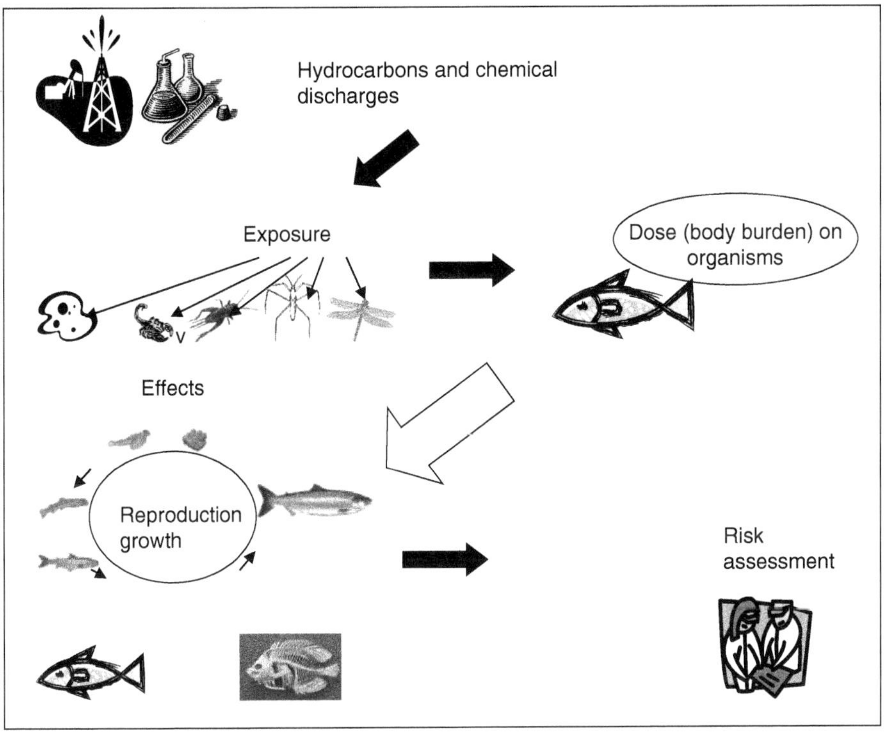

Figure 2.6 Framework for ecological risk assessment.

documents such as EPA's "Guidelines for Ecological Risk Assessment" and "Risk Characterization Handbook" (US EPA, https://www.epa.gov/risk/risk-characterization-handbook).

Typically, the ecological risk assessment procedure is a tiered assessment protocol involving the following:

i) A baseline assessment involving development of elements such as a low spatial and temporal resolution, general assessment endpoints, and identification of refinements needed. The next level would involve species/habitat-specific endpoints and probabilistic tools and methods to address magnitude and likelihood of risk, including reflection of measures of uncertainty.
ii) It is apparent that an ecological risk assessment involves much more depth than can be covered in this book, but indications include assessment of exposure and effects data, toxicity endpoints, and exposure modeling and field data/monitoring. Assessors must estimate the risk through integration of exposure and stressor–response profiles. The undertakings may include the evaluation of runoff and spray drift in an application for spraying of regions for pest control.

Incident data are used in risk assessments, as available, to provide evidence that modeled risk predictions are supported by actual effects in the field. An absence of evidence cannot be used as evidence of absence. An indication of the zone of coverage is depicted in Figure 2.7. It is important to consider how important data are to ecological risk assessment, as this represents a basic problem; the sparseness and uncertainty of available scientific data continues as a challenge. Hence, as apparent, there is need to incorporate a large number of features of the ecological system. In many respects, ecological risk assessment is very different from human

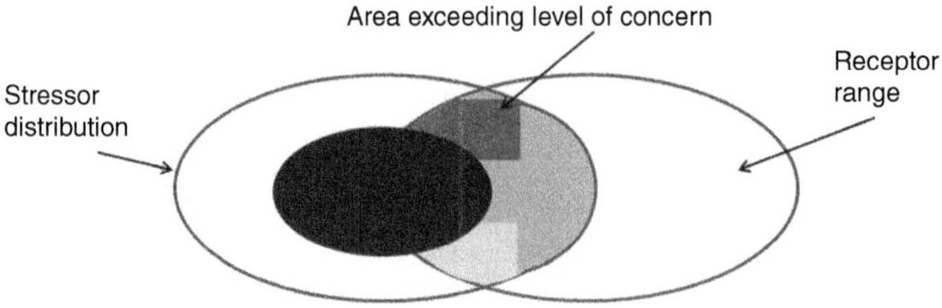

Figure 2.7 Schematic depiction of dose–response curve for noncarcinogenic effects.

risk assessment, although some of the dimensions are similar. The need is to integrate and synthesize effects and exposure characterizations into an overall conclusion and assess the risk. There are guidelines that have been published by the US EPA to improve the quality and consistency of ecological risk assessments, where these guidelines draw from a wide range of source documents including peer-reviewed issue papers and case studies. A major theme of the guidelines is the interaction among risk assessors, risk managers, and interested parties at the beginning and end of the risk assessment process (http://www.epa.gov/raf/publications/guidelines-ecological-risk-assessment).

2.10 Issues of Uncertainties in Risk

"Uncertainty" is implicit in all elements of risk assessment and management (data gathering, risk assessment, and risk management). If a very high degree of certainty is sought, costs of assembling and understanding the data are likely to be very high. Complete elimination of risk may mean foregoing the technology. As explained earlier in this chapter, reducing the risk to zero is generally very difficult, if not possible. In some circumstances this would correspond to foregoing the technology/activity. Hence, we typically define "acceptable risk" that is a compromise where risk is reduced to a level on the order of other familiar risks.

A recurring issue is that one of the concerns related to environmental risk is the ability to characterize low levels of risk. In many respects, the identification of many chemicals in our water supply is less related to an expanded use and abuse of chemicals but more a result of an ever-improving capability to quantify smaller and smaller concentrations using improved analytical procedures. Thus, we are not necessarily being exposed to larger risks than was the situation one or two decades ago; instead we are now more able to identify the chemicals that are present. We do, however, need data to allow quantification of the elements of exposure (which is the lead-in to Chapters 4 through 12).

A difficulty with assessments of many environmental risks is the availability of only minimal data. A major concern associated with the estimation of risk is the degree of confidence in the estimates of risk, given the paucity of available data. Alternative means of characterizing these dimensions are described in Chapters 4 through 12.

2.11 References

American Cancer Society (2013). Known and Probable Human Carcinogens. https://www.cancer.org/cancer/cancer-causes/general-info/known-and-probable-human-carcinogens.html (accessed 3 November 2016).

Asante-Duah, D.K. (1993). *Hazardous Waste Risk Assessment*. Boca Raton, FL: Lewis Publishers.
Bromley, A. (February 1991). *Chemecology*. Chemical Manufacturers Association.
Burgman, M.A. (2005). *Risks and Decisions for Conservation and Environmental Management*. Cambridge: Cambridge University Press.
California Department of Toxic Substances Control (CDTSC) (26 November 1991). Guidance for Risk Assessment of Town Gas Sites.
International Life Sciences Institute (ILSI) (2000). *Principles of Risk Assessment of Food & Drinking Water Related to Human Health*. Washington, DC: ILSI Press.
Islam, S., Lechner, A., Ferrari, U. et al. (2013). Social and economic impact of diabetics in Bangladesh: protocol for a case–control study. *BMC Public Health* 13: 1217.
Joseph, T., Dubey, B., and McBean, E. (September 2015a). A critical review of arsenic exposures for Bangladeshi adults. *Science of the Total Environment* 527–528: 540–551.
Joseph, T., Dubey, B., and McBean, E.A. (September 2015b). Human health risk assessment from arsenic exposures in Bangladesh. *Science of the Total Environment* 527–528: 552–560.
McBean, E. (2013). Risk characterization for arsenic-impacted water sources, including ground-truthing. *Stochastic Environmental Research and Risk Assessment* 27 (3): 705–711.
Ontario Ministry of Environment and Energy (1996). *Rationale for the Development and Application of Generic Soil, Groundwater, and Sediment Criteria for Use at Contaminated Sites in Ontario*. Calgary, AB: Ontario Ministry of Environment and Energy, Standards Development Branch.
Slovic, P., Fischhoff, B., and Lichtenstein, S. (April 1979). Rating the risks. *Environment* 21 (3): 14–20.
Sorensen, I., McBean, E., and Rahman, M. (2014). Retrofitting arsenic-iron removal plants for performance change in rural Bangladesh. *Journal of Water, Sanitation and Hygiene for Development* 0.4/3: 401–409.
Stallones, R.A. (1988). Epidemiology and environmental hazards. In: *Epidemiology and Health Risk Assessment* (ed. L. Gordis), 3. New York: Oxford University Press.
US EPA (1997). Integrated Risk Information System: Internet database system. https://www.epa.gov/iris (accessed 16 June 2016).
US EPA (2012). An Integrated Risk Management Model for Source Water Protection, doi: 10.3390/ijerph9103724.
US EPA https://www.epa.gov/sites/production/files/2014-11/documents/ecorisk-overview.pdf (accessed September 2016).

2.12 Problems

2.1 In a particular groundwater supply well, arsenic concentrations have been determined as 112 μg/l, which is considerably in excess of the Bangladeshi regulation of 50 μg/l. You are now faced with trying to estimate how much of the water that residents will consume, should come from a rainwater harvest system (so the resulting water intake) only reaches de minimis risk levels. The basic information for a typical resident in the village is as follows:

- Cancer slope factor for arsenic is 1.5 $(mg/kg\ day)^{-1}$.
- He/she consumes 3 l of water per day (it is a hot climate).
- Assume the residents weigh 60 kg.
- They live in the village 365 days/year (i.e. no vacations).
- They will likely live in the village for 30 years.
- Life expectancy of 63 years.

- Tip: Remember to have the correct units for all aspects of the intake equation and also that averaging time is the period of time for which the exposure is averaged (so in this case you are not using the typical AT = 70 years × 365 days/year, but rather AT = 63 years × 365 days/year).
  a) Specify five of the most important assumptions you need to make to answer this question.
  b) What is the risk of cancer, if the villagers were to consume the water from the well, for the time they live in the village?

2.2 What is the percentage of an individual's water consumption in Question 2.1 that must be consumed from a rainwater harvester to attain de minimis risk?

2.3 Workers are helicoptered in for work to remediate a former gold mine that is leaking cyanide into the air. Because of the dangerous conditions, the workers only work 6 months/year. It is expected that the remediation work at the mine site will take 6 years.
- It is noted that the climate is really hot and they are going to be working hard; hence they will breathe heavily (32 m^3/day).
- The cyanide concentration in the air at the mine site is expected to be 0.071 mg/m^3.
- Assume the cancer slope factor for cyanide is 0.61 $(mg/kg\ day)^{-1}$.
- Assume the worker weighs 70 kg and has an average life expectancy of 70 years.

Do you believe there is significant risk to the workers for this exposure scenario?

2.4 Visit the EPA website at https://www.epa.gov (the website is huge, so this represents only a small portion of the information on the website). In the document, reference is made to IRIS and HEAST. Relevant to these two databases are discussions of the capabilities of the databases that are used on a regular basis for risk assessments. Find on these sites the cancer slope factor for two chemicals (benzene, and vinyl chloride) and provide a brief paragraph that demonstrates the differences between EPA's IRIS and Tier 3, Other Toxicity Values.

2.5 An adult lives near one of the nation's largest nylon manufacturing plants and has been breathing air containing 0.005 mg/m^3 of benzene (a known carcinogen) daily for 40 years.
a) Estimate the dose of benzene using the appropriate intake equations and exposure parameters. Assume residential conditions and that the resident goes on vacation (i.e. leaves town) for 2 weeks every year.
b) Estimate the risk associated with the dose of benzene estimated in part (a), given that the inhalation cancer slope factor for benzene is 0.029 $(mg/kg\ day)^{-1}$.

2.6 Estimate the risk-based screening level (RBSL) of toluene (a noncarcinogen) in a drinking water well, assuming the water from the well is used for ingestion only. The reference dose for toluene is 0.2 mg/kg day. Assume residential conditions, an acceptable hazard quotient (HQ) of 1, exposure duration of 20 years, and an exposure frequency of 365 days/year.

2.7 IRIS is not the only source of toxicology information. What are some of the other sources of information related to emerging contaminants?

2.8 Dioxins and furans are notorious as sources of environmental risk to humans and the environment. List three examples of major source types that contain dioxin and furan emissions that cause significant concern to humans and the environment.

2.9 There are a number of countries in the world where there are major sources of nonanthropogenic arsenic in groundwater supplies. Name six of them.

2.10 That a carcinogenic risk for an individual may have arisen from having worked as a gas station/pumping of gas occurred over a 4-year period while the individual was a high school student. At the gas station, he worked 40 h a week. Ten years after his graduation at age 18, he then took a job in an industrial facility on a shift line also working 40 h a week making some industrial machinery. In both situations, he was breathing in benzene as per the concentrations as indicated in table below.

How many years can he continue to work on the industrial facility before he reaches 10^{-5} cancer risk?

Assume the CSF for benzene is 0.6 $(\text{mg/kg day})^{-1}$.

Benzene concentration (ug/m^3)	Activity
3×10^{-2}	Benzene concentration in air at gas station
4×10^{-1}	Benzene concentration in air at industrial facility

2.11 We typically refer to "de minimis" risk as acceptable. Why is that considered an acceptable appropriate action level? What is the magnitude of this "de minimis" risk?

2.12 Most people perceive of risk in one of three ways. The first of these is (i) an activity that is 100% safe. What are the other two ways? This is a simplistic representation of risk – why is it so frequently employed?

2.13 Is "alcohol drunkenness/intoxication" an acute or chronic risk. Indicate your reason for your answer.

2.14 Both toxicologists and epidemiologists really struggle with assessing whether a chemical is a carcinogen. Explain (briefly) indicating the most important reasons why you understand the challenges in their fields.

2.15 The figure below is a schematic of percentage response to different exposure concentrations. Label A, B, and C with the acronyms for (i) RfD, (ii) LOAEL, and (iii) NOAEL.

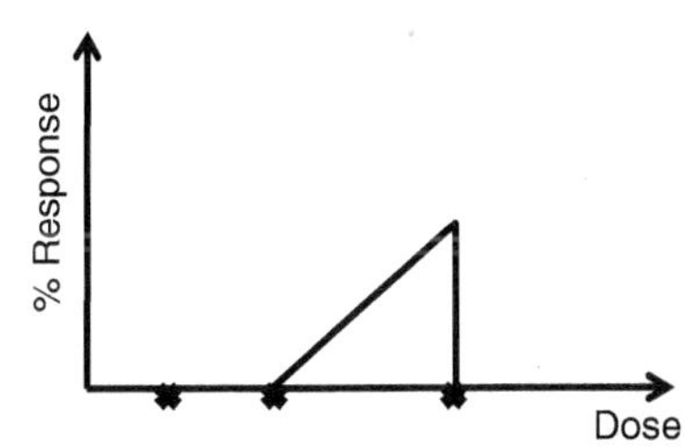

2.16 We frequently use the term "de minimis risk" as a guideline. List several examples (just approximate magnitudes are all that is necessary, mostly concept) of what constitutes de minimis risk and also indicate why we utilize this strategy for risk assessment.

2.17 Although human health risk assessment is difficult, it is not considered as challenging as an ecological risk assessment. What are the most important reasons for this statement?

2.18 In northern Canada, are trihalomethanes (THMs) real concerns – why? Explain your answer.

2.19 Should a landfill slope failure be considered a voluntary risk? Explain your answer.

2.20 Assessing water issues related to risk, most risk assessments include "multiple pathways." What does that mean, and why is this approach typically employed?

2.21 Approximately 80% of the time, there is an environmental factor contributing to waterborne disease outbreaks in North America. What is that environmental factor?

2.22 By relatively simple calculations, we realize that the water supply infrastructure in Toronto is aging virtually with calendar time. Why is this occurring?

2.23 Although human health risk assessment is difficult, it is not as difficult as ecological risk assessment. What are the most important reasons for this statement?

2.24 What is the major difference between epidemiology and toxicology results when they develop estimates of risk from data?

2.25 What does UF1, UF2, UF3, etc. mean, and why are these used?

2.26 Pharmaceuticals are being poorly treated in the wastewater treatment systems at three upstream communities, namely, A, B, and C, such that the drinking water supply at community D is potentially being impacted. What is the percent removal of pharmaceutical Ibuyxyz you need to ensure you remove in the water treatment plant at location D to protect the safety of the citizens? You need to make a series of assumptions to solve this questions – list them as you proceed.

CSF for Ibuyxyz = 0.61 mg/(kg day).

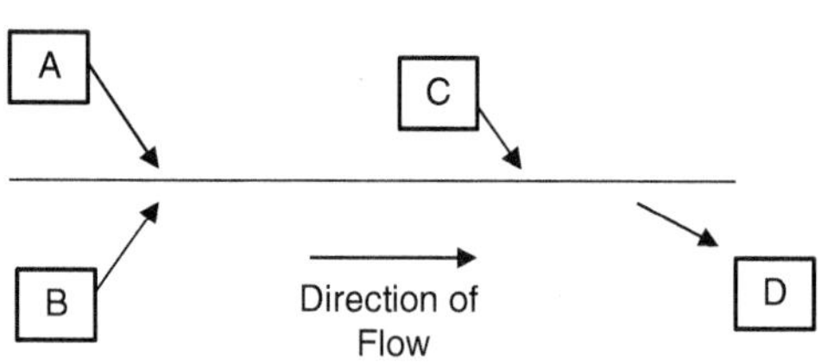

2.27 One of the more difficult aspects of risk assessment is in relation to chronic toxicity and the availability of data, as opposed to acute toxic levels. What is the reason there is much better understanding of toxic concentrations as opposed to chronic concentrations in terms of causing death?

2.28 You have been asked by your client to estimate the impact of their industrial activities on health in their workplace environment. In an assessment, you have estimated that the probability of cancer-causing death is 3×10^{-8}. Your client does not understand mathematics and asks you to prepare a brief statement as to what that really means.

2.29 A consultant has provided you with the following information: project A, involving the remediation of a site contaminated with benzene, a known carcinogen, has an environmental risk of 7×10^{-5} for remediation site workers. Indicate precisely what that means.

2.30 If 31 persons die each year in Canada due to lightning strikes, calculate the probability of any person dying due to lightning their lifetime? Briefly describe two of the major assumptions you are making in this calculation.

3

Factors Influencing the Assessment and Management of Risk

Everyone is entitled to his own opinion but not to his own facts.
—Daniel Patrick Moynihan

3.1 Background for Some of the Issues Influencing Risk Assessment and Management

Many dimensions associated with risk assessment and ultimately risk management cannot be readily quantified. This greatly increases the challenges associated with risk assessment and management of risks. These dimensions are many and varied, but three of the most critical issues are:

1) *Public perception*: Public perception of risk and the reality of the actual risk are not necessarily the same, but public consultation in many applications of risk assessment and management is an important consideration. Public perception of risk and its effect on risk assessment and management are described in Section 3.2.
2) *Challenges in quantification*: For events that happen frequently (e.g. motor vehicle accidents), we have sufficient data to quantify risks, but for other events, such as dam failures, we often lack the information necessary. As a result, it is sometimes necessary to employ more qualitative risk assessment practices than are described in Chapter 2. Qualitative risk assessment procedures as frequently used in practice are described in Section 3.3.
3) *Continued evolution of risk assessment*: Our understanding of environmental risks associated with many chemicals is reasonably well established, but assessment procedures for other types of risk (e.g. with microbial risk) procedures are still evolving as new information and procedures are developed. Section 3.4 briefly describes some of the new and evolving risk assessment procedures.

3.2 Issues of Perception Versus Reality in Risk Assessment

3.2.1 Influential Roles of the Public

Once a risk has been identified and characterized, political, social, economic, and engineering factors are considered. Together with risk characterization, these considerations are used to develop, analyze, and compare management options. However, "reality" and "perception" may be enormously different. Recall from the discussions in Chapter 2 that many aspects of risk

Risk Assessment: Procedures and Protocols, First Edition. Edward A. McBean.

Companion website: www.wiley.com/go/McBean/RiskAssessProcedure_1e

directly involve the public. Members of the public have legitimate rights and opportunities to be involved in decision-making on many aspects of risk.

The need for public involvement in this process exists because many environmental risks fall within the public sphere (e.g. the siting of a landfill). Hence, irrespective of reality, public perception about the risks involved is what "drives" the decision-making process to some degree, and, as will be demonstrated in this subsection of the book, the perception by the public about risk, and the reality of risk, may be very different. A useful characterization for some risk issues is depicted schematically below. The term "outrage" represents public sentiment and may be too extreme and hence inappropriate in some cases, but the magnitude of the difference between calculated and perceived risk is intended to be reflected by "outrage." It is very useful to keep in mind the concept of the "outrage component":

$$\text{Perceived risk} = \text{calculated risk} + \text{outrage}$$

As will be apparent in subsequent sections, risk communication is complex and often cross-disciplinary. Many challenges exist in communicating risk, especially to the public. One of the most effective approaches is to relate a particular risk to the risks of other activities/exposures with which members of the public are familiar.

3.2.2 Differences in Risk Characterization: Public Perception Versus the Reality of Risk

Although the technical aspects of risk assessment may be well based and defensible, the issues must include input from the public. Hence, risk assessment and management must involve assurance that the public accepts the decision-making (the issues are ones of decision-making with public input and, in many situations, decision-makers are politicians or bureaucrats). As will be seen in the paragraphs to follow, in many respects, it is the perception of the risk that governs, not necessarily the detailed science and engineering components. Basically, and generally, members of the public act semiautomatically to minimize their risks. We have already demonstrated (in Chapter 2) that we cannot make "risk" equal to zero, and a better understanding of the risks will assist in identifying strategies for its management. Much is gained through experience; knowledge of past mistakes and actions act to prevent new mistakes.

One way to address risk characterization is to consider the use of categories of risk. A means of classification of risks is as listed in Table 3.1.

The primary emphasis in this book is on the risks in the fourth category of Table 3.1. Some (large but infrequent events causing risk and alternative ways of assessing and managing risk) are considered in Section 3.3.

Table 3.1 Categories of risk.

Category	Description	Examples of risks
1	Familiar, high occurrence risks	Vehicular risks
2	Low probability events	Events whose consequences are so large that they must be taken seriously, e.g. earthquakes
3	Events with low probability that they have not occurred but whose consequences deserve attention	Nuclear generation station failure
4	Events that are hard to evaluate because they show up as increases in naturally occurring hazards	Environmental contamination

Consider the situation for siting a nuclear power plant or the smoking of cigarettes. A wide spectrum of members of the public will be interested in whether the human health exposure risks will be large with either type of human health exposure risk. Regardless, for both situations of exposure risks, there is a common element wherein the perception of risk by those involved is highly relevant to risk assessment. Quite possibly, the risk as perceived by the public is very different from the actual risk; however, given the necessity to involve the public in many aspects of decision-making on risk, the perceived risk by the public will be fundamentally important in influencing the decision-making.

The extent to which "perceived" and "actual" risks involved with many activities and concerns has been the subject of considerable research. When individuals in the public assess their perceived risk, they seldom have statistical evidence available. Slovic et al. (1979) have compiled interesting work in risk perception and have provided important insights into perceived risks as opposed to the reality of risks. Slovic et al. (1979) reported on perceptions from four sectors of people who were asked to rank the risk of dying (across US society as a whole) as a consequence of activities or technologies as per Table 3.2. Respondents were asked to order the items from least to most risky. Table 3.2 shows how three groups (the League of Women Voters [LOWV], college students, and experts) ranked the relative "riskiness" of 30 activities and technologies and, as apparent from the tabulated values, experts' judgments of risk differed markedly from the judgments of laypersons. This identifies the differences in perception of risk from one group to another and, assuming that the experts' opinions are relatively close to the truth, the differences between "perceived" and "real" risk. For example, the LOWV responses indicate their membership felt that the top five causes of death were, in order, (i) nuclear power, (ii) motor vehicles, (iii) handguns, (iv) smoking, and (v) motorcycles. Alternatively, the highest risk of death as ranked by a group of college students indicated that the highest death rates were attributed to (i) nuclear power, (ii) hand guns, (iii) smoking, (iv) pesticides, and (v) motor vehicles. Further, the rank ordering of the highest risks of death by experts is indicated in the fourth column of Table 3.2, and is evidently quite different from the other groups. Experts' judgments were sufficiently close to the calculated frequencies that it can be concluded that experts view the risk of an activity or technology as being synonymous with the annual number of fatalities.

As the results in Table 3.2 clearly demonstrate, there are substantial differences in the understanding of fatality rates arising from activities and technologies, from one group to another. Perception and reality are substantially different since different groups perceive the death rate from individual sources of risk as substantially different. There are many similarities between groups (e.g. motorcycles, cars, and handguns are highly risky), but there are differences as well. Students view bicycles as being less risky than the LOWV, whereas they view surgery and X-rays and nuclear power and police work as much less risky. It is likely that people consider the low death rate associated with nuclear power but consider it to be a high-risk technology because of the potential for disaster.

Risk judgments of laypeople are different in that there is only a moderate relationship to the annual number of fatalities. As a result, for laypeople, risk may not be synonymous with fatalities. As an example, the perceived risk of nuclear power was disproportionately higher than the annual number of fatalities. Both groups substantially underestimated the risk of high probability events and overestimated the risk of low probability events.

As an alternative means of illustrating the differences between perceived and real risk, Slovic et al. (1979) developed a figure depicting the differences between perceived or judged rates of death and actual rates of death arising from a number of causes in the United States for 41 common causes of death as included herein as Figure 3.1 (note that Figure 3.1 has been plotted using logarithmic axes). The differences between perceived and real risks are demonstrated by the deviation of the plotted points from the 45° line on Figure 3.1.

Table 3.2 Rank ordering of risks of death in the United States as a consequence of activity or technology.

	LOWV	College students	Experts
Nuclear power	1	1	20
Motor vehicles	2	5	1
Handguns	3	2	4
Smoking	4	3	2
Motorcycles	5	6	6
Alcohol beverages	6	7	3
General (private) aviation	7	15	12
Police work	8	8	17
Pesticides	9	4	8
Surgery	10	11	5
Firefighting	11	10	18
Large construction	12	14	18
Hunting	13	18	13
Spray cans	14	13	23
Mountain climbing	15	22	29
Bicycles	16	24	15
Commercial aviation	17	16	16
Electric power	18	19	16
Swimming	19	30	10
Contraceptives	20	9	11
Skiing	21	25	30
X-rays	22	17	7
High school and college football	23	26	27
Railroads	24	23	27
Food preservatives	25	12	14
Food coloring	26	20	21
Power mowers	27	28	28
Prescription antibiotics	28	21	24
Home appliances	29	27	22
Vaccinations	30	29	25

Source: This information is from a study appearing in Slovic (2000).

If perception of the rate of death or fatality was accurate, all of the plotted points would fall along the 45° line; however, as apparent from Figure 3.1, the individual points are scattered about a curved line. In general, rare causes of death were overestimated and common causes of death were underestimated. Further, very interesting findings include the following: accidents were judged to cause as many deaths as diseases, whereas diseases actually take about 15 times as many lives. Homicides were incorrectly judged to be more frequent than diabetes and stomach cancer. Homicides were also judged to be about as frequent in causing death as strokes, although strokes claim about 11 times as many lives.

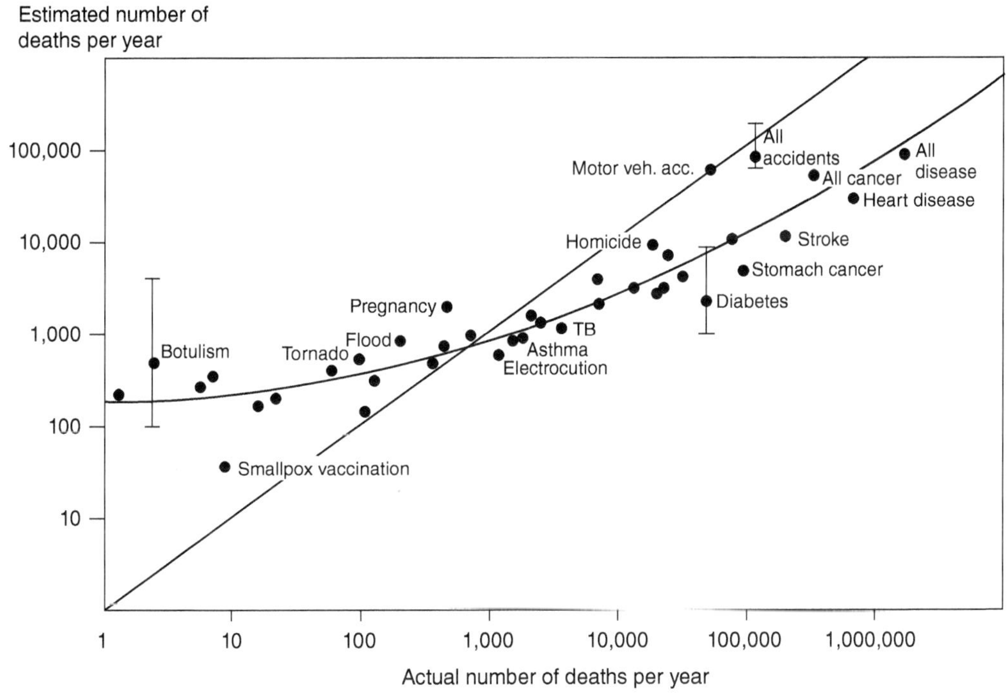

Figure 3.1 Estimated versus actual deaths per year for 41 causes of death. *Source:* From Slovic (2000).

Basically, overestimated items were those that tend to be dramatic and sensational (likely influenced by the media), whereas underestimated items tended to be unspectacular events that claim one victim at a time but cause many overall (e.g. disease). The findings as derived from Figure 3.1 indicate that relatively rare causes of death tend to be overestimated in frequency and more frequent or common causes of death tend to be underestimated. Specifically, the frequencies of death due to botulism, tornadoes, and pregnancy are greatly overestimated since the perceived rate of death is substantially above the 45° line. It is likely that the notorieties associated with media coverage of these causes of death receive considerable attention in the news media and hence are more in the forefront of the public's mind. As well, although not considered in the work of Slovic et al., a recent disaster can distort public judgment.

In general, rare causes of death were overestimated and common causes of death were underestimated (Slovic et al., 1979). Deaths caused by cancer, for example, are likely considered by most respondents as "old people are going to die of something"; these types of deaths do not generally cause headlines, and hence the deaths from diseases are underestimated. People tend to understand rather effectively the risks such as motor vehicles; this is likely as a result of media coverage, which occurs on a regular basis, allowing people to be relatively aware of the risks.

Another relevant dimension of the data plotted in Figure 3.1 is that cancer causes approximately 25% of deaths. You will recall that this was the subject of attention in Chapter 2 and the validity of this extent of death is relevant in that it shows how incremental death rates are of great interest, but they are incremental to a large number; in environmental risk assessment, we are dealing with the tails of the distribution as described in Chapter 2.

All of these findings indicate that there are substantial differences between the perceived risk of death and the actual risk of death. Hence, while the actual risk of death is clearly important, when the public is involved (and this is true in many circumstances), equally important is the perceived risk of death.

While the above deals with overarching concerns of risk, it is important to understand in the widest sense of the problem that the public perception of risk is important and influential in issues involving public perception of risk and decision-making involving the public. Although the technical aspects of risk assessment may be well based and defensible, for many situations, risk must deal with the public, and hence, in many respects, it is the perception of the risk that will influence decision-making, not the real risk. In part this occurs because people's response to the hazards they perceive, who is imposing the risks, and their willingness to accept the risks influence how the risks must be assessed as part of the decision-making process.

3.2.3 Characteristics of Risk Which Influence Risk Perception

Before the calculations are actually completed, and still with a view to understanding public perception of risks, we must first deal with the differences between public perception of risks and some of the problematic features that influence the perception by members of the public.

As Slovic et al. (1979) indicated, "people respond to the hazards they perceive. If their perceptions are faulty, efforts at public and environmental protection are likely to be misdirected." The causes of death that are relatively infrequent tend to be the type that evoke big press coverage (e.g. the impact of tornadoes or the Ebola virus) and/or sympathetic responses (e.g. pregnancy) and hence are likely substantially overestimated by the public because they receive notoriety due to press/media coverage, whereas the incidence of death of children due to waterborne illness is far in excess of death rates caused by AIDs and malaria combined.

An important element to learn from the above is that when dealing with risk characteristics, there are a number of characteristics of the risk that will influence public perception. Some of those characteristics are listed in Table 3.3.

As outlined in the previous section, although the technical aspects of risk assessment may be well based and defensible, for many situations, aspects of issues that must be decided upon involve input from the public. When laypeople are asked to evaluate risks, they do not have the evidence from which to necessarily make decisions. Instead, they rely on inferences based on what they have read, experienced, or heard. Hence, for many risk assessments, the execution of risk assessments must deal with ensuring that the public accepts the decision-making (the issues are ones of decision-making with public input and, in many situations, decision-makers are politicians or bureaucrats). In many respects, it is the perception of the risk that governs, not the detailed science and engineering analyses. Basically, as technical people, we prepare the information for decision-makers, but we do not make the decisions.

Part of the challenge of risk assessment is whether the public and/or decision-makers understand the risks and the implications of probability. For example, corporate managers become

Table 3.3 Characteristics of risk that influence public perceptions.

1	Voluntariness of the risk
2	Immediacy of the effect
3	Knowledge about risk
4	Control over risk
5	Newness of the technology
6	Chronic/catastrophic
7	Common/dread
8	Severity of consequences

annoyed with consultants who give them the probabilities of possible events instead of telling them exactly what will happen (Slovic et al., 1979). Another example of criticisms occurred where the results of a blue-ribbon panel of scientists reported that they were 95% certain that cyclamates do not cause cancer. Food and Drug Administration commissioner in the United States indicated, "I'm looking for a clear bill of health, not a wishy-washy, iffy answer on cyclamates." Senator Muskie indicated that he did not want "one-armed" scientists who do not respond "on the one hand, the evidence is so, but on the other hand..." when asked about the health effects of pollutants. There is a misunderstanding of the ability of risk assessment to be understood, even by highly educated individuals.

Further complicating risk assessment procedures when dealing with public perception is the fact that people's beliefs change slowly and are extraordinarily persistent even in the face of contrary evidence. Once formed, initial impressions tend to structure the way that subsequent evidence is interpreted. New evidence appears reliable and informative if it is consistent with a person's initial belief; contrary evidence is dismissed as unreliable, erroneous, or unrepresentative.

If a very high level of certainty is sought, the resulting costs are likely to be high; eliminating the uncertainty may mean eliminating the technology and foregoing its benefits.

3.2.3.1 People's Behavior

As demonstrated in Chapter 2, we cannot make risks equal to zero, and a better understanding of the risks assists in the management. Much is gained through experience. From knowledge of past mistakes and actions, people look to prevent new mistakes.

People tend to respond to the hazards they perceive. Even for the circumstances where the perception of risk is in error, attempts to persuade the public that they are incorrect may be viewed with suspicion. People's willingness to accept risks depends on the hazards they perceive, who are imposing the risks, and their willingness to accept the risks. Further, a recent disaster may greatly distort judgment.

This means that decisions are driven by public perceptions and governmental regulations. Hence, regardless of the realities of risks, the public perception is what may be driving the means by which risk management occurs.

These dimensions demonstrate that perceptions are highly variable – variable, but certainly relevant – since perception will influence what risks are considered as acceptable and must be considered in completing a risk assessment. It is important to acknowledge that while there may be differences between the real risk level and the perceived risk level, whether the two are the same or not, if differences exist they cannot be ignored because they influence the people's attitudes.

3.2.4 Magnitudes and Consequences of Risk Influence People's Willingness to Accept Risk

Individuals choose to voluntarily accept some risks, and while they do not necessarily understand the magnitudes of the risk, they are presumably aware there are risks involved. Examples of the risk of death per million hours of exposure include:

Mountain climbing	40 000
Cigarette smoke	3 000
Swimming	2 560
Air travel	500
Struck by lightning	100
Living near nuclear power plant	0.5

While there are many ways to characterize risk, it is useful to keep in mind that most people perceive of risks in one of three ways, namely,

1) An activity that is 100% safe.
2) An activity that is 100% dangerous, and hence to be avoided.
3) Everything else is 50/50.

Features that influence the members of the public's willingness to accept risk include the magnitudes and consequences of the risk. Examples of the relevant dimensions include:

1) New technologies are more suspect than familiar technologies.
2) If the individual feels they have personal control, then the individual is more likely to accept the risk if they have minimal to no control (e.g. driving a vehicle as opposed to being a passenger in an airplane).
3) People tend to respond to the hazards they perceive. Even for the circumstances where their perception of risk is in error, attempts to persuade the public that they are incorrect are generally viewed with suspicion.
4) Generally, if an individual is confronted with both small and large/infrequent risks, he/she will be more willing to accept the former.

Dimensions of risks that influence people's willingness to accept risks are indicated in Table 3.4.

Risks with a large potential loss and low probability of occurrence are considered differently from one with a low potential loss and high likelihood of occurring. Further, a desire for certainty is typical. While every technology is a gamble, one way to reduce the anxiety generated by confronting uncertainty is to deny that uncertainty. This type of denial is illustrated by the case of people faced with material hazards, who often view their world as either perfectly safe or sufficiently predictable to preclude worry. The search for certainty comes at a high price; if a very high level of certainty is sought, the costs are likely to be high. Eliminating the uncertainty may mean eliminating the technology and foregoing its benefits.

3.2.5 Examples of Trade-Offs Between Contributing Factors

The concept of risk trade-off may be familiar from economics and finance, where the term "risk return trade-off" describes the principle that risk increases with the potential return and that every investor must recognize their own tolerance for risk and desire for return and weigh these factors when selecting investments. Likewise, trade-offs may exist in environmental risk. For example, a wastewater treatment that reliably removes contaminants to a very low level may be very expensive to install and operate, while an alternative technology that is far less expensive may not provide the same degree of treatment. Stakeholders must weigh the costs

Table 3.4 Examples of dimensions of risks that influence people's willingness to understand and accept risks.

Communication issues can be very real – 10^{-9} versus 10^{-12}; the 3 orders of magnitude may not be correctly interpreted by nontechnical people
Technical aspects are typically not understood by the public; this may arise due to a lack of trust
People do not feel they have control over water resources and environmental risks
People's beliefs change slowly and tend to be highly persistent in the face of contrary evidence
Once formed, initial impressions tend to structure how subsequent evidence is interpreted
New evidence appears reliable and informative if it is in agreement with a person's initial belief, while contrary evidence is frequently interpreted as unreliable, prone to error, or unrepresentative

and benefits of both options against the project objectives in order to make a decision. With respect to environmental risk in general, the same is usually true; uncertainty is a source of discomfort, but certainty comes at a price. In many cases, due to prohibitively high costs, eliminating uncertainty means eschewing the activity and foregoing its benefits.

3.2.5.1 Underestimation of Risk

We saw in Chapter 2 and earlier in this chapter that people tend to overestimate the risks of low probability events, especially if those risks are involuntary, uncontrollable, and unfamiliar, and may have severe consequences, and tend to underestimate the risk of familiar, high probability events. The underestimation of risks may also occur for reasons including:

- Failure to consider the ways in which human errors can affect technological solutions (e.g. a disastrous fire at Browns' Ferry Nuclear Plant caused by a technician who used a candle to look for a gas leak);
- Overconfidence in current scientific knowledge (e.g. failure to recognize the harmful effects of X-rays until widespread use had occurred).
- Failure to anticipate human response to safety measures (e.g. an improved highway may increase the number of motor vehicle accidents because the road is used more frequently and cars travel faster).

The following are intended as simple examples of risk trade-offs.

Example 3.1 Having a glass of beer contributes to the possible cirrhosis of the liver and development of oral cancer. On the other hand, the glass of beer may induce relaxing effect and assist in obtaining a good night's sleep, which would assist in lowering risk.

Example 3.2 As a different example, we may cook a meal in a microwave. Even at low levels, microwaves result in adverse nerve impacts on people and hence elevate risks. If instead a gas stove is used, gas combustion may fill the kitchen with carbon dioxides and nitrogen oxides, thereby elevating risk.

Example 3.3 Beer in a green glass may contain chromium, of which a small amount enters the beer. Chromium is a known carcinogen but must not be totally avoided because it is essential to life. If instead beer were to be consumed from plastic bottles, bisphenol A may dissolve into the beer from the plastic, creating an alternative risk.

As per Table 2.4, activities such as smoking 1.4 cigarettes, travelling 10 miles by bicycle, etc. increase chances of death by one-in-a-million. These are activities that people may be exposed to on a daily basis and provides a point of comparison, which can be useful in facilitating people's understanding of risks. EPA generally assumes that approximately one-in-a-million is an acceptable risk for environmental hazards.

As indicated in Chapter 2, risks of one-in-a-million are in the tails of the distribution, as schematically depicted in Table 3.5. However, the politics of risk management frequently focus on the adequacy of supporting scientific evidence, and it is very challenging for nontechnical people to understand risks at the tails of a distribution. Defining the extent of an exposure risk, and utilizing evidence of comparable risks with which members of the public are familiar, can be very helpful in making members of the public aware of the magnitudes of risks.

Many factors may influence the magnitude of risks and complications may influence the magnitude of the risks. Examples of reasons for underestimating risks in terms of estimating the percentile for events to occur are indicated in Table 3.6.

Table 3.5 Risk as characterized using probability and magnitudes.

↑	Impact		Likelihood after Control			↑
			Impact After Control →			
	Insignificant	Minor	Moderate	Major	Extreme	
Almost Certain	Moderate Risk	Moderate Risk	High Risk	High Risk	High Risk	Almost Certain
Likely	Low Risk	Moderate Risk	High Risk	High Risk	High Risk	Likely
Possible	Low Risk	Moderate Risk	Moderate Risk	High Risk	High Risk	Possible
Unlikely	Low Risk	Low Risk	Moderate Risk	Moderate Risk	High Risk	Unlikely
Rare	Low Risk	Low Risk	Low Risk	Moderate Risk	Moderate Risk	Rare
	insignificant	Minor	Moderate	Major	Extreme	
				Impact →		

Table 3.6 Examples of reasons for underestimation of risk.

Failure to consider the ways in which human errors can affect technological solutions (e.g. the disastrous fire at Browns' Ferry Nuclear Plant as a result of a technician looking for an air leak using a candle)
There may be overconfidence in current scientific knowledge (e.g. there was a failure to recognize the harmful effects of X-rays until widespread use had occurred)
Failure to anticipate human response to safety measures (e.g. the flood that overtops a dam would have less impact if the dam had never been constructed)
An improved highway may increase the number of deaths on a highway because this improved highway increases the number of miles driven

Nuclear power had the dubious distinction of scoring at, or near, the extreme on all of the characteristics associated with high risk (Slovic et al., 1979). The public typically perceives environmental risks as one or more of the following: risks are involuntary, delayed, unknown, uncontrollable, unfamiliar, potentially catastrophic, dreaded, and severe. Perceptions of nuclear power are likely the result of the belief that if the risk occurs, then the consequences are very high. Hence, perception is as important as reality. It is likely that people consider the low death rate associated with nuclear power, but consider it to be a high risk technology because of its potential for a huge disaster.

It is noteworthy that social and political contexts influence how we see the issues. The public in the United States are highly concerned about nuclear power (20% of US power is nuclear). In France, the public does not even care (78% of France's power is nuclear generated). New evidence appears reliable and informative if it is consistent with one's initial belief; contrary evidence is dismissed as unreliable, erroneous, or unrepresentative.

Consider cigarette smoking as an example; smoking cigarettes at the rate of one pack per day has a loss of life expectancy of 6.4 years for men and 2.3 years for women.

Complicating the assignment if risks when setting standards for environmental toxins, consideration needs to be given to:

Combined effects of different exposures
Individual sensitivities within population
Effects of chronic and acute exposures

3.2.5.2 The Influence of Voluntary and Involuntary Aspects of Risks

Peoples' willingness to accept risk is highly dependent upon whether the risks are being imposed on them or they are willingly accepting the risk. People tolerate risks when they have a choice (e.g. cigarette smoking), but people are generally opposed to risks imposed on them (e.g. deteriorating environmental quality).

In many situations, alternatives involved in relation to risks are that they are involuntary. There is no question that there are risks that are real; many of the situations of environmental risk are involuntary risks. The risks are exposure risks, which people do not really understand and wish to avoid, e.g. a landfill adjacent to their property. People often feel water resources and environmental risks are forced upon them and are therefore more adverse to accepting those risks.

3.2.5.3 Dreadfulness of the Outcome

Tolerance for risk is often directly related by how terrible the potential consequences might be, should they materialize. Tolerance is particularly sensitive to impacts on "innocent people" such as children. Sometimes the very mentioning of low probability hazard may increase the judged probability of the hazard (regardless of what the evidence indicates, e.g. nuclear power).

Public dread and actual deaths are caused by common sources of energy.

3.2.5.4 Visibility of the Hazard

The profile of any potential hazard is important. Although deaths in cars are much more frequent than deaths resulting from nuclear power, the latter is viewed as being catastrophic and receives much more attention in the media. In this way, the media has a strong influence on public perception of risk.

3.2.5.5 Media Influences on Perception of Risks

We cannot minimize our risks by simply avoiding those risks about which we are aware because awareness is heavily influenced by the media. However, the media coverage may be very poor and a poor guide to actual dangers. This issue is best demonstrated by example.

A recent disaster can distort judgment. The official death toll from the 11 September 2001 terrorist attacks was 2974. But in 2002, the American death toll on the roads grew by more than 1500. As well, a swan washes up on a British shore, dead from bird flu, and the press panics, while the 3000 people who die every year on British roads (13 times the number of people throughout the world who have ever died from bird flu) go largely unnoticed. The media influence can be highly influential.

Tornadoes and floods are estimated as higher than actual. As judged by deaths across North America, are these at de minimis levels? To consider this, assume the population of North America is approximately 300 million. Large events receive large coverage. Consider cancer that causes death at the rate of approximately 500 000 deaths/year. Tornadoes/floods cause approximately 1100 deaths/year. The number of deaths per year 500 000/0.25 = 2 000 000. Prob. of death (tornadoes and floods) = 100/year/2 × 10^6 = 0.5 × 10^{-4} = 100 deaths as opposed to 500 000 deaths/year.

The differences between risk perception and actual hazards reveal the discrepancy between the scenarios that we fear and those that are seriously harmful to us.

3.3 Qualitative Risk Characterization and Probability–Impact Matrix Procedures

3.3.1 Introduction to Probability–Impact Matrix Procedures

Risk is the amount of harm that can be expected to occur during a given time period and as a result of a specific harmful event. The statistical terms and the level of risk were the focus of Chapter 2 through the calculation of the product of the probability that harm occurs (e.g. that an accident happens) and multiplied by the severity of that harm (i.e. the average amount of harm that occurs). However, for many circumstances and issues, the calculations cannot be completed, but neither the probability nor the harm or severity can be estimated with both accuracy and precision. Examples of issues include those of dam failure or snow avalanche conditions – in most respects, quantification of the probability and the harm of various types of failure scenarios are not easily quantified. The risks are real, but the quantitative and semiquantitative data necessary to do the calculations are not. To provide quantification of likelihood and/or consequence, data which are needed typically do not exist, requiring a different approach.

As an alternative, a strategy that may be effective is to represent the two components of risk (likelihood and impact) that are characterized by "words" or perhaps a scale from 1 to 10. The risk calculation is undertaken using a "probability impact or risk matrix." A risk matrix is a matrix that is used during risk assessment to define the various levels of risk as the product of the harm probability categories and harm severity categories. An example of this approach is depicted in Tables 3.7 and 3.8; the concept is simple, but deducing specific values for entry into the matrix can be very challenging. For example, the harm severity can be categorized into catastrophic, critical, marginal, and negligible. As an example, an event that has "rare" likelihood and negligible consequence would have a low risk (Figure 3.2). Alternatively, an event that has a "likely" likelihood and a "critical" consequence would have a high associated risk.

The procedure is very helpful in that it is a simple portrayal and hence easy for individuals to understand and be traceable. Many standard risk matrices exist for different applications where differentiation between the examples is the wording that is used and/or scaling such as illustrated in Tables 3.7 and 3.8.

Table 3.7 Example of a risk matrix – likelihood, consequence, and risk scoring.

	Negligible	Marginal	Critical	Catastrophic
Rare	Low	Low	Moderate	High
Unlikely	Low	Low	Moderate	Extreme
Possible	Low	Moderate	High	Extreme
Likely	Moderate	High	High	Extreme
Certain	High	High	Extreme	Extreme

Table 3.8 Likelihood, consequence, and risk scoring based on Alberta template.

		Consequence descriptor					
	Score	Not applicable	Insignificant	Minor	Moderate	Severe	Catastrophic
	Not applicable	0	1	2	4	8	14
Likelihood descriptor	Most unlikely	1	1	2	4	8	16
	Unlikely	2	2	4	8	16	32
	Medium	4	4	8	16	32	64
	Probable	8	8	16	32	64	128
	Almost certain	16	16	32	64	128	256

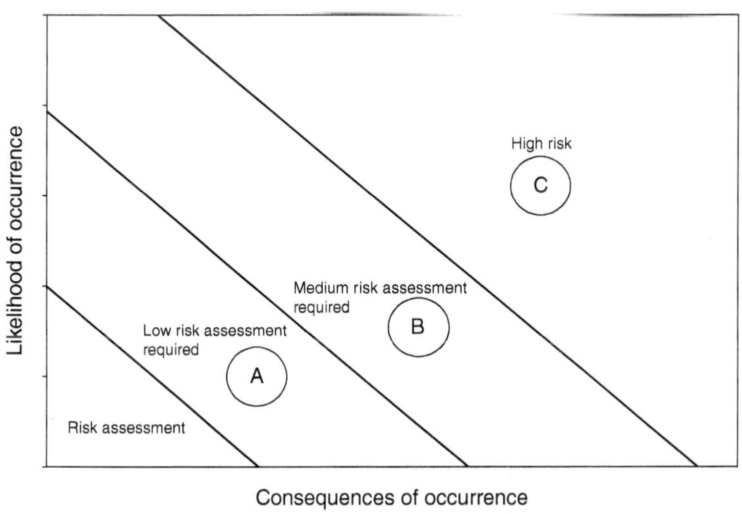

Figure 3.2 Schematic characterization of a likelihood and consequence trade-offs. A, do nothing; B, monitor; C, high risk and hence something must be done.

Many other examples of such matrices exist. There is value in describing alternative approaches as the potential exists to select alternatives, depending upon individual circumstances. Alternatives include use of Table 3.6 which indicates a simplified.

Table 3.9 indicates a simplified model of the probability and impact classification.

After assigning the total (scores) for likelihood and impact of risk categories, the values can be assigned by multiplying the two variables, indicating the degree of risk.

These might be expanded, using the information in Table 3.10, for the likelihood score risk.

The second step is to set the impact on a scale of 1–5 (as an example), as illustrated in Table 3.11.

The calculation of exposure risk is as demonstrated in Table 3.12.

The last step can be to build the risk matrix as shown in Table 3.13.

Table 3.9 A simplified model of the probability and impact classification.

Classification probability	Score	Impact classification	Score
Low	1	Major	3
Medium	2	Medium	2
High	3	Easy	1

Table 3.10 Probability or likelihood score risk.

Likelihood level	Score
Very low	0–20
Low	21–40
Medium	41–60
High	61–80
Very high	81–100

Table 3.11 Impact value assignments.

Magnitude of impact	Impact definition	Score	Rating
High impact/high probability	**Very high** They are large risks	5	A
High impact/medium probability Medium impact/high probability	**High** These risks have either a high probability of occurrence or a significant impact	4	B
Medium impact/medium probability	**Medium** There is a medium chance that the risks appear noticeable impact	3	C
Medium impact/low probability Low impact/medium probability	**Low** These risks can occur in some situations and have a low to medium risk	2	D
Low impact/low probability	**Insignificant** There are risks with low probability of occurrence and low impact. Likely able to be neglected	1	E

Table 3.12 Calculation of exposure risk.

		Impact		
		Low (insignificant)	Medium (reasonable impact)	High (have a significant impact)
Likelihood	Low (unlikely)	E	D	C
	Medium (may occur)	D	C	B
	High (likely to occur)	C	B	A

Table 3.13 Example of a risk matrix.

		Occurrence likelihood		Impact		Degree of risk	
	Risk	Probability	Score	Probability	Score	Rating	Score
1	Without	Very low	20	Very high	5	E	12.5
2	A small number	Low	40	High	4	D	22
3	Reasonable number	Medium	60	Medium	3	C	31.5
4	Large number	High	80	Low	2	B	41
5	Very large	Very high	100	Very low	1	A	50.5

3.3.2 Issues with the Risk Matrix Approach

While helpful, several problematic features associated with use of the risk matrix approach include the following:

1) *Poor resolution of entries in the matrix*: It is possible to correctly and unambiguously compare only a small fraction of hazards. The results may assign identical ratings to quantitatively very different risks.
2) *Susceptibility to errors*: Assignments of values may mistakenly involve higher qualitative rates to quantitatively smaller risks.
3) *Suboptimal resource allocation*: The effective allocation of resources to risk-reducing countermeasures cannot be based on the categories provided by risk matrices.
4) *Ambiguous inputs and outputs may exist*: Categorizations of severity cannot be made objectively for uncertain consequences. Inputs to risk matrices (frequency and severity categorizations) and resulting outputs (risk ratings) require subjective interpretation, and different users may obtain opposite ratings of the same quantitative risks.

3.4 Microbial Risk Assessment

Chapter 2 provided a general development of carcinogenic and noncarcinogenic risk assessment methodology and dealt with the more established types of risk assessments. However, risk assessments associated with waterborne pathogens are very real, but challenging, in terms of the characterization of risks.

Waterborne fecal pathogens are microorganisms that may occur in water as a result of contamination from human or animal feces. Throughout the last 150 years, consumption of drinking water of poor sanitary quality has been linked to human illness, most notably gastrointestinal-related illness such as diarrhea. The classification of waterborne pathogens include bacteria, protozoans, helminthes, and viruses, and many scholarly articles describe in detail the attributes, health concerns, etc. associated with these and will not be further elucidated herein – suffice it to indicate that the issues of illness and death on a world scale is enormous.

Microbial risk assessment involves the steps to understand, reduce, and prevent risks caused by waterborne pathogens. Although microbial risk assessment evolved from chemical risk assessment, there are fundamental differences. For example, the exposure of chemical risk assessment is potentially chronic, whereas in microbial, the exposure is acute.

Quantitative microbial risk assessment (QMRA) is an effective way to estimate the risk of infection by specific microbes (Austin et al., 2012; Besner et al., 2011).

Attention herein is restricted to a methodology called quantitative microbial risk assessment. QMRA is the application of the principles of risk assessment to estimate the potential effects of an exposure to infectious microorganisms (Haas et al., 1999). The WHO (2006) describes QMRA as "... a rapidly evolving field that systematically combines available information on exposure and dose-response to produce estimates of the disease burden associated with exposure to pathogens." The WHO (2006) states that QMRA can be valuable "... when the fraction of disease that can be attributed to drinking water is ...difficult to measure directly through public health surveillance or analytical epidemiological studies." Epidemiological data are frequently not well reported due to the accessibility and structure of healthcare systems, and the data are difficult to obtain from individuals. Given the variety of sources of infection and the difficulties in attaining high quality epidemiological data, QMRA is a useful tool to describe the potential risks from water supply and for informing water safety management.

Risk assessment and management has become an increasingly important tool in the control of water safety (WHO, 2006). The WHO (2006) has incorporated QMRA into its third edition of the Guidelines for Drinking-Water Quality (GDWQ). The use of QMRA to estimate disease burdens to inform management decisions had been demonstrated for use in developed countries (e.g. Haas et al., 1999); however QMRA had only begun to be explored for use in developing countries where data may be scarce (Howard et al., 2006). Howard et al. (2006) demonstrated that the QMRA procedure can be simplified to produce what may be a useful tool in developing countries with limited data, for the use in making water supply management decisions. Hunter et al. (2009) used a similar methodology to examine the effect of poor reliability of drinking water supply systems in developing countries.

Protection of the general public from harmful pathogens within the environment is a concern of regional health units. A complicating feature is that the health risks posed to humans by pathogens are subject to both exposure time and exposure pathways (Figure 3.3). One of the higher-risk activities associated with pathogens that the general public engages in is spending time in the water at a beach.

Of the known protozoans, *Cryptosporidium parvum* and *Giardia lamblia* are of greatest concern because of their significant impact on individuals at low infectious doses: 1–10 for *C. parvum* and 20 for *G. lamblia* (Metcalf et al., 2003). These protozoans may cause cryptosporidiosis and giardiasis, which are diseases accompanied with mild to severe diarrhea, stomach cramps, nausea, and vomiting that can last for extended periods of time. *Cryptosporidium* and *Giardia* are also of considerable concern because of their resistance to conventional wastewater treatment processes, resulting in exposure hazards.

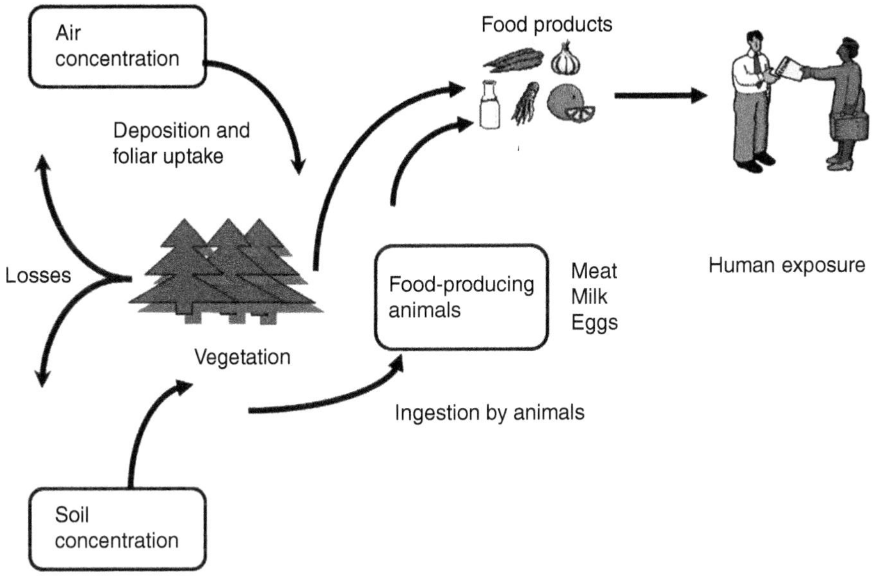

Figure 3.3 Example depiction of pathways of exposure for human health risk assessment.

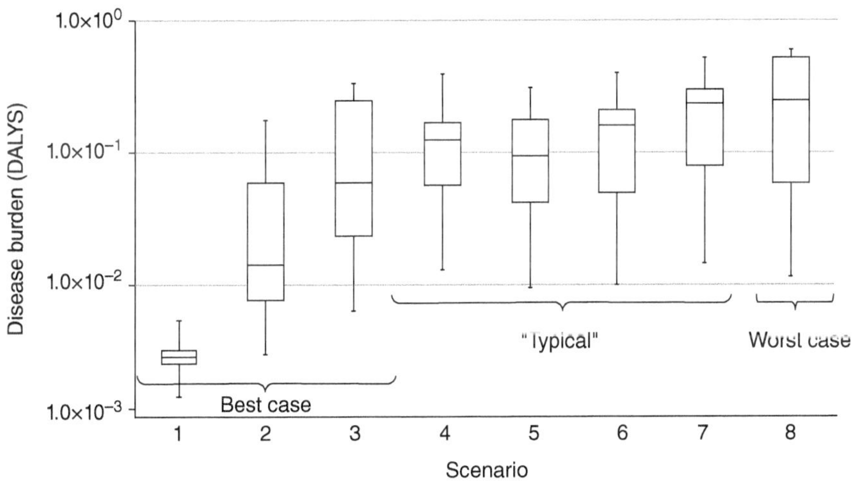

Figure 3.4 Box and whisker plot representing the median, 25th and 75th percentile, maximum and minimum estimates of risk for different water use scenarios.

The reference level of risk for waterborne diseases recommended by WHO (2006) is 1×10^{-6} disability-adjusted life years (DALYs) per person per year. An indication of how DALYS may be interpreted as related to varying levels of risk, in the application to different water use scenarios, is shown in Figure 3.4. However, WHO (2006) also states that the reference level of risk can be adapted to local circumstances on the basis of a risk–benefit approach.

A key element in QMRA is the selection of reference pathogens. WHO (2006) recommends the selection of representative organisms that "… if controlled, would ensure the control of all pathogens of concern. Typically this implies inclusion of at least one bacterial pathogen, virus, and protozoan." *Escherichia coli* O157:H7 or enterotoxigenic *E. coli* (ETEC), *C. parvum*, and rotavirus may serve as reference pathogens for bacterial, protozoan, and viral pathogens, respectively (Howard et al., 2006; Hunter et al., 2009).

Both *E. coli* O157:H7 (Howard et al., 2006) and ETEC (Hunter et al., 2009) have been used as reference pathogens in QMRA in developing countries. Thomas et al. (2015) used ETEC rather than *E. coli* O157:H7 because it is a "... far more significant pathogen in developing countries that is largely by food and water with little or no person to person spread" (Hunter et al., 2009).

Exposure assessment requires the estimation of the number of pathogenic microbes to which an individual is exposed. The standard QMRA consists of four basic steps (after Ahmed et al., 2010):

1) *Hazard identification* defines the potential microbes (e.g. protozoan, virus, and bacteria) that are harmful to the system and also investigates the effect of the microbes. Most drinking water-related QMRA address enteric viruses (*hepatitis A, Norwalk virus, rotavirus*), parasites (*Ascaris, Elmeria, Cyclospora, Toxoplasma*), protozoa (*Cryptosporidium* and *Giardia*), and bacteria (*E. coli, Shigella, Salmonella, Vibrio cholerae*) (after Haas et al., 1999). Unlike microbial water quality assessment, *E. coli* O157:H7 can be considered for simplicity.
2) *Exposure assessment* identifies the exposure route (e.g. inhalation, ingestion, and dermal absorption), the time period for the exposure, and the level of the exposure. The focus of the exposure assessment is the amount of oral consumption per day and depends on the place and the type of consumer.
3) *Dose–response assessment* is typically conducted by using a model to define the response as the probability of infection. Microbial concentration is required in colony-forming units for bacteria, plaque-forming units for viruses, and cysts or oocyst counts for protozoa. Many studies have been conducted on different hosts to observe the response over the dose and aggregated in a QMRA database (e.g. see Haas et al., 2014). The database assumes the response is either beta-Poisson or exponential model with the following equations:

$$P_{in} - 1 - e^{-(kd)}$$

$$P_{in} = 1 - \left[1 + d(2^{1/\alpha} - 1)/N_{50}\right]^{-\alpha},$$

where P_{in} is the response (probability of infection), d is the dose (concentration multiplied by the volume of ingestion = Conc × Volume), kd is the probability of survival (depends on the organism and test), N_{50} is the the experimental dose at which 50% of the population is expected to be affected, and α is the beta-Poisson distribution parameters. Table 3.14 has the distribution parameter values for commonly available microbes.

The calculation of the disease burden in DALYs requires two key pieces of information (Havelaar and Melse, 2003):

1) Average duration of the adverse health response, including loss of life (D).
2) Weights for severity to the unfavorable health conditions (S), and DALY values are calculated for each negative health consequence expected to result from infection using Equation (3.1):

$$\text{DALY} = \text{D} \times \text{S} \tag{3.1}$$

DALY values are calculated for each negative health consequence expected to result from infection, and summed together, to determine the maximum disease burden.

Table 3.14 Related QMRA studies.

Organisms	Reference	Best fit model	Parameters for the best fitted model			Host type and route	Dose units	Response
			K	μ	N50			
Salmonella	McCullough and Elsele (1951)	Beta-Poisson		3.18×10^{-01}	3.71×10^{04}	Human, oral	CFU	Positive stool culture
Shigella	DuPont et al. (1972)	Beta-Poisson		2.65×10^{-01}	1.48×10^{03}	Human, oral	CFU	Positive stool sample culture isolation
Vibrio cholerae	Hornick et al. (1971)	Beta-Poisson		2.50×10^{-01}	2.43×10^{02}	Human, oral	CFU	Infection
E. coli O157:H7	DuPont et al. (1971)	Beta-Poisson		1.55×10^{-01}	2.11×10^{06}	Human, oral	CFU	Positive stool isolation
Legionella pneumophila	Muller et al. (1983)	Exponential	5.9×10^{-02}			Guinea pig, inhalation	CFU	Infection
Cryptosporidium parvum	Messner et al. (2001)	Exponential	5.7×10^{-02}			Human, oral	Oocysts	Infection
Giardia duodenalis	Rendtorff (1954)	Exponential	1.99×10^{-02}			Human, oral	Cysts	Infection

CFU, colony-forming unit.

As an example, Thomas et al. (2015) used QMRA analyses for the various scenarios for comparison to piped water usage, and the "best," "worst," and "typical" water source combinations are identified in Figure 3.1.

The QMRA analysis revealed that the use of filtered water or boiled water stored in a kettle 100% of the year represented the "best" case scenario, while 100% use of untreated shallow well water represented the "worst" case scenario for residents in the resettled communities studied. Four "typical" water use scenarios were identified. The results showed that these annual water usage combinations provided some risk reduction over exclusive use of surface water ("worst" case); however, these "typical" scenarios did not reach the lower levels of risk associated with the exclusive use of filtered water or boiled water ("best" case).

QMRA is a useful tool to compare the risks associated with alternative water supply scenarios. In this example, QMRA was used to examine the differing risks between "best" and "worst" case scenario water sources as well as "typical" water source combinations used by periurban residents of Phnom Penh. Using QMRA in this way allows one to examine the risk due to existing water consumption practices and to compare the risk to possible intervention strategies. QMRA can be used to inform and better target interventions and be used as an advocacy tool with policy-makers to illustrate the increased health risk to households when resettled in a community without piped water supply.

3.5 References

Ahmed, W., Vieritz, A., Goonetilleke, A., and Gardner, T. (2010). Health risk from the use of roof-harvested rainwater in Southeast Queensland, Australia, as potable or nonpotable water, determined using quantitative microbial risk assessment. *Applied and Environmental Microbiology* 76: 7382–7391.

Austin, R., Choi, C. Balagopal, T., and Gurian, P. (2012). Event-based microbial risk assessment and response analysis of cryptosporidium in municipal water distribution networks. *12th Annual International Conference on Water Distribution Systems Analysis 2010, WDSA 2010*, Tucson, AZ, pp. 459–470.

Besner, M.-C., Prevost, M., and Regli, S. (2011). Assessing the public health risk of microbial intrusion events in distribution systems: conceptual systems, conceptual model, available data and challenges. *Water Research* 45 (3): 961–979.

DuPont, H.L., Formal, S.B., Hornick, R.B. et al. (1971). Pathogenesis of *Escherichia coli* diarrhea. *The New England Journal of Medicine* 285 (1): 1–9.

DuPont, H.L., Hornick, R.B., Snyder, M.J. et al. (1972). Immunity in shigellosis. II. Protection induced by oral live vaccine or primary infection. *The Journal of Infectious Diseases* 125 (1): 12–16.

Haas, C.N., Rose, J.B., and Gerba, C.P. (1999). *Quantitative Microbial Risk Assessment*. New York: John Wiley & Sons, Inc.

Haas, C., Rose, J., and Gerba, C. (2014). *Quantitative Microbial Risk Assessment*. New York: John Wiley & Sons, Inc.

Havelaar, A. and Melse, J. (2003). *Quantifying Public Health Risk in the WHO Guidelines for Drinking-Water Quality: A Burden of Disease Approach*. Bilthoven, The Netherlands: National Institute for Public Health and the Environment.

Hornick, R.B., Music, S.I., Wenzel, R. et al. (1971). The broad street pump revisited: response of volunteers to ingested *Cholera vibrios*. *Bulletin New York Academy of Medicine* 47 (10): 1181–1191.

Howard, G., Pedley, S., and Tibatemwa, S. (2006). Quantitative microbial risk assessment to estimate health risks attributed to water supply: can the technique be applied in developing countries with limited data? *Journal of Water and Health* 4 (1): 49.

Hunter, P.R., Zmirou-Navier, D., and Hartemann, P. (2009). Estimating the impact on health of poor realiability of drinking water interventions in developing countries. *Science of the Total Environment* 407: 2621–2624.

McCullough, N.B. and Elsele, C.W. (1951). Experimental human salmonellosis: I. Pathogenicity of strains of *Salmonella meleagridis* and *Salmonella anatum* obtained from spray-dried whole egg. *The Journal of Infectious Diseases* 88 (3): 278–289.

Metcalf & Eddy, Tchobanoglous, G., Burton, F.L., and Stensel, H.D. (2003). *Wastewater Engineering: Treatment and Reuse*, 4e. Boston, MA: McGraw-Hill.

Messner, M.J., Chappell, C.L., and Okhuysen, P.C. (2001). Risk assessment for *Cryptosporidium*: a hierarchical Bayesian analysis of human dose response data. *Water Research* 35 (16): 3934–3940.

Muller, D., Edwards, M.L., and Smith, D.W. (1983). Changes in iron and transferrin levels and body temperature in experimental airborne legionellosis. *Journal of Infectious Diseases* 147: 302–307.

Rendtorff, R.C. (1954). The experimental transmission of human intestinal protozoan parasites. II. *Giardia lamblia* cysts given in capsules. *American Journal of Epidemiology* 59 (2): 209–220.

Slovic, P., Fischhoff, B., and Lichtenstein, S. (April 1979). Rating the risks. *Environment* 21 (3): 14–20.

Slovic, P. (2000). *The Perception of Risk*. Earthscan Publishing.

Thomas, K., McBean, E., Schantz, A., and Murphy, H. (2015). Comparing the microbial risks associated with household drinking water supplies used in peri-urban communities of Phnom Penh, Cambodia. *Journal of Water and Health* 13 (1): 243–258.

World Heath Organization (2006). Guidelines for Drinking-Water Quality: First addendum to the 3rd edition, volume 1 recommendations. Geneva: World Health Organization.

3.6 Problems

3.1 As fossil fuels become more expensive, there is increased interest in green energy, of which wind turbines is one. However, one of the issues associated with wind turbines is bird kill. Rank in order of annual bird kill, the bird kill rate occurs from the following sources: wind turbines, high-rise buildings, eagles, and swine flu.

3.2 Issues associated with the site-specific risk assessment procedures:

a) In many areas, there are site-specific risk assessment methodologies. Assume you have just completed a site-specific risk assessment (SSRA). Who is able to sign a record of site conditions (RSC) (list a few types of individuals who can sign the RSC) and explain why the SSRA regulation is likely set up this way

b) Under what circumstances, is it possible that completion of a risk assessment may save a property owners a considerable amount of money

3.3 In Figure 3.1, the best fit line is above the diagonal line for low fatality numbers. Is this suggesting that all events in this area of the figure are above the diagonal line or are likely also ones below the diagonal line? What is the reason the curve is below the diagonal line in the high magnitudes?

3.4 Cancer is responsible for approximately 25% of deaths in North America. Motor vehicle accidents account for only approximately 10% of deaths. See the Figure 3.1 for reference.

i) Which of these two causes of death are better understood insofar as magnitude by the public? Explain how you know this.

ii) Select one of the two (motor vehicles and cancer) of the rates of causes of death indicated in Figure 3.1 that are the least understood by the public.

3.5 The rates of discovery of the importance of microorganisms to human health are only now starting to become substantial. What is the most important reason?

3.6 The official death toll from the 11 September 2001 terrorist attacks was 2974. However, in 2002, the American death toll on the roads grew by more than 1500 in the single year alone to greatly surpass the death in the terrorist attack. As well, a swan washes up on a British shore, dead from bird flu, and the public panics staying away from the shoreline for months on end. Meanwhile, the 3000 people who die every year on the British roads (13 times the number of people throughout the world who have ever died from bird flu) go largely unnoticed. What is the fundamental reason for these types of responses and what if anything should be done about this type of reporting?

3.7 We approach risk management very differently for environmental concerns for residents versus construction workers. What are two of the most important reasons we do this?

3.8 An individual really likes spicy food (the spicier the better). The individual purchases, on a regular basis, an imported salsa, and consumes some of the salsa every second day within the year such that a 250 ml bottle is consumed in 15 servings. This has been ongoing for 10 years. Unfortunately, the individual has not been aware that there are traces of toluene (mg/l) in the salsa. Is the individual likely to become ill from consumption from the salsa?

Toluene reference dosages

Acute reference dose	1.02 mg/(kg_{BW} day)
Chronic reference dose	0.8 mg/(kg_{BW} day)

Sample no.	Arsenic concentration (ug/m^3)	Ln transform of arsenic
1	1.43	0.358
2	12.1	2.493
3	4.8	1.569
4	39.2	3.669
5	1.77	0.571
6	1.01	0.010
7	5.1	1.629
Mean	9.34	1.471
Std. dev.	13.71	1.297
Skew	2.262	0.690

3.9 Is it suspected there is atrazine in the groundwater, and hence are you, as a risk professional, considering managing the risk via point-of-use water treatment?

What does it mean in terms of managing the risk?

3.10 You are tasked with the completion of an inhalation risk for the workers in the restaurant. Prepare a flowchart of the steps you would take to compete the inhalation risk for the people that work in the restaurant. In this flowchart, identify the key data you would assemble to complete this task.

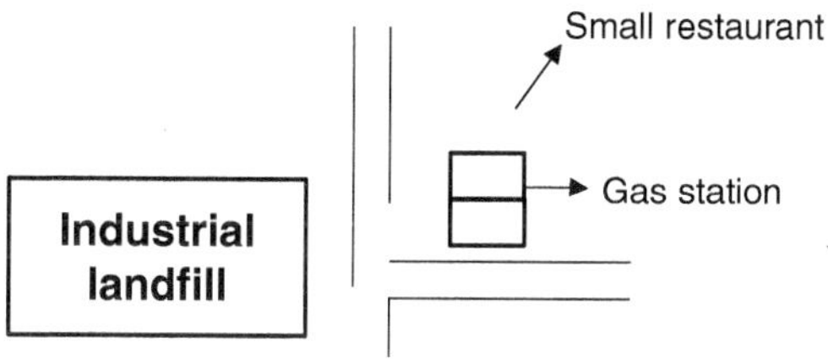

3.11 What is the reason you might develop a record of site conditions under the Site-Specific Risk Assessment (SSRA) Act of Ontario? There is a very specific reason for the value of this step.

3.12 You may find that when preparing the SSRA, you exceed the generic criteria or tables. Does that mean you may conclude your analyses and report to the client, or is there merit in undertaking a next step (and what is that step, if you think there is one?)

3.13 An individual purchases lettuce at the grocery store and takes 3 days to consume the lettuce. Specifically, on 5 January, the individual purchases lettuce (i.e. there is some consumed each day, such that over the next 3 days, the entire head of lettuce is consumed). Therefore, the individual is exposed for three successive days to the salmonella. If you do the analyses to evaluate whether the individual is likely to become ill, what dose–response rate would you employ, and why? Note that this is not an easy question, so indicate how you would approach the evaluation.

3.14 The figure in Question 3.3 depicts actual risk versus perceived risk of death in the United States per year. Briefly explain two features (i) why is the curved line, through the plotted points, below the 45° line, and (ii) what are the causes of death that are in the northeast portion of the figure and why people's perception is so different than actual death rates.

3.15 The town of Walkerton, Ontario, in 2000 suffered a tragedy when the pathogens in the water made one-half of the people in the town ill, and seven people died. List the sequences of issues (briefly) that resulted in the event to cause such a catastrophe.

3.16 Indicate the regions on the figure below, where you would expect to plot a project evaluation that has a low risk of failure. This approach to risk assessment is not as formal as the procedures that rely upon, for example, ingestion, body weight, averaging time, etc. What are the primary reasons for this rather simple approach?

4

Characteristics of Environmental Quality Data

Never tell me the odds.

—Hans Solo – Star Wars Episode V: The Empire Strikes Back

4.1 Background to Data

Risk assessment has become an increasingly important requirement for environmental decision-making. As described in Chapters 2 and 3, risk assessment is intended to be an objective and transparent procedure. However, of equal importance is that risk assessment must necessarily consider assumptions and uncertainties such as may exist, for example, in the ability to assign parameters for use in the risk calculations. We are looking to identify the underlying patterns contained within a dataset, as part of the risk assessment process.

Concern with the quality of the environment is pervasive. Members of the public, politicians, the media, lawyers, scientists, engineers, and so on are all watchful of current environmental quality levels and the perceived trends in these levels. Much of the concern with environmental quality is both real and appropriate and arises from a legacy of situations that represent (in hindsight) inadequate environmental protection. On the other hand, the legacy of some examples is overblown by the media and a lack of understanding. However, whether the concern by the public is warranted or not is in many cases immaterial. Attention to, and concern about, environmental quality is a phenomenon that must be addressed by professionals throughout the environmental industry.

During the past several decades, the public has been increasingly demanding that the risks associated with the exposure to environmental contaminants be reduced. The demands have been directed in part toward government agencies responsible for soil, water, and air quality for the protection of human health and ecosystems. This has resulted in dramatic increases in the requirements for professionals to analyze and interpret environmental quality data. The collection of samples from the field, the need for laboratory analyses of these samples, and the time requirements for interpreting the resulting information can represent enormous expenditures for government agencies, corporations, and citizens. Consequently, when analyzing the resulting data, diligence must be used to ensure that the correct interpretation of the data is obtained.

There are many reasons for the increased focus on environmental quality. Examples might include the population increases and urban densification that result in more locally concentrated pollutant discharges and deteriorated environmental quality. The reasons also include enhanced laboratory technologies (in today's context, we are able to know, analyze for, and report concentrations at levels that historically were not available [e.g. the LC-MS-MS that can measure down to ng/l]). The new technologies have resulted in the ability to monitor emerging contaminants such that measurement of chemical concentrations and pharmaceuticals are at

Risk Assessment: Procedures and Protocols, First Edition. Edward A. McBean.

Companion website: www.wiley.com/go/McBean/RiskAssessProcedure_1e

levels that previously were not quantifiable. With the intensive monitoring efforts now being carried out, instances of deteriorated environmental quality are now identified, which might otherwise have remained undetected. Equally important is that because the concentrations can now be able to be monitored, it does not mean there is actually an environmental risk.

The focus on environmental quality has resulted in greatly enhanced needs to understand the evolution of environmental quality conditions. Part of the understanding of the evolution is gained from the enhanced statistical interpretation of environmental quality data capabilities that we now have. However, by no means it is sufficient in isolation. Other needs may include the mathematical modeling of air quality and surface and subsurface water and soil quality. Thus, in addressing the statistical analyses of environmental quality data, the statistical analyses described in this book must be tempered with numerous other considerations. The use of statistics is only one of the tools we employ when interpreting environmental quality data. Statistical analyses are not an interpretation of the facts, but when properly used, these analyses make the facts easier to see and allow other evidence to enter into judgments about environmental phenomena (Unwin et al., 1985).

As will be seen in Chapters 4 through 12, an extensive theoretical basis exists for statistical methodologies. However, much of the traditional statistics literature is premised on the availability of very lengthy data records, a circumstance that seldom exists for environmental phenomena. In interpreting the data, we must therefore understand the limitations and assumptions of the statistical tests and how assumptions implicit in these tests may affect the interpretation. In addition to brief datasets, the difficulties of dealing with environmental data include the presence of numerous parameters, high degrees of variability in some constituents, and "censored" data where censored indicates that the magnitude of a constituent is known only as being less than some specified magnitude. As well, analyses must reflect the frequent occurrence of incompatible data due to different sampling methods, different laboratory analytical methods, and/or different times or spatial intensities of monitoring. Frequently, some of the information available is anecdotal. For example, some data are the result of nonstatistical sampling, grab samples, and vagueness in terms of where, how, and when the samples were collected. Conceptually, environmental processes are characterized as complex, multifaceted, chaotic, and dynamic, and in using statistics, we are trying to assist in the assessment regarding environmental quality phenomena. The result is that we cannot use many of the procedures described by statisticians in the available texts (or, at a minimum, we have to be inventive in how the tests are applied). One of the reasons for preparation of this book is to respond by using examples and discussions of the advantages and disadvantages of the various statistical analysis procedures and how statistics can be utilized to assist in these endeavors. Clear, complete, statistically defensible, and understandable information is essential in making informed decisions. Study of statistics attempts to model "order" in "disorder," such as the quantification of equation error or measurement error. One must always temper statistics with a clear understanding of the problem, so that spurious information is not introduced, nor valid information omitted. Statistical analyses of data are not an interpretation of the facts; statistical analyses are just another way of making the facts easier to see and therefore interpret (McBean and Rovers, 1990).

4.2 Characteristics of Environmental Quality Data

4.2.1 Indications of the Sources of Variability in Environmental Data

The nature of the variability of environmental quality data greatly influences the statistical analysis of the data. The specifics of statistical analyses will depend upon the ways the phenomena of interest are defined and sampled. In general, the ability of a sample of environmental quality data to characterize the population from which it is drawn is related to such aspects as (i) the size of the sample, (ii) the degree to which it was selected at random, (iii) the degree of independence between the observations comprising the sample, and (iv) how the data can be employed.

If extensive monitoring requirements are specified in governmental legislation, this will translate into sizable expenditures in terms of time and dollars to collect the data. Nevertheless, samples of environmental quality are just that, namely, subsets of the populations that are of interest. To clarify this somewhat, some of the features that contribute to the variability and problems of data analysis in groundwater quality phenomena are briefly described in Table 4.1. The result is that in many assessments of environmental phenomena, estimates of groundwater quality must be developed from only very brief data records, brief being with respect to both temporal and spatial considerations. The result is that there may be substantial quantities of data, but there is only a finite amount that is usable for specific applications.

These features stress the difficulty with statistical interpretation of environmental data in relation to risk assessments. There is a need for the analysis procedure to be sensitive to small changes (e.g. early detection of contamination is desirable), and yet a point of diminishing marginal returns also occurs. There may be more data collected than is necessary to make a decision, and thus, some of the money spent in data collection may be unnecessary. It is prudent to assess when the situation of diminishing marginal returns is being reached, such that additional data collection effort may be an inappropriate expenditure of money.

The result is that on many occasions we must be inventive in terms of how statistical analyses for risk assessment should proceed. Many standard statistical analysis techniques that are valid for long records have little credibility in situations possessing only a brief record or, at a minimum, the techniques must be modified to correct for biases created because of the brevity of the record. A further complicating factor is that many of the data records are highly variable or "noisy" due to, for example, seasonal phenomena. An additional consideration arising in part with increasing instrument technology is that we can measure features that previously were reported only as censored (less than) data. Further, a number of chemicals have maximum concentration levels (MCLs) to which humans and the environment are exposed while not incurring injury, where these MCLs are very close to the technological instrumentation capability. The result is that problems associated with statistical analyses of censored datasets are increasing.

For the variety of reasons previously indicated, there is not a single approach to statistical analyses. Instead, what is frequently needed is a series of approaches, each of which possess useful attributes that may be appropriate in addressing a particular question.

4.2.2 Independence of Successive Data Values

Time-series analyses are pertinent to the problem of estimating trends and cycles (e.g. seasonal variations). For example, consider the situation where there is a tendency for a groundwater monitoring well that is yielding high chloride concentrations today to also yield similar values

Table 4.1 Indications of the features that contribute to the variability and problems of data analysis in groundwater quality.

- Sample collection may involve drilling, sampling, and laboratory analysis for many water quality constituents. The expense argues for utilization of brief records
- Sample collection and laboratory analyses have inherent difficulties, resulting in uncertainties in subsequent interpretation of the findings
- Many important groundwater phenomena take years to evolve, making the available timeframe for sampling programs only statistically "windows" of temporally varying processes

tomorrow and for nearby monitoring wells to also yield sampling aliquots with elevated chloride concentrations. Hence, these types of sampling results are not necessarily independent, one from another. Similarly, replicate sampling (e.g. the splitting of field or laboratory samples into several samples) does not create independent samples. As a result, there are differing degrees of independence of monitoring results, and these aspects must be assessed during the statistical analyses of the resulting data.

Many statistical analysis procedures assume independence of data. Dependent samples exhibit less variability, and statistics determined from dependent data will therefore have underestimated sample variances. Dependence can severely influence the results of testing any hypothesis. See as an indication, Example 4.1, for an indication of hypothesis testing.

Concerns with hypothesis testing include those dealing with independence/dependence of successive samples. One approach to minimize dependence in samples is to allow sufficient time between sampling times to allow the "real-time memory" of the system to be exceeded. For example, the statistical analysis of annual peak flows in a river (the highest flow in a river in an individual year) involves statistical analysis of independent events since the peak flow in 1 year is unlikely to be related to the peak flow in a later year.

For some situations, we can avoid the problem of dependent data. We might "deseasonalize" the data by removing the periodic characteristic(s) associated with the seasons. However, the ability to remove seasonality in environmental quality data records may be constrained by the brief length of environmental data, which limits our ability to isolate the seasonal variability from the other sources of variability, which are present in the dataset.

Example 4.1 Example of Statistical Hypothesis Testing

Consider the question of whether a landfill is leaking leachate, which will contaminate the underlying groundwater, as depicted in Figure 4.1.

One way to consider this question is to monitor the groundwater quality both upgradient (Point A) and downgradient (Point B) from the landfill. We might then compare the quality at B and A to determine if there is a difference. Thus, a hypothesis might be the following.

Hypothesis: There is no statistically significant difference in the quality at the two locations.

Outcome 1: If we accept the hypothesis then, we are concluding there is insufficient information to indicate the landfill is leaking.

Outcome 2: If we reject the hypothesis, evidence exists to indicate that the landfill is leaking.

The question of hypothesis testing in environmental phenomena is a recurring one. The details of hypothesis testing will be a recurring question throughout this book, and the quantitative aspects of hypothesis testing will be left to later chapters.

4.2.3 Uncertainties and Errors in Environmental Quality Data

There are different levels of "observational" data. For example, "proxy" data are observations of one variable that have a high probability of being indicative of levels of another variable. Such data are indirect "observations." Another example is remotely sensed data by which many "indices" are derived from satellite data including temperature and vegetation. Image classification techniques may allow patterns in "data" to be recognized as signatures of certain environmental features.

In the strictest sense, there are virtually no data that are direct "observations" of a variable. Data on animal (or human) demography are collected by observing various signs, such as spores, tracks, nests, houses, income tax reports, and so on. Even something as obviously "observable" as the digital elevation on a mountain is subject to interpretation, according to the methods used to produce the measurements. The validity of these "data" depends on such features as one's

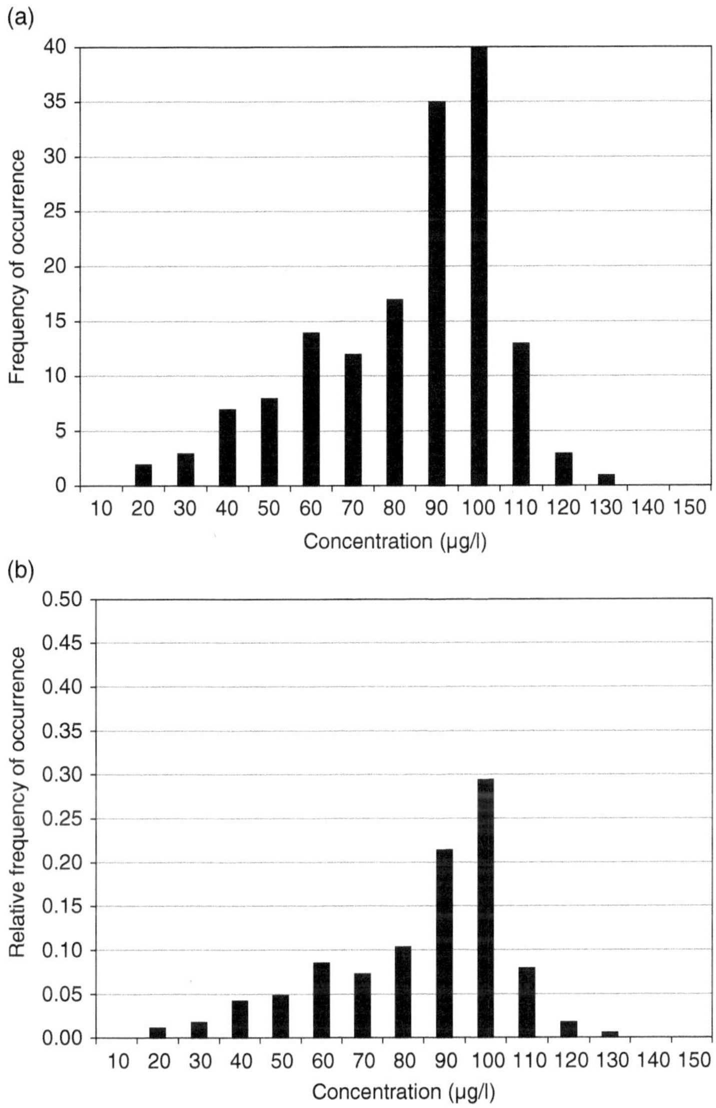

Figure 4.1 (a) Frequency-of-occurrence histogram of nickel concentrations. (b) Relative frequency of occurrence of nickel concentrations.

confidence in a particular measurement method and the reliability of calibration techniques of the instruments, all of which must then be considered when interpreting the data.

Quantifying observations involves the employment of sensory techniques, indirect measurements of related variables, and various levels of processing. It can then be argued that data are observed only in the context of the experimental design in which they are produced. It becomes a matter of interpretation as to what degree of processing is appropriate to produce a quantified observation that will then be called "data."

Errors in sampling procedures, inadequate sample storage and preservation techniques, and laboratory analytical errors are examples of errors in environmental datasets. As a demonstration of the multifaceted initiation points for such errors to exist in a dataset, further examples of the sources of error in the collection and analysis of groundwater quality data are listed in Table 4.2 and for air quality listed in Table 4.3.

Table 4.2 Examples of sources of error in the sampling and analysis of groundwater quality data.

- Sampling of a nonhomogeneous region in which wells and springs intersect more than one chemical type in water
- Piezometers and wells that are inadequately flushed out prior to sampling of groundwater so that a sample of the groundwater is not representative of conditions in the adjacent soil environment
- Sampling stations that are subject to temporal variations of chemical concentrations can exhibit significant sampling error
- Cross-examination of a sample may occur at the time of sampling as a result of an unclean container into which the sample is placed
- An error in the laboratory protocol of the experiment during the laboratory samples
- Improper preservation techniques. For example, groundwater samples are often particularly susceptible to changes in the pressure of oxygen and carbon dioxide and improper sample storage. Improper preservation techniques can result in a chemical alteration of the sample as it adjusts to new equilibrium conditions. For example, the pH levels in groundwater samples have been noted as increasing up to 1.0 pH units due to CO_2 escape to the atmosphere during storage

Table 4.3 Examples of possible sources of error in the collection of air quality data.

- Instrumentation error due to poor calibration or "drift" of the calibration of the instrument with time
- Channel error incurred during transmission from monitoring locations to a central data processing unit
- Fluctuations in meteorological conditions or quantities being released by the emission source resulting in nonrepresentative air quality conditions

There is always a degree of uncertainty associated with each discrete measurement of environmental quality. In interpreting data, each discrete measurement is really a range of statistically probable values instead of a single value. There are two subdivisions of reproducibility criteria, namely, replication and repeatability. Replication is when two or more results are obtained by the same operator in a given laboratory using the same apparatus of successive determinations on identical test material, within a short period of time. Often this is done during quality assurance and quality control testing of a laboratory to assure that the lab results are trustworthy. Alternatively, repeatability is a quantitative expression of the random error associated in the long run with a single operator in a given laboratory obtaining successive results with the same apparatus under constant operating conditions on identical test material. Obviously, the requirements for quality assurance and quality control can be substantial.

Many of the statistical analyses described in this book are concerned with sampling errors and the estimation of *population* characteristics from *samples* of data. The fact that sampling errors are inherent in random data does not mean, however, that statistical manipulations and sophistication can in any way overcome faulty data. The quality of any statistical analysis is no better than the quality of the data utilized. Furthermore, statistical considerations should not be used to replace judgment and careful thought in analyzing data. Statistics must be regarded as a tool or an aid to understanding, but never as a replacement for careful thought.

4.3 Some Summary Indications of Approaches for Statistical Analyses

The concern with the statistical interpretation of data is widespread. However, the variability encountered in one circumstance is quite possibly very different from that encountered in another circumstance. There is no convenient "recipe" that can be utilized to stipulate an approach that can be universally applied. Instead, statistical analyses of environmental data have become the science of collecting, analyzing, and interpreting data with the findings at each decision point assisting in identifying the next stage of analysis. Statistics is concerned with scientific methods for collecting, organizing, analyzing, summarizing, and presenting data as well as with drawing valid conclusions and making reasonable decisions. Statistical analyses do not consist of a standard set of rules. Instead, analyses involve successive tests and refinements, with each test an improvement in understanding of the data (and its information content) being gained. The findings of the tests may well be that additional statistical analyses are needed.

Since there is no convenient recipe, the practitioner needs to have available at his or her disposal a set of approaches, with each approach having utility in application to a class of problems. In selecting the procedure for use in a particular application, there are no absolute rules, only guidelines. To a large extent, the selection of the best procedure involves careful scrutiny of the characteristics of the problem at hand and the assumptions implicit in the particular statistical interpretation.

An analyst must still understand the basics in terms of both how to characterize a problem and the method by which the results may be interpreted. With the automated calculation procedures available in today's software, it is all too easy to employ a computer package without understanding the basis for the statistical procedures. The intent herein is to explain the features of the various techniques in terms understandable to the nonstatistician and to provide some order to the available procedures by presenting the advantages and disadvantages of the limitations of the procedures. The discussion included describes some of the sources of error and how these error sources should be reflected in the statistical analyses and in their interpretation. Proper application of statistical methods by someone who understands the utility and limitations of these methods can be most helpful in revealing the information that the data hold.

The emphasis in the chapters dealing with the statistical analyses of data is on examples to develop the equations and relationships and then to demonstrate the usefulness and application of the various procedures. References are provided for those wishing to pursue specific features in greater depth. Much of the theoretical background is omitted, with the focus being on the engineering applications to environmental quality data.

The various statistical methods are tools for data analysis that, like any tools, have proper and improper applications. The person using a statistical method is responsible to see that it is applied properly since the value of the results obtained depends on it. Even in the best circumstance, however, statistical data analysis can provide only evidence, never proof. Inferential statistics are aimed at distinguishing between random noise in the data and the real effects that are of interest. Only through careful consideration and interpretation of all the evidence can one hope to even begin finding answers to questions about causes for, and effects on, environmental quality.

4.4 Samples and Populations

As emphasized in previous chapters, for many environmental quality data, only brief data record lengths are available. The brevity of data on environmental quality bases is the result of numerous causal factors, including:

1) The expense of collecting and analyzing of field data
2) Limited sampling opportunities may exist due to the lengthy evolution over time of environmental phenomena

3) Heterogeneities in data exist arising from phenomena such as seasonality, trends, missing data, and the vagaries of nature, all of which limit the record length available for quantification of individual features.

Factors such as these often create a considerable challenge for the extraction of information from the available environmental data record.

The brevity of most environmental data records means that the statistical analysis procedures described in many books have only limited applicability to the analyses of environmental quality data. It must always be kept in mind having only samples, not population statistics, indicating that careful consideration must be considered as a result of the records "brevity."

Observed data represent only a sample from the population. Thus, the statistical procedures utilized must be sensitive to the fact that available data are only samples. As the sample size becomes very large, the summary descriptors of the samples approach those of the population values. Only a sample is collected because it is too expensive and time consuming to collect large numbers of measurements.

Statistical analyses are based on the idea that the sample will be "typical" in some way and that it will enable us to make predictions about the entire population. A sample statistic is then a value that expresses the property of the dataset and the sample information is utilized to estimate specific characteristics of the population. Because of the inherent variability in environmental quality data, a sample statistic will usually differ from the corresponding parameter of the population, but the sample information is the only information currently available. The process of estimating the characteristics of a population through sampling is referred to as inference, and questions of inference will be a focus throughout much of this book.

In mathematical terms, we deal with a sample of "n" individual values:

$$\{x\} = x_1, x_2, x_3, \ldots, x_n \tag{4.1}$$

where the {} will be used throughout this book to indicate a vector of individual values. The "n" individual observations are a sample from the set of possible outcomes of the experiment (population).

4.5 Probability and Statistics

Probability and statistics are two different concepts. Probability is a measure of the likelihood or chance that something will happen. It is a number between zero and one, with zero representing the impossible event and one representing the certain event. Alternatively, statistics are aspects that are measured or counted. The interesting part of statistics begins when attempts are made to find the relationship between one statistic and another.

4.6 Graphical Data Descriptors

The first step in data analysis is often a graphical study to improve our understanding of the characteristics of a data sample. These data descriptors are graphical and illustrate the character of the dataset, but do not summarize it. Often it is of interest to identify the probability distribution function (e.g. as will be of interest in Chapters 5 through 7); to assist in this, it is often helpful to develop a frequency histogram from the data as described below, which can be thought of as an estimate of the underlying probability distribution.

4.6.1 Histograms of Data

A histogram is a plot that indicates the frequency at which data occur within discrete intervals. Specifically, the relative number of occurrences of an event, for example, within the class interval or concentration interval 5–10 μg/l, is denoted as $P(E_1)$ and is the probability of event E_1, which equals the fraction n_1/n if the event occurred n_1 times in n sampled results. The frequency distribution is often represented in the form of a histogram, the principle of which is based on the fact that the area of each rectangle represents the proportion of observations falling in that interval.

A histogram or frequency histogram thus consists of a set of rectangles having (i) bases on a horizontal axis (the X axis) with centers at the designated class marks and lengths equal to the class interval sizes and (ii) areas proportional to the class frequencies. Boundaries of intervals are chosen so that no observation falls on a boundary.

Example 4.2 A frequency histogram of nickel concentration data as monitored in the groundwater is depicted in Figure 4.1a. For example, 11 reported values out of the total of 174 values are within the interval between 60 and 70 μg/l.

Alternatively, the relative frequency of occurrence histogram is presented in Figure 4.1b. Thus, for example, the ordinate frequency value of concentration in the interval between 60 and 70 ug/l is 11/174 or approximately 0.063. The sum of the ordinates over the range of individual class intervals equals unity.

According to Shaw (1964), a class interval is best chosen between one-quarter and one-half the standard deviation of the data (see Section 4.7.2 for details on how to calculate the standard deviation). If the class interval is selected as too large, the true form of the distribution is masked or blurred. If the class interval is selected as too small, then too many gaps appear in the resulting histogram. As the class interval of a histogram decreases for large samples, it becomes easier to pass a smooth curve through the classes. Thus, the effectiveness of a graphical analysis in characterizing the data (and for identifying the probability density function) depends on the sample size and the interval selected to plot the abscissa. For small samples, it is difficult to separate the data into groups that will provide a reliable indication of the frequency of occurrence.

Histograms are a familiar method of displaying numerical information and can be utilized to gain a rapid visual characterization of the data. As indications of this, concentration data for three possible situations are as depicted in Figure 4.2a–c where Figure 4.2a illustrates data that would be referred to as negatively skewed, meaning the long tail of the histogram is to the left; Figure 4.2b indicates data that are relatively symmetric; and Figure 4.2c illustrates data that would be referred to as positively skewed.

Obvious advantages of histograms as a means of visual representation of the data include the following:

i) The total range of the data in a sample is readily apparent.
ii) The most frequently occurring value, or mode, can be easily recognized.
iii) The range of greatest abundance can be readily identified.
iv) The general shape or character of the density distribution of data is apparent.

4.6.2 Probability Density Functions

As the number of samples "n" becomes large and the interval discretization small, the histogram will tend to a smooth continuous curve, referred to as the probability density function. Only a very large number of samples will permit a complete delineation of the probability density.

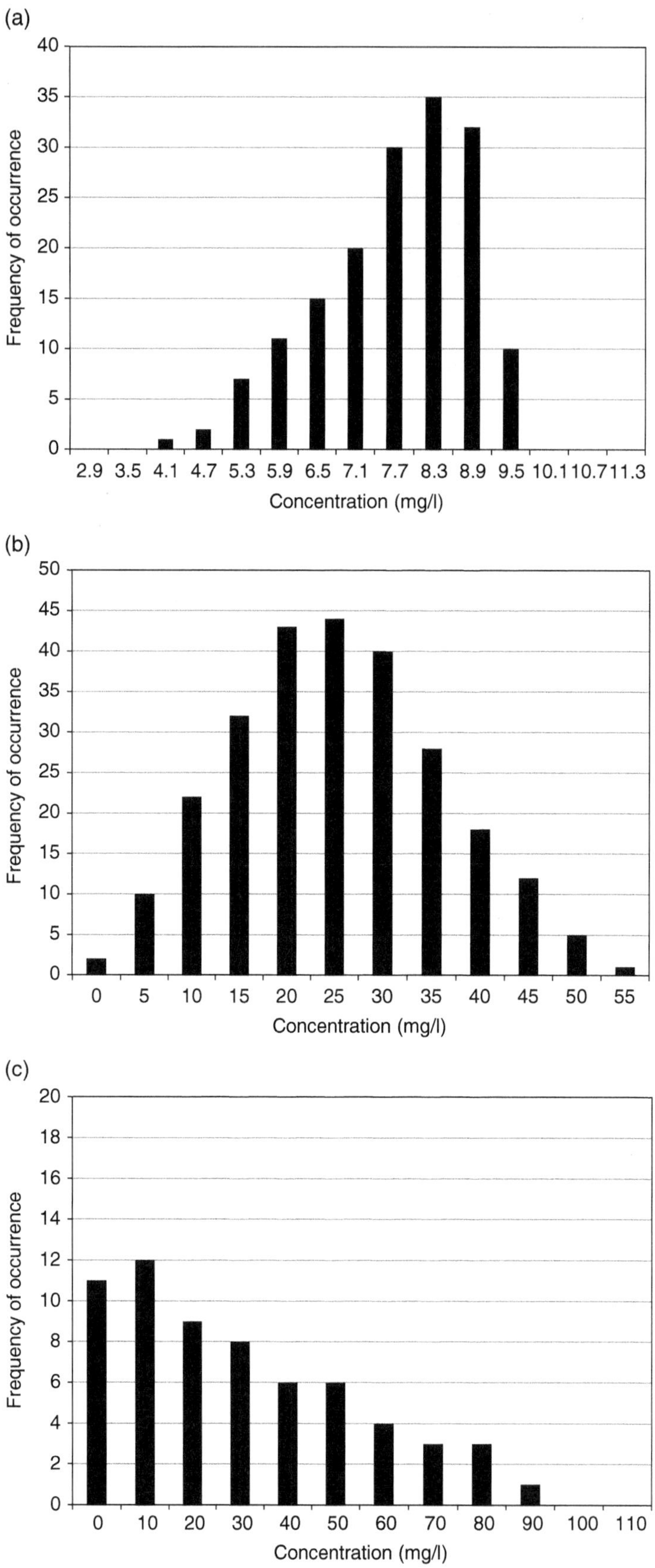

Figure 4.2 (a) Histogram of dissolved oxygen concentrations as an example of negatively skewed data. (b) Histogram of chloride concentrations as an example of symmetric data. (c) Histogram of lead concentrations as an example of positively skewed data.

A probability distribution is a plot of the probability density (i.e. relative frequency) versus the data values to describe the behavior of a random variable. Often the resulting plot of the probability distribution can be represented in equation form. This mathematical representation is called a probability density function and is denoted as f(x). Examples of candidate distributions relevant to environmental quality data include the normal distribution, the lognormal distribution, the Poisson distribution, and so on, the details of which will be examined in considerable extent in Chapters 5 through 7. For environmental engineering problems, we frequently assume a distribution and use the available data to "fit" a distribution to the data such as depicted in Figure 4.3. Since an array of distributions exist, we select the distribution to utilize for a particular application based on the assumption that the distribution appears to be a reasonable descriptor of the data. There are tests that evaluate the "goodness of fit" of the distribution to the data.

There is no requirement for the probability density to be unimodal or single peaked, as is the case for the example plotted in Figure 4.3. Generally, the probability density functions of data are single peaked, but for particular applications, the distribution may be bimodal, such as depicted in Figure 4.4. The distribution of data may be bimodal due to, for example, seasonal loading patterns where higher wastewater lagoon loadings are discharged to the receiving waterbody during the winter, as opposed to during the summer. However, for the majority of environmental quality data concerns, the distribution of the data may be assumed to be unimodal, and, equally important, most statistical procedures assume the distribution of data is unimodal.

It should be noted that there are environmental quality monitoring data and there are assumed probability density distributions. Of interest is the probability distribution that represents a good descriptor of the monitoring data; to ensure that this is the situation, one or more of a number of alternative tests will be utilized to ensure that an assumed distribution is a reasonable descriptor of the data. However, demonstrations of use of these tests are developed in Chapter 5.

4.6.3 Cumulative Distribution Functions

A probability distribution function can also be represented by its cumulative probability distribution function, F(x). The cumulative probability function is obtained by adding or

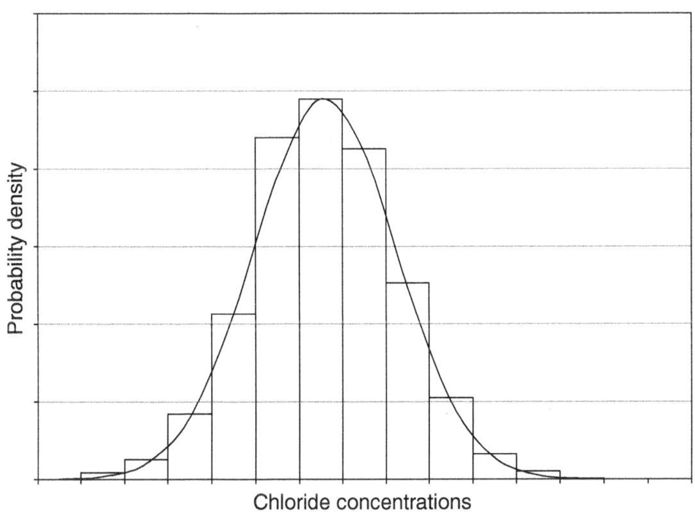

Figure 4.3 Probability distribution function fit to historic chloride data.

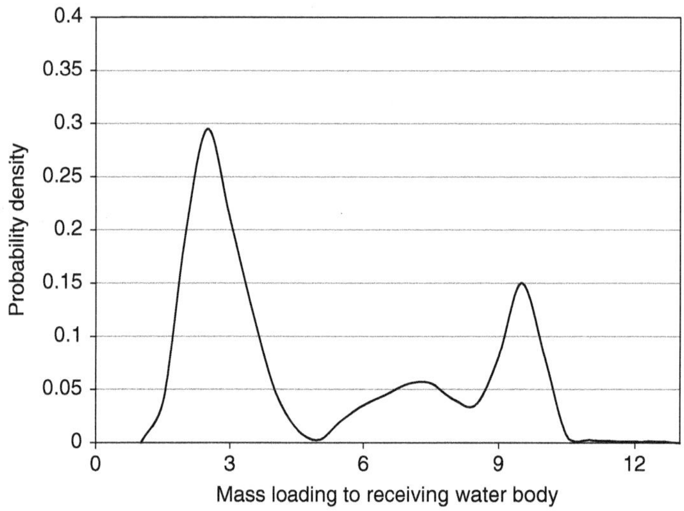

Figure 4.4 Example of bimodal distribution function.

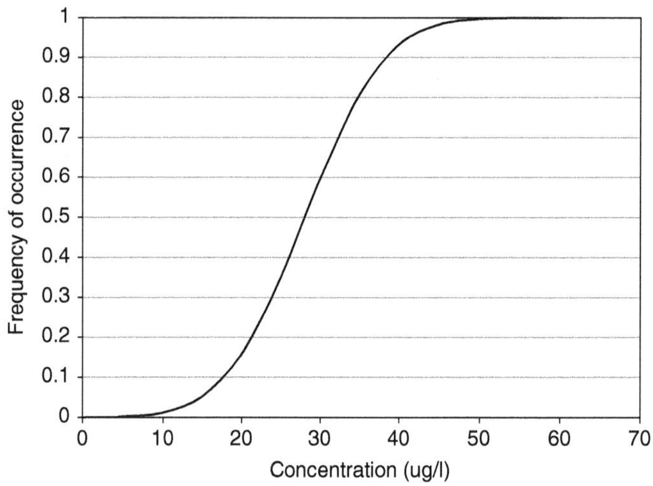

Figure 4.5 Cumulative distribution function of chloride concentrations.

accumulating the individual increments of the probability distribution function. The cumulative probability distribution function is defined as the probability that any outcome in X is less than or equal to a stated, limiting value x. In mathematical terms,

$$F(x) = \text{Prob}[X \le x] = \int_{-\infty}^{x} f(x)\,dx \tag{4.2}$$

As x approaches infinity, the area under the curve approaches unity (since by definition, the area under the probability distribution function is unity).

The probability distribution function is the slope of the cumulative probability function. An example of the cumulative distribution function is depicted in Figure 4.5, as derived from the data presented in probability distribution format in Figure 4.3.

4.7 Summary Measures of the Distribution of Data

There are many mathematical measures that may be used to objectively summarize the nature of a dataset. This is required for several important reasons including:

i) To help identify an appropriate probability distribution to characterize the dataset
ii) To characterize the data in a manner that will allow comparisons between datasets

For example, we might be interested whether the mean concentration of the river water quality changes as a result of a wastewater discharge. For this problem, we are interested in determining whether the summary statistics of the water quality are different for monitoring results both upstream and downstream of the point of discharge. The parameters that are utilized for these purposes are generally summary measurements of distribution properties; one is the central tendency or typical value of the data, and the other is the degree of spread of the distribution.

In keeping with the earlier comments that environmental quality records are normally brief, the emphasis in all of the discussion that follows is on sample statistics, as opposed to population statistics.

4.7.1 Measures of Central Tendency

Measures of central tendency are concerned with identifying the "average" value of a set of data. However, "average" is a very imprecise term. There are numerous ways to calculate a measure of central tendency, each of which has a different interpretation:

a) *Mean*: The arithmetic mean is the first moment about the origin (or as sometimes referred to as the center of gravity of the data), calculated by the following equation:

$$\bar{x} = \frac{\sum_{i=1}^{n} x_i}{n} \tag{4.3}$$

As the number of samples, n, increases, the sample mean $\bar{x}$ approaches the population mean, μ. The mean is the most commonly utilized measure of central tendency.

Example 4.3 Values of air quality monitoring results for benzo(a)pyrene were recorded as follows: 5.2, 7.3, 3.4, 4.4, 7.4, 4.3, 2.3, 1.2, 5.6, 4.5, 7.5, 6.1, 4.4 μg/m^3.

The mean of the monitoring results is

$$\bar{x} = \frac{63.6}{13} = 4.89 \text{ mg}/\text{m}^3$$

It is important to distinguish carefully between the mean $\bar{x}$ of a set of observations or a frequency distribution and the mean, μ, of the corresponding probability distribution function toward which $\bar{x}$ tends as the number of observations increases. Thus, $\bar{x}$ is the sample mean, while μ is the population mean. A measure such as the mean $\bar{x}$ that can be calculated from an observed frequency distribution is called a statistic, while the corresponding measure of the probability distribution, μ, is referred to as a parameter.

b) *Median*: The median is the middle value of the set of data, or the median has the property that one-half the values exceed it and one-half are less than the median. If the sample size, n, is an odd number, the sample median is the middle term when the samples are arranged in increasing order. When n is even, there is no middle term, and the median is defined as the average of the term just below the middle and the term just above the middle.

Example 4.4 Determine the median of the air quality measures of benzo(a)pyrene as reported in Example 4.3.

First, the values are ranked in order of magnitude as 7.5, 7.4, 7.3, 6.1, 5.6, 5.2, 4.5, 4.4, 4.4, 4.3, 3.4, 2.3, 1.2. Since the number of samples is odd, the median is 4.5. If the number of samples had been even, then the average of the middle two numbers is utilized.

Example 4.5 Determine the median and the mean of the hydraulic conductivity of a clay soil, given the three measurements.

The monitoring values (all in cm/s) are as follows:

$$3.8\times10^{-8},\quad 5.0\times10^{-8},\quad 4.8\times10^{-6}$$

The mean value $\overline{x}$ is

$$\overline{x} = 1.63\times10^{-6}$$

whereas the median, m, is

$$m = 5.0\times10^{-8}.$$

Note the difference in magnitude between the mean and the median values. This occurs because the mean is strongly influenced by the single value, which is two orders of magnitude higher than the second highest value. The result is that the mean, when there are a few very high values, is not necessarily a very good representative value of the central tendency of the dataset (since the mean is virtually the largest value divided by the number of samples and is insensitive to the magnitude of the two other values). The relative magnitudes are depicted in Figure 4.6.

As apparent from the values listed as part of Example 4.6, when there are outliers, the mean is not a good representation of the central tendency of the data. The median is superior to the mean as a measure of central tendency when the data are skewed.

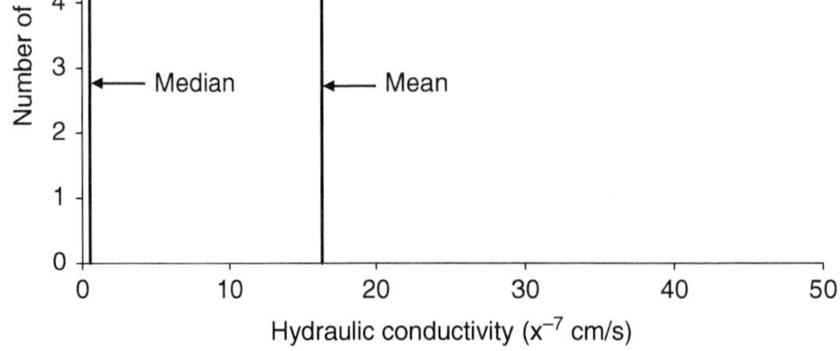

Figure 4.6 Comparison of mean and median values from Example 4.5.

c) *Mode*: The third measure of central tendency is the mode, the most commonly occurring value. The mode is not widely employed since it forms a poor basis for any further calculations of an arithmetical nature.

Example 4.6 The mode of the data illustrated in Figure 4.7 is 1.00 mg/l, the median is 2.71 mg/l, and the mean is 4.45 mg/l.

For a unimodal distribution (i.e. a distribution possessing a single peak), which is moderately skewed (asymmetrical), we have the approximate empirical relation

$$\text{Mean} - \text{mode} = 3(\text{mean} - \text{median}) \tag{4.4}$$

d) *Harmonic mean*: The harmonic mean H of a set of n numbers x_1, x_2, x_3 is defined as the reciprocal of the arithmetic mean of the reciprocals of the numbers, i.e.

$$H = \frac{1}{1/n\sum_{i=1}^{n} 1/x_i} = \frac{n}{\sum_{i=1}^{n} 1/x_i} \tag{4.5}$$

Example 4.7 Determine the harmonic mean of the monitoring values of 2, 4, and 8.

Using Equation (4.5) the harmonic mean is found to be 3.43.

e) *Geometric mean*: The geometric mean, G, of a set of n positive numbers $x_1, x_2, x_3, \ldots, x_n$ is the nth root of the product of numbers

$$G = \sqrt[n]{x_1 x_2 x_3 \ldots, x_n} \tag{4.6}$$

Example 4.8 Determine the geometric mean of the numbers 2, 4, and 8. Using Equation (4.6), the geometric mean is

$$G = \sqrt[3]{2 \cdot 4 \cdot 8} = 4$$

The geometric mean of a set of positive numbers $x_1, x_2, x_3, \ldots, x_n$ is less than or equal to their arithmetic mean but is greater than or equal to their harmonic mean:

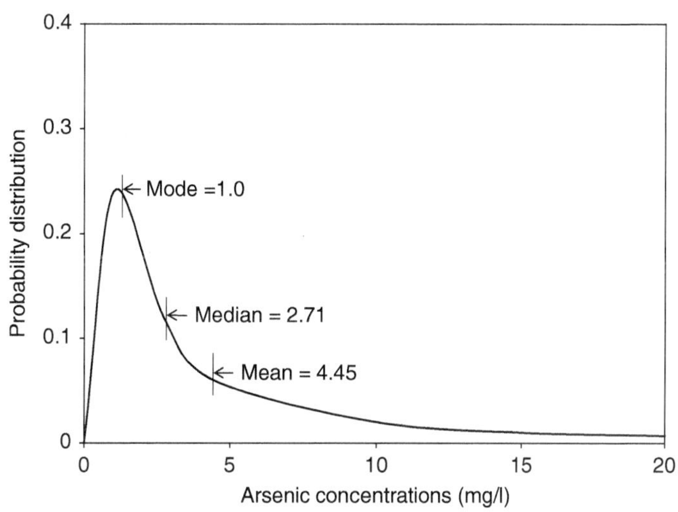

Figure 4.7 Schematic depiction of three measures of central tendency of data.

$$H \leq G < \bar{x} \tag{4.7}$$

The equality signs in Equation (4.7) apply only if all the numbers x_1, x_2, x_3, …, x_n are identical.

Example 4.9 The set of numbers 2, 4, 8 has arithmetic mean 4.67, geometric mean 4, and harmonic mean 3.43.

f) *Summary comments on measures of central tendency*: When the distribution of the data is symmetrical and unimodal, the values of the mean, median, and mode are identical. For example, this is true for the normal distribution. When the distribution of data is skewed, the relative locations of each of the measures of the central tendency are as schematically depicted in Figure 4.7.

4.7.2 Measures of the Dispersion of Data: Variance, Standard Deviation, and Range

The most common measures of dispersion or spread of the data around the mean are the variance, standard deviation, and range.

a) *Variance and standard deviation*: The more representative of the measures of dispersion of the data (in that they reflect the array of data) are the variance and standard deviation. The variance and standard deviation are defined as

$$S^2 = \text{Sample variance} = \frac{\sum_{i=1}^{n} (x_i - \bar{x})^2}{n-1} \tag{4.8}$$

and
Standard deviation of sample, S

$$S = \sqrt{\frac{\sum_{i=1}^{n} (x_i - \bar{x})^2}{n-1}} \tag{4.9}$$

The dimensions for the standard deviation are the same as the dimensions of the mean; this is one of the primary reasons that the standard deviation is utilized so extensively.

The simple variance estimates that are often used in risk assessment require that the data are independent or uncorrelated.

As characterized by both Equations (4.8) and (4.9), these measures are the sample estimates of the population parameters. The calculation of the population values would proceed in accord with the equations as follows:
Population variance

$$V = \sigma^2 = \frac{\sum_{i=1}^{N} (x_i - \mu)^2}{N} \tag{4.10}$$

and
Population standard deviation

$$\sigma = \sqrt{\frac{\sum_{i=1}^{N}(x_i - \mu)^2}{N}} \tag{4.11}$$

where (4.10) and (4.11) apply only when N is the number of elements in the population. As noted in Section 4.7, when the distribution of a random variable is not known, but a set of observations $(x_1, \ldots, x_n)$ is available, the moments of the distribution of n can be estimated with the sample values. However, two quantities are often considered when estimating these summary descriptors of the data, namely, the bias and the variance. The bias measures the difference between the average value of an estimate and the quantity being estimated, whereas the variance measures the spread or width. The bias is especially marked in small samples. A bias correction involving $((n-1)/n)$ is referred to as Bessel's correction function. The correction function is a defined approximation that is intended to reflect the uncertainty associated with including estimated parameters in the calculation of moments. For example, a correction function shows up in the variance and standard deviation calculation (but not in the calculation of the mean) because the calculation sequence for the variance and standard deviation includes the estimated mean, $\bar{x}$. In words, the variance is increased by the ratio $n/(n-1)$ to reflect the fact that implicit in the calculation sequence is the sample-estimated parameter, $\bar{x}$. Generally, with $n > 15$, the sample estimates of the standard deviation and the population value are similar. This is evident by examination of $n/(n-1)$; when n is large, the ratio of $n/(n-1)$ is very close to unity.

Example 4.10 Estimate the variance and the standard deviation of the lead concentration data listed below.

Lead concentrations (μg/l)				
37.0	34.9	28.4	27.3	31.8
28.3	26.9	19.9	31.8	32.6

In summary, $n = 10$,

$\bar{x} = 29.9$ μg/l and $S = 4.84$ μg/l

Degrees of freedom

The number of "degrees of freedom" of a statistic is defined as the number of independent observations, n, in the sample (i.e. the sample size) minus the number k of population parameters, which must be estimated from sample observations. In symbols, the number of degrees of freedom, df, is defined as

$$df = n - k \tag{4.12}$$

The term "degrees of freedom" will be referred to on numerous occasions throughout the book. It is defined here at its first utilization.

b) *Measures of the range of data*: An additional measure of dispersion or spread of data is the range. The range of a dataset is simply the difference between the largest and the smallest value. Since only two of the values from the entire dataset are utilized, the range is not a very useful parameter because it is sensitive to only these two values.

A measure similar to the range is the "quartile" and "10–90" percentile range (the range that contains 80% of the data). Quartiles divide the data into four parts, each containing the same number of values. The difference between the upper quartile and the lower quartile is referred to as the interquartile range.

A variation on the range, but one that is not so sensitive to the extreme values, is the 10–90 percentile range referred to above, defined as

$$P_{90} - P_{10} \tag{4.13}$$

c) *Mean deviation of data*: The mean deviation, or average deviation, of a set of n numbers x_1, x_2, ..., x_n is

$$\text{Mean deviation} = \frac{\sum_{i=1}^{n} |x_i - \bar{x}|}{n} \tag{4.14}$$

d) *Modified interquartile range* (*MIQR*): The MIQR is the difference between the 75th and 25th percentiles of the data divided by 1.34. The MIQR has a divisor of 1.34 because for data distributed in accord with the normal distribution (see Chapter 3 for further description of the normal distribution), the standard deviation (SD) = MIQR. If the SD and MIQR are quite different, the distribution of data may be substantially skewed, since data that are skewed have a high standard deviation. The advantage of the MIQR is that it is not influenced by any outliers.
e) *Box–whisker plots*: Turkey invented the box–whisker plot as depicted in Figure 4.8. The box's ends are the quartiles, with the median drawn inside the box. The "whiskers" extend out to the farthest points that are not outliers. Box–whisker plots are especially used in illustrating the differences between groups of data. Box–whisker plots are potentially useful ways of summarizing the various measures of the spread or dispersion of the data. Box–whisker plots are effective for exploratory data analysis as a way to visualize the "spread" of the data. Specifics of box–whisker plots vary from one application to another, but the essence involves depiction of the mean, a measure of the spread (such as the quartile) and the range (highest to lowest). The spread is the "box" and the range is the "whiskers." Figure 4.8 provides an example of a box–whisker plot and demonstrates disease burden for

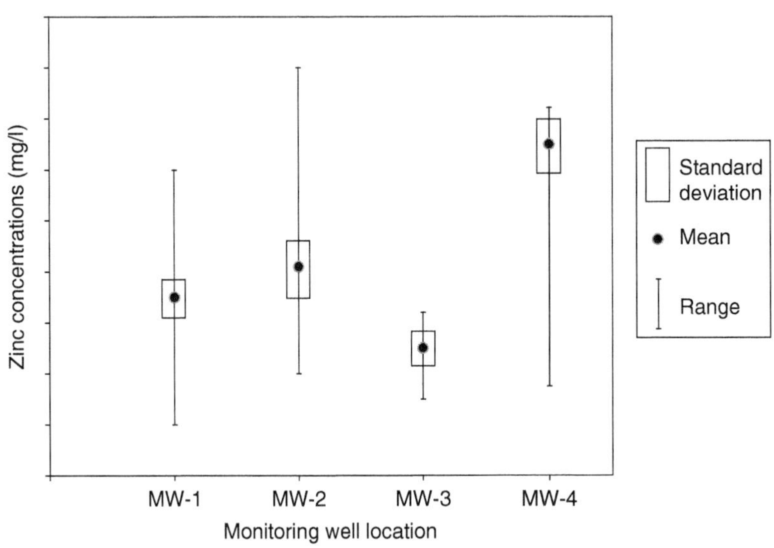

Figure 4.8 Box–whisker plot as summaries of environmental quality data.

a number of water supply sources (after Thomas et al., 2015). The box–whisker plots present a useful and quick graphical summary of data for different locations or from one time period to another.

4.7.3 Skewness

The mean and standard deviation are useful for summarizing the properties of "reasonable" symmetrical distributions. To understand the characteristics more fully, another common descriptor of a dataset is the skewness, a measure of the symmetry. The sample estimate of the skewness "a," is determined from

$$a = \frac{n\sum_{i=1}^{n}(x_i - \bar{x})^3}{(n-1)(n-2)} \tag{4.15}$$

The population measure of the skewness, α, is calculated from

$$\alpha = \frac{\sum_{i=1}^{N}(x_i - \mu)^3}{N} \tag{4.16}$$

The sample estimate differs from the population estimate as a result of the bias correction, hence the basis for the n/((n − 1)(n − 2)) term in Equation (4.14).

Because the magnitude of the skewness as determined using Equation (4.15) is dependent upon the units of the variable employed, a common procedure is to calculate a normalized skewness, C_s. C_s is calculated by dividing Equation (4.15) by the cube of the standard deviation, or

$$C_s = \frac{n\sum_{i=1}^{n}(x_i - \bar{x})^3}{(n-1)(n-2)S^3} \tag{4.17}$$

where S is the standard deviation defined in Equation (4.9). C_s is referred to as the coefficient of skewness and is dimensionless. Since Equation (4.17) is normalized (or unitless), it is effective at comparing environmental quality records over time, space, and/or different constituents.

As most distributions are unimodal (i.e. have a single peak), a primary difference between distributions is their degree of symmetry. Distributions of data that are symmetric will have counterbalancing effects in terms of positive and negative deviations from the mean, with the result that the calculation of skewness using Equation (4.17) will approach zero. Distributions whose right-hand tail is longer than its left-hand tail (e.g. Figure 4.2c) are referred to as skew to the right (i.e. C_s is positive or is said to be positive skew), and those with a longer left-hand tail (e.g. Figure 4.2a) as skew to the left (i.e. C_s is negative or is said to be negative skew). The skewness of a symmetric distribution such as depicted in Figure 4.2b will approach zero.

As a guideline of magnitudes, a distribution is considered to be highly skewed if the absolute value of the coefficient of skewness is >1. Distributions with C_s from 0.5 to 1 are considered as moderate skew, and those with C_s values from 0 to 0.5 as essentially symmetrical. It is noteworthy that the skewness coefficient value is very sensitive to the number of data points because the calculations involve the cube of deviations from the mean. A robust estimate (i.e. an estimate that is relatively insensitive to individual data values) of the skewness needs approximately 50 data points, but this is just a general guideline and should be assessed for specific applications.

Example 4.11 Calculate the skewness of the data from Example 4.9.
Using Equation (4.14)

$$a = -74.4.$$

The negative value of "a" indicates that the data are skew to the left. In fact, the single value of 19.9 contributes the majority of the numerator.

The normalized skewness is determined using Equation (4.17) as

$$C_s = -0.66.$$

This magnitude of skewness indicates the data are moderately skewed.

4.7.4 Kurtosis

The fourth moment about the mean is the kurtosis, which is a measure of the "peakedness" of the distribution. The sample estimate of the kurtosis is obtained from the following calculation:

$$K = \frac{n^2 \sum_{i=1}^{n} (x_i - \bar{x})^4}{(n-1)(n-2)(n-3)} \tag{4.18}$$

Similar to the skewness, kurtosis is also dimension specific and thus is usually made dimensionless by dividing by the standard deviation to the power 4, i.e.

$$C_K = \frac{K}{S^4}. \tag{4.19}$$

An additional related transformation usually employed is to subtract "3" from the coefficient of kurtosis:

$$C_{K'} = \frac{K}{S^4} - 3. \tag{4.20}$$

The basis for Equation (4.20) is that C_K' will equal zero when the data are normally distributed. The normal distribution will be described fully in Chapter 5 and thus will not be fully described at this point – it is mentioned here for the sake of completeness.

If the data being characterized has a relatively greater concentration of probability mass near the mean than does the normal distribution, the kurtosis is greater than 3 and the distribution is referred to as leptokurtic, as depicted in Figure 4.9. If the data being characterized has a relatively smaller concentration of probability near the mean than does the normal distribution, the kurtosis will be less than 3 and the distribution is referred to as platykurtic, as depicted in Figure 4.10. C_K' is positive for leptokurtic, negative for platykurtic, and zero for normal (mesokurtic).

Example 4.12 Calculate the kurtosis and the coefficient of kurtosis for the data listed in Example 4.9.

The sample estimate of the kurtosis is calculated using Equation (4.18) as

$$K = 2655.$$

The coefficient of kurtosis is calculated using Equation (4.20) as

$$C_K = 4.85$$

and

$$C_K = 1.85.$$

Hence, the data would be described as leptokurtic.

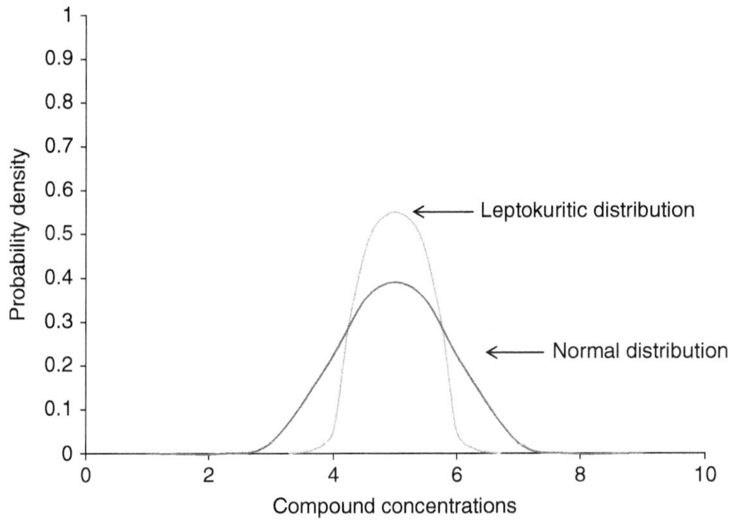

Figure 4.9 Leptokurtic distribution.

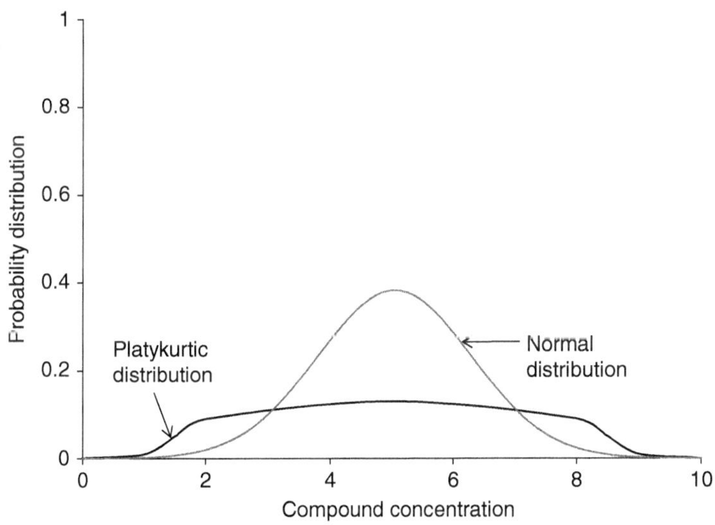

Figure 4.10 Platykurtic distribution.

4.7.5 Some Summary Comments

Apparent from the preceding sections is the acknowledgment of sample statistics being potentially different from population statistics. A summary of the equations for calculating the sample statistics and the population statistics is provided in Table 4.4.

If only brief data records are available, there is little merit in going beyond the fourth moment (kurtosis) as a means of characterizing the data, and thus moments higher than kurtosis are infrequently utilized in environmental quality assessments. Since the calculation of the moments involves $(x_i - \bar{x})^m$, where "m" refers to the moment about the mean, when m is large, the resulting moment is very sensitive to individual values.

Table 4.4 Summary of equations for samples and populations.

	Sample equations	Population ("N" very large)
Mean	$\bar{x} = \frac{\sum_{i=1}^{n} x_i}{n}$	$\mu = \frac{\sum_{i=1}^{N} x_i}{N}$
Variance	$S^2 = \frac{\sum_{i=1}^{n}(x_i - \mu)^2}{n-1}$	$V = \sigma^2 = \frac{\sum_{i=1}^{N}(x_i - \mu)^2}{N}$
Standard deviation	$S = \sqrt{\frac{\sum_{i=1}^{n}(x_i - \bar{x})^2}{n-1}}$	$\sigma = \sqrt{\frac{\sum_{i=1}^{N}(x_i - \bar{x})^2}{N}}$
Skewness	$a = \frac{n\sum_{i=1}^{n}(x_i - \bar{x})^3}{(n-1)(n-2)}$	$\alpha = \frac{\sum_{i=1}^{N}(x_i - \mu)^3}{N}$
Kurtosis	$K = \frac{n^2\sum_{i=1}^{n}(x_i - \bar{x})^4}{(n-1)(n-2)(n-3)}$	$k = \frac{\sum_{i=1}^{N}(x_i - \mu)^4}{N}$

4.8 Further Summary Measures of the Distribution of Data

Section 4.7 developed the equations for the moments of the distribution of data. Employing these equations, several other summary measures exist for which there is significant utility and they will be described in the section to follow.

4.8.1 Coefficient of Variation

The coefficient of variation (COV) is often used to describe the relative amount of variation in a population. The sample estimate for the COV is defined as

$$COV = \frac{S}{\bar{x}} \tag{4.21}$$

The COV is dimensionless and thus is effective at characterizing the relative variability of a distribution. The magnitude can be used to compare the extent of variability at alternative sampling locations. The utility of the COV as a measure arises since the data from different populations or sources, the mean and standard deviation, tend to change together so that the COV tends to remain constant. Being dimensionless, it represents a normalizing parameter to characterize the scatter of the data.

A small change from Equation (4.21) is the COV expressed as a percent, simply as

$$COV' = \frac{100S}{\bar{x}} \tag{4.22}$$

Example 4.13 Monitoring results at three sampling locations are listed below. Compare the sampling records in terms of the variability.

		Sampling location		
		A	B	C
		40.0	19.9	37.0
		29.2	24.1	33.4
		18.6	22.1	36.1
		29.3	19.8	40.2
$\overline{x}$	=	29.28	21.48	36.68
S	=	8.74	2.05	2.80
COV	=	0.30	0.10	0.07

In terms of variability (as quantified by the coefficient of variation), the concentrations at A vary the most, followed by B and then C.

The COV is also useful and simple as an indicator of skewness. If the data are likely skewed (either positively or negatively), the value of the COV will be large (>1) since the standard deviation is larger than the mean.

4.8.2 Standard Error of the Mean

Statistics like the mean are derived from individual random variables and so are also random in nature; they are random variables. As one of these random variables, the standard error of the mean, $SEM_{\overline{x}}$, is defined as

$$SEM_{\overline{x}} = \frac{S}{\sqrt{n}} \tag{4.23}$$

The standard deviation and the standard error of the mean $(SEM_{\overline{x}})$ are related to each other and yet are quite different. The $SEM_{\overline{x}}$ is smaller than the standard deviation, S. The size difference occurs because the $SEM_{\overline{x}}$ is an estimate of the error (or uncertainty) involved in estimating the mean of a sample and not an estimate of the error (or variability) involved in measuring the data from which the mean is calculated. Variability of the sample mean is much less than that of any single observation, allowing us to better pinpoint the true population mean from the sample average than from any single measurement. The $SEM_{\overline{x}}$ is used when the uncertainty of the estimate of the mean is of concern. The uncertainty decreases as the sample size increases.

In practice the confidence interval (CI) of the mean is expressed as

$$CI = \overline{x} \pm t\frac{S}{\sqrt{n}} \tag{4.24}$$

where $\overline{x}$ is the mean, S is the standard deviation, n is the number of samples, and t is the t-distribution value. The value of "t" is a function of the number of degrees of freedom, df (see Section 4.7.2), and a level of significance.

For the 95% CI, the value of t is as noted in Table 4.2.

The use of "t" compensates for the tendency of a small number of values to underestimate the uncertainty. A more complete discussion of the basis for the value of "t" is contained in Section 5.4. The CI information is presented here only for the sake of completeness and will be dealt with more extensively in Chapter 8.

Table 4.5 Table of t-values for different sample sizes.

Number of samples n	$t_{0.05,df}$
2	12.71
3	4.30
4	3.18
5	2.78
10	2.26
∞	1.96

Example 4.14 For the data listed below, compute the 95% confidence interval estimate of the mean

$$25.0, 36.1, 45.2, 19.6, 21.8 \text{mg/l}.$$

where mean $\bar{x}$ = 29.4 mg/l; standard deviation S = 10.8 mg/l and number of samples, n = 5. The t value from Table 4.5 is 2.78. Hence, the 95% confidence interval is

$$\text{Confidence interval} = 29.4 \pm \frac{2.78(10.8)}{\sqrt{5}}$$

$$= 17.26 - 45.70 \text{ mg/l} \quad \text{for the 95\% confidence interval for the mean.}$$

4.8.3 Standard Errors

The standard deviation of a sampling distribution of a statistic is often referred to as its standard error. Summarized in Table 4.6 are the standard errors for a number of statistics. The

Table 4.6 Equations for the standard error of alternative statistics.

Sampling distribution	Standard error	Remarks
Means	$\sigma_{\bar{x}} = \frac{\sigma}{\sqrt{n}}$	This is true for large or small samples. The sampling distribution of means is very nearly normal for $n \ge 30$ even when the population is nonnormal
Standard deviation	$\sigma_s = \frac{\sigma}{\sqrt{2n}}$	For $n \ge 100$, the sampling distribution of S is very nearly normal, σ_s is given by this formula only if the population is normal (or approximately normal)
Median	$\sigma_{med} = \sigma\sqrt{\frac{1}{2n}}$	For $n \ge 30$ the sampling distribution of the median is very nearly normal. The given result holds only if the population is normal (or approximately normal)
Coefficient of variation	$\sigma_v = \frac{v\sqrt{1+2v^2}}{\sqrt{2n}}$	Here $v = \sigma/\mu$ is the population coefficient of variation. The given result holds for normal (or nearly normal) populations and $n \ge 100$

Table 4.7 Summary of equations for additional descriptions of data.

Coefficient of variation	$COV = S / \bar{x}$
Standard error of the mean	$SEM_{\bar{x}} = S / \sqrt{n}$
Standard error of the standard deviation	$SE_s = S / \sqrt{2n}$

standard error of the mean from Section 4.8.2 is noted as the first row of Table 4.6, and the other rows relate to other less frequently used statistics.

4.8.4 Summary Descriptors

The summary equations for the additional descriptors are listed in Table 4.7. These measures are utilized for subsequent analyses and will be referred to at various points in the book.

4.9 Conditional Probability and Bayes Theorem

4.9.1 Basic Probability Concepts

Developing an understanding of conditional probability is one of the most important concepts for risk assessment. It will be done in the following segment using a series of examples.

Example 4.15 Consider the outcome of rolling of a die. What is more is likely:

i) Rolling at least one 6, in four throws of a single die? Or,
ii) Rolling at least one double 6 in 24 throws of a pair of dice?

For (i), the chance of one 6 = 1/6. The average number of 6's in four rolls = 4 × 1/6 = 2/3.

For (ii), the assessment is not so simple. For one of the dice, there is 1/6 of getting a 6, but then probability of getting a second 6 (and hence two 6's) is a conditional probability. To address this, it is important to assess the characteristics of probability.

The total probability of all outcomes is unity

$$P(O_i) \geq 0.$$

$$P(O_1) + P(O_2) + \cdots + P(O_n) = 1.$$

The array of possible outcomes includes those listed in Table 4.8.

Combining events with logical operations involves "AND," "OR," and "NOT." If we have two events, X and Y, we can assess the probabilities of additional events, namely,

X and Y: Both events X and Y occur.
X OR Y: The event X or the event Y occur.
NOT X": The event X does not occur.

These events can now be used to estimate probabilities of alternatives. It is noted that if two events have nothing in common, they are referred to as "mutually exclusive." Note that

Table 4.8 Basic outcomes from dice roles.

Event description	Possible outcomes	Probability
A: Dice add to 3	(1,2), (2,1)	P(A) = 2/36
B: Dice add to 6	(1,5), (2,4), (3,3), (4,2), (5,1)	P(B) = 5/36
C: One of the dice shows 1	(1,1), (1,2), (1,3), (1,4), (1,5), (1,6)	P(C) = 6/36
D: The other of the dice shows 1	(1,1), (2,1), (3,1), (4,1), (5,1), (6,1)	P(D) = 6/36

$$P(X) = 1 - P(\text{NOT } X).$$

And this is valuable as it is easier to characterize P(NOT X) than it is to characterize X.
Event X: a double "6" not rolled.

$$P(X) = 1 - P(\text{NOT } X) = (1 - 36)/36 = 35/36.$$

The multiplication rule is

$$P(X|Y) = \frac{P(X \text{ and } Y)}{P(Y)}.$$

Or rearranging

$$P(X \text{ AND } Y) = P(X)P(Y).$$

Then the probability of X is the event of getting at least one 6 in four rolls of a single die:

$$\begin{aligned} P(\text{NOT } X) &= P(A_1 \text{ AND } A_2 \text{ AND } A_3 \text{ AND} A_4) \\ &= (5/6)^4 = 0.482. \end{aligned}$$

And hence

$$P(X) = 1 - P(\text{NOT } X) = 1 - 0.482 = 0.518.$$

Next is the task of determining the probability of getting at least one double 6 in 24 throws.
Then P(NOT Y) = P(A_1 AND A_2 AND…A_{24})
where the probability of each A (no double 6 in the ith roll) is

$$P(A_i) = \frac{35}{36}.$$

Therefore, P(NOT Y) = $(35/36)^{24}$ = 0.509.
So the probability

$$\begin{aligned} P(Y) &= 1 - P(\text{NOT } Y) \\ &= 1 - 0.509 = 0.498 \end{aligned}$$

The probability of getting at least one 6 in four throws = 0.518 and the probability of at least one double 6 = 0.498.

4.9.2 Bayes' Theorem

It is now useful to carry it to the next step to lead to hypothesis testing. Again, this will be demonstrated by an example.

Example 4.16 What is the probability that if a coin is flipped 20 times, one gets heads 20 times?

P(X) = 20 tails in 20 flips of a coin
P(NOT X) = no tails in 20 flips of a coin

Assign A = event of getting a tail given each toss of the coin is independent:

$$\begin{aligned} P(\text{NOT X}) &= P(A_1 \text{ AND } A_2 \text{ AND} \ldots \text{AND } A_{20}) \\ &= (0.50)^{20} = 9.5 \times 10^{-7}. \end{aligned}$$

The result is that it is highly unlikely but it is possible to get heads 20 times.

The utility of these relationships is that it demonstrates the pathway to get to Bayes' theorem. This is important in making good decisions when there is imperfect information.

Example 4.17 Assume a rare disease infects 1 out of every 1000 people. There is a good (but not perfect) test to determine if a person has the disease. If a person has the disease, the test comes back positive 99% of the time. However, the test also produces some false positives where about 2% of uninfected patients also test positive.

There are two events.

Event X: Patient has the disease.
Event Y: Patient tests positive.

P(X) = 0.001	1 patient in 1000 has the disease
P(Y|X) = 0.99	Probability of a positive test, given infection is 99%
P(Y|NOT X) = 0.02	Probability of a false positive, given no infection is 2%
P(X|Y)	Probability of having the disease, given the test was positive

	X	NOT X	
Y	X and Y	NOT X and Y	
NOT Y	X and NOT Y	NOT X and NOT Y	
	X	NOT X	SUM
Y	P(X and Y)	P(NOT X and Y)	P(Y)
NOT Y	P(X and NOT Y)	P(NOT X and NOT Y)	P(NOT Y)

Test result:

$$P(\text{X and Y}) = P(Y|X)P(X) = 0.99(0.01) = 0.00099$$

$$\begin{aligned} P(\text{NOT X and Y}) &= P(Y|\text{NOT X})P(\text{NOT X}) \\ &= 0.02(0.999) = 0.019\ 98. \end{aligned}$$

	X	NOT X	SUM
Y	0.000 99	0.019 98	0.020 97
NOT Y	P(X and NOT Y)	P(NOT X and NOT Y)	P(NOT Y)

$$P(X) = 0.001 \quad P(\text{NOT X}) = 0.99$$

$$P(X|Y) = \frac{P(X \text{ and } Y)}{P(Y)} = \frac{0.000\,99}{0.020\,97} = 0.0472$$

Hence, despite the high accuracy of the tests, less than 5% of those who test positive actually have the disease.

Of a group of 1000 people tested, on average, only 21 people will test positive and only 1 of them has the disease. Twenty false positives come from the much larger unaffected group. Issues of false positives are frequent.

	Disease	No disease	Sum
Tests positive	1	20	21
Tests negative	0	979	979
Sum	1	999	1000

4.10 Summary

One of the best ways to describe statistically quantitative uncertainty is by using probability distributions. This will be the focus of discussion in Chapters 5 through 7. We will use the summary statistics noted in this chapter to fit the distributions to the data.

Descriptions herein of alternative procedures to summarize sets of data into some useful statistics have been provided. These summary statistics will be utilized extensively throughout the remaining chapters of this book.

4.11 References

McBean, E. and Rovers, F. (1990). Flexible selection of statistical discrimination tests for field-monitored data. In: *Groundwater and Vadose Zone Monitoring* (ed. D.M. Nielsen and A.I. Johnson), 256–265. Philadelphia, PA: ASTM.

Shaw, D.M. (1964). *Interpretation Geochimique des elements en trace dans les roches cristallines.* Paris, France: Masisson et Cie.

Thomas, K., McBean, E., Schantz, A., and Murphy, H. (2015). Comparing the microbial risks associated with household drinking water supplies used in Peri-urban communities of Phnom Penh, Cambodia. *Journal of Water and Health* 13 (1): 243–258.

Unwin, J., Miner, R.A., Srevers, G., and McBean, E. (1985). Groundwater quality data analysis. *NCASI Technical Bulletin No. 462.*

4.12 Problems

4.1 Measurements of concentration from 200 samples show a mean of 0.839 mg/l and a standard deviation of 0.050 mg/l. Find the 95% confidence limits for the mean concentration. Does this allow the data to be assumed to be described by a normal distribution?

4.2 The background measurements for a groundwater monitoring well are 0.8, 3.1, 1.7, 0.6, 1.1, 2.8, 1.8, and 0.9 for magnesium. Calculate the 95% limits on the mean of the background concentrations using these data.

4.3 Using the data from the above problem (i.e. data before you log-transformed everything), calculate the arithmetic mean and the geometric mean. Which one is a better indicator of the central tendency of the data?

4.4 You are informed that the risk of illness being caused by a single visit to a hospital (any one hospital of many possible) is 0.000 06 per visit. Your task is to assess the risk potential for society. Given this, what will be the units of the standard deviation of illness potential?

4.5 Explain why Bessel's correction function is employed in the estimate of the variance.

4.6 We flip a coin 36 times and for each of the coin tosses, the result is tails.
a) What is the chance of obtaining a head on the 37th toss?
b) If we say the coin is fair, what evidence can we use to support the claim?

4.7 In testing whether the consumption of fish is hazardous to human health, you arrange to have collected a sample set of six individual fish to have analyzed for chemical concentrations. However, there are two dangers you have to face when analyzing the data – one is determining a false positive and one determining a false negative. Explain what both of these terms mean in this type of application.

4.8 Explain in technical terms why we need to employ Bessel's correction function for calculating the variance. What is the implication of Bessel's correction function?

4.9 Your technician has been analyzing water quality samples using a chloride meter. At the end of the assembly of data, you realize that the calibration curve that the technician was employing was incorrect. Each individual measurement should have been two times as large as the technician was reporting. Would the variance of resulting data change, and if so, how (approximately)?

4.10 The following indicate concentration levels of TCE in a monitoring well downstream from an industrial sewage lagoon. Assuming that the data are lognormally distributed, calculate the arithmetic mean of the data transformation.

After log transformation, which of the two measures of central tendency is a better representation of the data, and why?

6.323	4.71	5.616	4.191	4.836	6.322	3.516	
4.106	2.056	5.973	5.157	3.609	6.042	3.013	5.354

Part II

Characterization of Common Distributions

5

The Normal or Gaussian Distribution

The probable is usually what happens.

—Aristotle

5.1 Introduction

One of the objectives of data analysis is to use available sample observations to identify the most appropriate population distribution function and to describe the problem of interest. This is important because if we can select a probability distribution function to describe the distribution of the data, then we can make inferences based on the known statistical properties of the selected distribution. Thus, we are interested in selecting one distribution from many possible population distribution functions that exist (e.g. normal, lognormal, beta, gamma, and so on), which can be used to make probability statements.

Although a number of alternative distributions may, in theory, be necessary to describe the wide variety of features that exist in environmental quality concerns, in practice, a majority of environmental quality variables exhibit frequency distribution shapes that can be approximated by a relatively small number of distributions. In fact, two distributions, namely, the normal distribution and the lognormal distribution, are applicable to a widespread set of concerns. There is an important supporting rationale for the widespread utility of the normal and lognormal distributions as will be described in this and the following chapter. Specifically, this chapter focuses on the normal distribution or, as it is also frequently referred to, the Gaussian distribution, and Chapter 6 focuses on the lognormal distribution.

The terms "Gaussian" and "normal" refer to precisely the same distribution and the two terms are frequently used interchangeably. As reported by Bulmer (1965), the normal distribution was discovered in 1733 by Huguenot refugee Abraham de Moivre, although this distribution is more frequently associated with the name of Gauss who derived it in 1809. The term "normal" is utilized because so many outcomes may be described by the distribution. Henceforth, in this book, the distribution will be referred to by the term "normal" distribution.

The normal distribution applies to many applications, including those within the environmental quality field, in large part because of a mathematical property known as the central limit theorem. The central limit theorem is one of the most significant theorems in probability and pertains to the limiting distribution of a sum of random variables. The theorem states that "under very general conditions," as the number of variables in a sum becomes large, none of which is dominant, the distribution of the sum of random variables approaches the normal distribution, almost regardless of the distribution of the individual variables. Thus, any variable

Risk Assessment: Procedures and Protocols, First Edition. Edward A. McBean.

Companion website: www.wiley.com/go/McBean/RiskAssessProcedure_1e

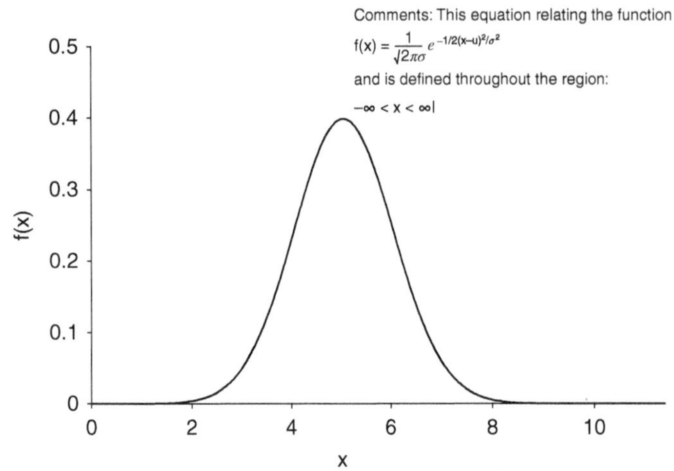

Figure 5.1 Schematic of the normal or Gaussian distribution.

that can be regarded as the sum of a large number of small independent contributions is likely to follow the normal distribution. If a physical or chemical process in environmental phenomena is the result of the totality of a large number of individual effects, then according to the central limit theorem, the sum of the individual effects will tend to be normally distributed; the existence of some of the individual effects on the high side is counterbalanced by an (approximately) equivalent number of individual effects on the low side such that the resulting shape is symmetrical, as depicted in Figure 5.1.

As will be apparent in the chapters to follow, the normal distribution is the assumed underlying distribution for numerous statistical tests. The importance of the normal distribution will be evident and consequently will be dealt with extensively in this chapter. The normal distribution will be the assumed underlying distribution for tolerance intervals and confidence intervals (Chapter 8), the t-test (Chapter 9), and parametric ANOVA (Chapter 12), among others. The failure of data to follow a normal distribution (at least approximately) precludes subsequent tests from being utilized, for otherwise, incorrect conclusions may be reached. Thus, a sizable percentage of this chapter is focused on various procedures to determine the extent to which environmental quality data can be characterized by the normal distribution.

5.2 The Mathematics of the Normal Distribution

The first step of data analysis is to identify the probability distribution function that best describes the random nature of a particular variable. A primary purpose for examining the features of a distribution such as the normal distribution is to characterize a set of data. The intent is to use the individual data to "fit" a distribution. If a probability distribution function possesses the general shape of the phenomena of interest (e.g. bell shaped as per Figure 5.1), the individual values from the data record are used to fit the probability distribution as a means of estimating properties of the data such as exceedance probabilities of specified events. If the data record was very lengthy, one could use the data record itself to assign probability exceedances, but the availability of an appropriate probability distribution function lessens the need for the lengthy data record.

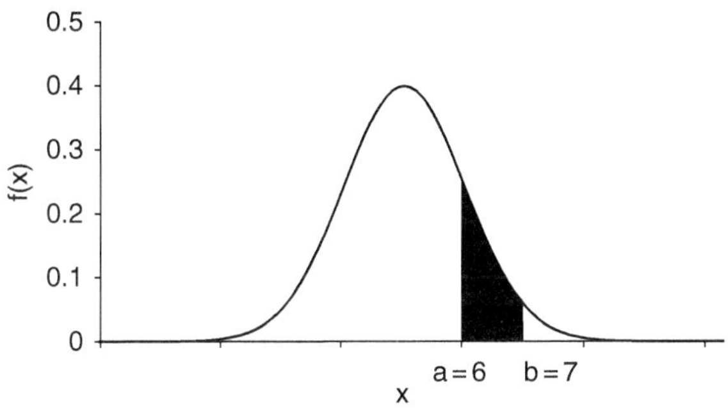

Figure 5.2 Probability density of normal distribution.

As indicated in Section 5.1, the normal distribution is the most common descriptor of data. Mathematically, the normal distribution is defined as

$$f(x) = \frac{1}{\sqrt{2\pi}\sigma} e^{-\frac{1}{2}\left(\frac{x-\mu}{\sigma}\right)^2} \tag{5.1}$$

where (5.1) is defined for the region $-\infty < x < \infty$ and where μ is the mean (a location parameter), s is the standard deviation (SD) (a scale parameter), and σ^2 is the variance. A summary notational form for this is N(μ, σ^2) where "N" refers to the normal distribution. Hence, for example, N(0,1) refers to a normal distribution with a mean of zero and a variance (and SD) of unity.

In mathematical terms, Equation (5.1) gives the height of the normal probability distribution at any point along the scale of x as depicted by the f(x) ordinate in Figure 5.2. Equation (5.1) extends to infinity in both directions as indicated above by the defined region. The existence of nonzero values over the range of $-\infty$ to $+\infty$ is one of the major problems associated with applying the normal distribution to environmental quality issues; it allows, for example, nonzero probabilities of negative values, which implies there is a possibility of having negative concentrations, something that is not possible.

It is significant to note that the probability distribution described by Equation (5.1) is, in fact, a probability density. To estimate the probability of occurrence, the probability density function (Equation [5.1]) must be integrated over a range. For example, the probability that a sample value x is less than or equal to a specified value "X" is evaluated by integrating from $-\infty$ up to the level X using

$$\text{Prob}[x \le X] = \int_{-\infty}^{X} \frac{1}{\sqrt{2\pi}\sigma} e^{-1/2\left(\frac{x-\mu}{\sigma}\right)^2} dx \tag{5.2}$$

This corresponds to the area under the curve as depicted in Figure 5.3. A property of the normal distribution (and all probability distributions) is that the total area bounded by the curve and the x-axis is unity. Hence the area under the curve between two ordinates x = a and x = b where a < b represents the probability that x lies between a and b. This relationship is shown in Figure 5.2, i.e.

$$\text{Prob}\{a \le x \le b\} = \int_{a}^{x} \frac{1}{\sqrt{2\pi}\sigma} e^{-1/2\left(\frac{x-\mu}{\sigma}\right)^2} dx \tag{5.3}$$

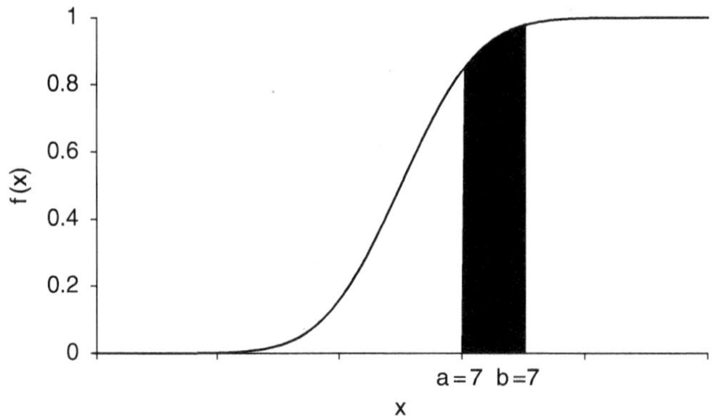

Figure 5.3 Cumulative density function for normal distribution.

Because of the extensive use of the normal distribution and the infinite number of possible values of μ and σ, probabilities for the normal distribution are usually evaluated using the so-called "standard" normal distribution. The standard normal distribution is derived by transforming the random variable x to the random variable z using $z = (x - \mu)/\sigma$. When the variable x is expressed in terms of standard units, Equation (5.1) becomes

$$f(z) = \frac{1}{\sqrt{2\pi}} e^{-\left(\frac{z^2}{2}\right)} \tag{5.4}$$

Equation (5.3) is the probability density function for the standard normal curve and is the form of the information provided in the normal distribution tables. In this case, we say that z is normally distributed with a mean of zero and a variance of unity, or N(0,1).

As can be seen from Equation (5.1), the normal distribution is a two-parameter distribution. In other words, all aspects of the distribution are known once the mean (μ) and variance (σ^2) are specified. Both of these parameters characterize the central tendencies of the distribution (the mean being the centroid or center of gravity and the variance being the dispersion or spread of the distribution) around the centroid. Therefore, when an issue relates to environmental quality near the "tails" of the distribution (on the very low and/or the very high end of the data record), the normal distribution assumption is not necessarily the best distribution to employ.

The area under any probability distribution curve is equal to unity by definition (i.e. the integration from – to + infinity). However, for the normal distribution, the full range is seldom of interest in concerns dealing with environmental quality. Of greater interest are features such as how often one would expect values to be within various proximities to the mean. Three commonly employed guidelines are indicated in Table 5.1 and shown graphically in Figure 5.4. In words, the results from Table 5.1 indicate that given data can be described by the normal distribution:

i) You would expect 68% of monitoring values to lie within ±1 SD from the mean. The normal distribution is symmetric, and therefore 34% of the values should be within the mean and the mean +1 SD, and 34% of the values should be within the mean and the mean −1 SD.
ii) You would expect 95% of the monitoring values to lie within ±2 SD of the mean. Because the distribution is symmetrical about the mean, you would expect (approximately) 2.5% above the mean +2 SD, and 2.5% below the mean −2 SD.
iii) You would expect 99.73% of the monitoring values to lie within ±3 SD of the mean.

The normal distribution function (from Figure 5.1 and from the values listed in Table 5.1) illustrates that random fluctuations around the mean (or average) are to be expected but that

Table 5.1 Indications of percentages of monitoring values that would be expected to be in the vicinity of the mean value, assuming the normal distribution.

Number of standard deviations from the mean	Percentage of values in vicinity of mean
±1	68.26
±2	95.45
±3	99.73

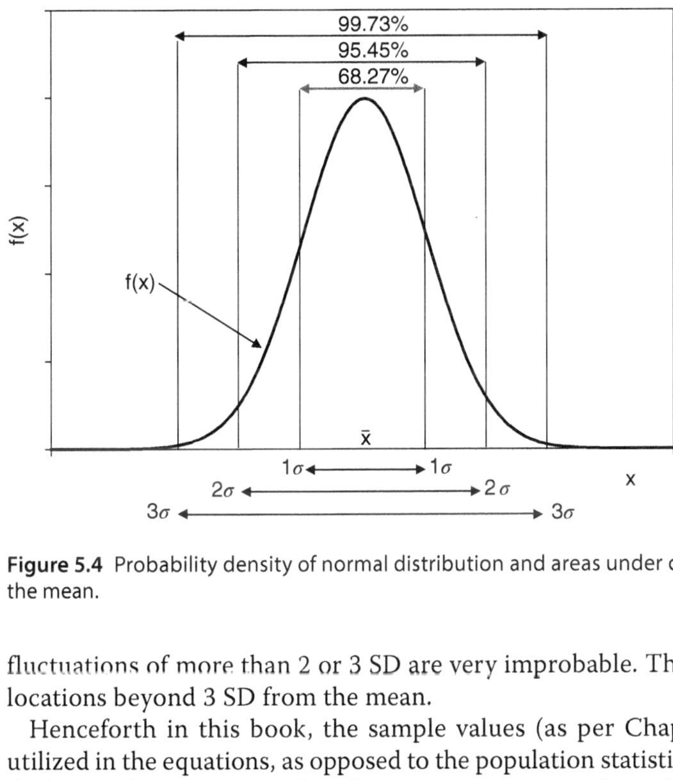

Figure 5.4 Probability density of normal distribution and areas under curve for different distances from the mean.

fluctuations of more than 2 or 3 SD are very improbable. The curve is practically at zero for locations beyond 3 SD from the mean.

Henceforth in this book, the sample values (as per Chapter 4 characterization) will be utilized in the equations, as opposed to the population statistics, since in practical applications dealing with environmental quality concerns we never have the population statistics.

5.3 Tests for Normality

As indicated in Section 5.2, we utilize the data to fit the normal distribution to, in turn, allow us to quantify various aspects of the probabilities. However, implicit in this undertaking is the assumption that the normal distribution is a reasonable descriptor of the data. Other distributions may be appropriate for specific circumstances, and thus it is of interest to establish criteria that can be employed to determine that the data are, in fact, adequately described by the normal distribution. In other words, data must be subjected to goodness-of-fit tests to determine the appropriateness of different distributions. To accomplish this, a series of alternative tests of varying complexity are available. As will be seen, several approaches exist to identify whether a probability distribution function adequately describes the random nature of the random variable. A number of these tests will be considered in the sections to follow.

5.3.1 Coefficient of Variation Test for Normality

The coefficient of variation (COV) test is the simplest (and least rigorous) of the available tests to determine whether the data can be characterized by a normal distribution. Specifically, the COV is defined as

$$COV = \frac{S}{\overline{X}} \tag{5.5}$$

Given that the normal distribution is a symmetrical distribution and that environmental quality data cannot be negative, then if S is very large in relation to the mean, it is not possible for the data to be effectively described by the normal distribution; for this situation, the data distribution must be highly skewed. For environmental quality data (e.g. where individual concentration measurements cannot be less than zero), an indication of the failure of the data to be described by the normal distribution is sometimes adopted as occurring when the COV is greater than 0.5 and/or 1.0. If the COV exceeds 1.0 there is strong evidence that the data are not normally distributed. If the data are from a normal distribution, then less than 10% of all sample COV will exceed 1.0 by random chance. If the data are nonnormal, then the sample SD will be large relative to the sample mean and, therefore, make the probability of a sample coefficient exceeding 1.0 quite large.

Example 5.1 The results from the monitoring of chloride levels in the groundwater are summarized in Table 5.2. Utilize the coefficient of variation (COV) test to determine if the data can be characterized by the normal distribution.

From the data in Table 5.2, the statistical parameters are determined as

Mean	$\overline{X}$	= 18.3 mg/l
Standard deviation	S	= 4.06 mg/l
Coefficient of variation	COV	= 0.222

In summary, the COV is less than unity, which means that (at least by this test) the data can be described by the normal distribution. The general bell-shaped distribution of the data is depicted in the histogram of chloride concentrations in Figure 5.5a and confirms that the data are distributed in a relatively symmetric fashion.

Example 5.2 The results from the monitoring of benzene concentrations in water in a roadside ditch are as listed in Table 5.2, column II. Utilize the coefficient of variation (COV) test to determine if the data may be characterized by the normal distribution.

From the data in Table 5.2, the statistical parameters were determined specifically:

Mean	$\overline{X}$	= 4.47 mg/l
Standard deviation	S	= 4.98 mg/l
Coefficient of variation	COV	= 1.11

In summary, the COV is greater than unity, which indicates that the lead concentration data are not well described by a symmetric distribution. The histogram of data as illustrated in Figure 5.5b indicates that the data are not well described by the normal distribution.

Table 5.2 Chemical concentrations in monitoring records.

	Column I	Column II
Number of samples	Chloride concentrations (mg/l)	Benzene concentrations (μg/l)
1	25.25	1.1
2	13.32	1.4
3	15.78	1.8
4	22.63	2.05
5	20.36	2.11
6	21.43	2.15
7	11.74	2.23
8	22.27	2.54
9	18.11	2.58
10	27.50	2.61
11	18.12	2.85
12	16.16	3.15
13	16.39	3.50
14	16.66	4.80
15	18.71	5.12
16	14.90	7.40
17	14.09	10.15
18	15.64	5.35
19	16.09	2.94
20	20.47	23.5

Example 5.3 Monitoring results at two sampling locations are listed below. Comment on whether the data are symmetrically distributed and/or skewed using the coefficient of variation test.

Sampling location	
A	B
37.0	2.56
34.9	1.00
28.4	0.25
27.3	0.60
31.8	0.40
32.6	0.85
28.3	0.38
26.9	2.25
19.9	0.41
31.8	1.75
$\bar{x} = 29.9$	1.05
S = 4.84	0.84
COV = 0.16	0.80

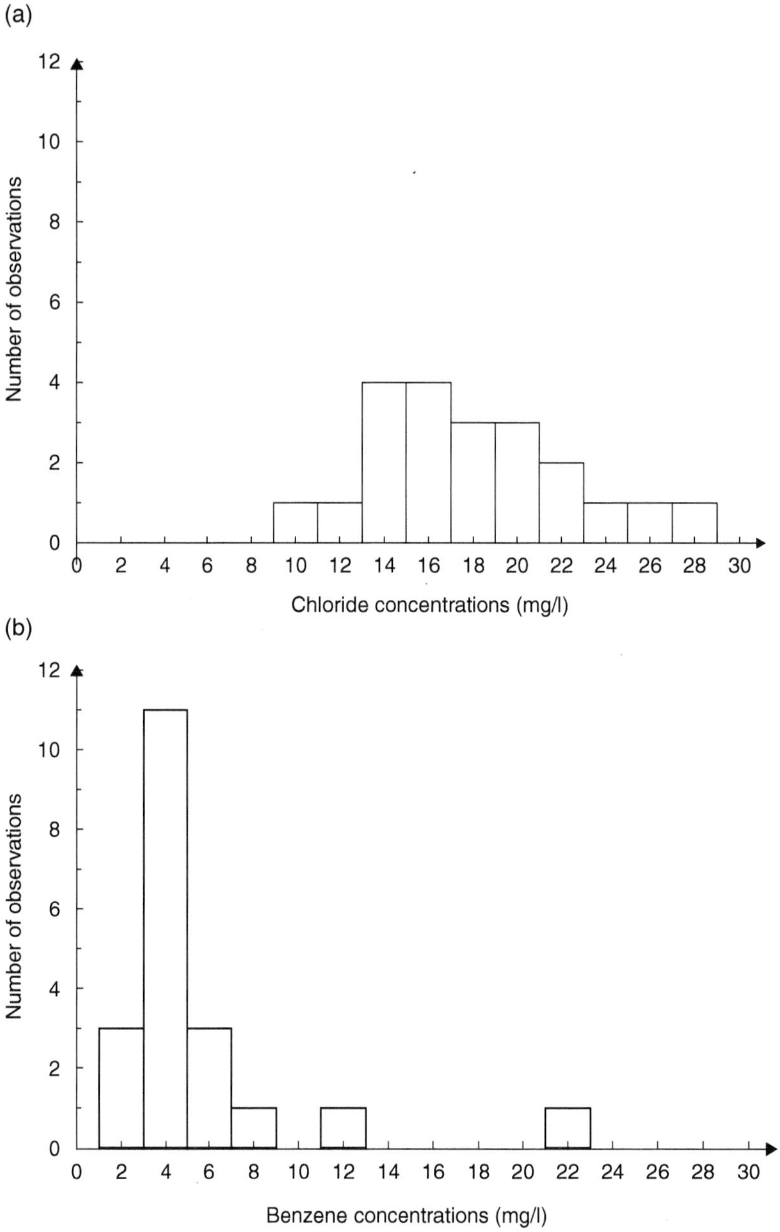

Figure 5.5 Histogram of chloride (a) and benzene concentrations (b).

The coefficient of variation test indicates that the monitoring results at location A are relatively uniform, whereas those at B are skewed. The histograms of Figure 5.5a and b indicate the above comments are true.

The coefficient of variation test is an easy, but only an approximate measure, or indicator of skewness.

5.3.2 Skewness and Kurtosis Coefficient Tests for Normality

As described in Sections 4.7.3 and 4.7.4, the normal distribution has zero skewness and a kurtosis coefficient value of 3. Some deviations from these magnitudes for samples (as opposed to populations) must be expected. These calculation procedures are computationally simple, utilizing spreadsheet procedures. The skewness coefficient, C_S, for the data should approach zero since the normal distribution is a symmetric distribution and the positive deviations should cancel the negative deviations. Although not a firm criterion, when the absolute value of C_S is >1, the distribution is not likely a normal distribution. It is to be noted that this test using skewness is not robust when the dataset is small since the cube of deviations makes the calculations very sensitive to individual members of the dataset. In summary then, for "large" datasets, the skewness coefficient must be small to support the assumption that the data are adequately described by the normal distribution. For "small" datasets, the utility of the skewness coefficient as an indicator that the data may be adequately described by the normal distribution is less certain since it is highly sensitive to individual values within the dataset. It is to be noted that strict definitions on what comprises "large" and "small" do not exist; however, 5–10 samples is definitely small and more than 50 is generally large.

Note that just because a distribution is symmetric, it does not indicate normality. For example, it will be shown in Section 5.5 that the student's t-distribution is symmetric but not normal (except as the sample size becomes large). The t-distribution has fatter tails, and hence a greater proportion of the random values from the t-distribution are farther from the mean than with the normal distribution. If the absolute value of the skewness coefficient is less than unity, the data do not show evidence of significant skewness (i.e. may be accepted as normally distributed).

Example 5.4 Using the chloride concentration data listed in Example 5.1, the skewness coefficient is $C_S = 0.65$. As characterized by the skewness test, the distribution of data is moderately skewed. The chloride concentrations are plotted in histogram form in Figure 5.5a.

The kurtosis coefficient test may also be used to consider whether a dataset is normally distributed.

Example 5.5 Using the benzene concentration data from Example 5.2, the coefficient of skewness of data is 3.27. This magnitude of skewness coefficient indicates the data are highly skewed. The benzene concentrations are plotted in histogram form in Figure 5.5b and indicate the right skew of the data. The coefficient of kurtosis = 12.0 (3 is magnitude for the normal distribution) – correct for the $C_K = 12 - 3 = 9$.

5.3.3 Probability Plots

Another procedure useful in assessing the appropriateness of a particular distribution for a particular data test utilizes probability graph paper. Probability plots are procedures routinely used for making quantitative and subjective statistical analyses of datasets by plotting the magnitude versus a cumulative distribution function. If the sample of data is from the distribution function that was used to scale the probability paper, the data will appear as a straight line when data are plotted. The use of probability plots requires only a general understanding of simple statistical concepts.

Cumulative probability plots were first developed by Hazen (1914) as a means of simplifying the interpretation of reservoir storage data. The method has since found widespread application in graphical studies of numeric data in many fields.

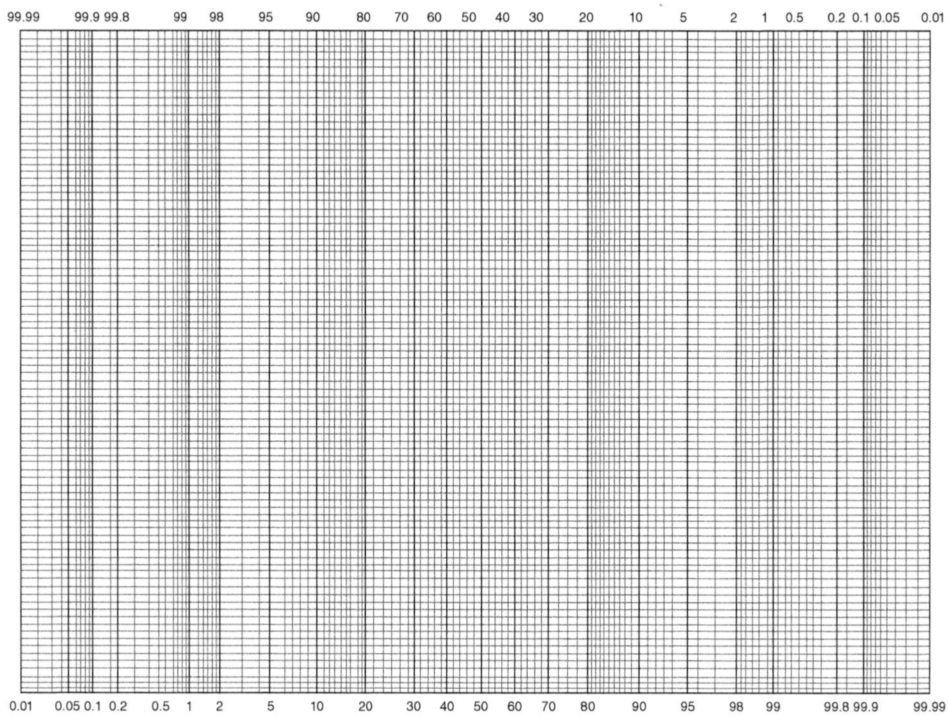

Figure 5.6 Normal probability paper.

The horizontal axis along the top of probability paper characterizes the probability that the value of the random variable will be exceeded (i.e. the exceedance probability). The horizontal axis on the bottom provides the nonexceedance probability as depicted in Figure 5.6. The abscissa scale of normal probability paper is an arithmetic scale. The ordinates are percentages, descending from 99.99 to 0.01 as indicated in Figure 5.6. The ordinate scale is stretched at both ends; the distance from 50 to 60 is smaller than that from 60 to 70 and so on. The probability (cumulative percentage) scale is arranged such that a cumulative normal density distribution will plot as a straight line. The probability plot is intended to transform a curvilinear distribution into a straight line. Utilizing results plotted on probability paper, the probability of the concentration exceeding some X $\mu g/m^3$ is found by entering the value of the random variable, moving horizontally to the frequency curve, and then moving vertically to the probability scale.

Probability paper is specifically designed paper where "normal probability paper" is a particular type, while other forms of probability paper exist. The objective of attempts to identify the underlying population for a dataset is to obtain a model that can be used to make probability statements about the occurrence of values of a random variable. By plotting the data and determining if the data plot as a straight line, we can determine the appropriateness of the distribution as a descriptor of the data.

The plotting of the individual data points requires the determination of "plotting position" for the probability. When successive values of the dataset are independent, the exceedance probability can be estimated from one of a number of formulas, examples of which are listed in Table 5.3 where "m" refers to the rank of values, with the largest value equal to 1, and "n" is the total number of values.

Kimball (1960) reviewed the problem of choosing a plotting position for the ranked observations in a series of n independent identically distributed random variables (nine alternatives).

Table 5.3 Examples of plotting position formulas.

Formula name	Formula
California	$\frac{m}{n}$
Hazen	$\frac{2m-1}{2n}$
Beard	$1-(0.5)^{1/n}$
Weibull	$\frac{m}{n+1}$
Chegadayev	$\frac{m-0.3}{n+0.4}$
Blom	$\frac{m-3/8}{n+1/4}$
Tukey	$\frac{3m-1}{3n+1}$

Source: Adapted from Viessman et al. (1989).

Although the nine different formulas in Kimball (1960) give similar plotting positions near the middle of the data ($i \approx n/2$), they can lead to quite different plotting positions at the extreme values, near $i = 1$ and $i = n$. In general, the plotting formulas have been described by Bloom (1958) as

$$P_m = \frac{m-c}{n+1-2c} \tag{5.6}$$

where P_m is the plotting position of the mth point and c is a function of the record length, n, and the distribution being sampled, and $0 \leq c \leq 1$.

The probability, P_m, is an estimate of the probability of values being equal to or less than the ranked value $p(x \leq X)$. Thus, the probability of a concentration being exceeded can be obtained using a plotting position formula. For example, the expected number of exceedances, x, for a given number of future measurements, N, using the Weibull plotting formula is

$$x = N\frac{m}{n+1} \tag{5.7}$$

where m and n are as defined previously.

To calculate the recurrence interval, R (the number of measurements before an exceedance is expected to occur), a value of $\overline{x} = 1$ can be used in Equation (5.6). Rearranging, the recurrence interval or return period can be estimated using

$$R = \frac{n+1}{m} \tag{5.8}$$

Thus, the recurrence interval based on the Weibull plotting formula of a concentration to be equaled or exceeded is equal to the number of measurements plus 1 divided by the rank of that magnitude of observation.

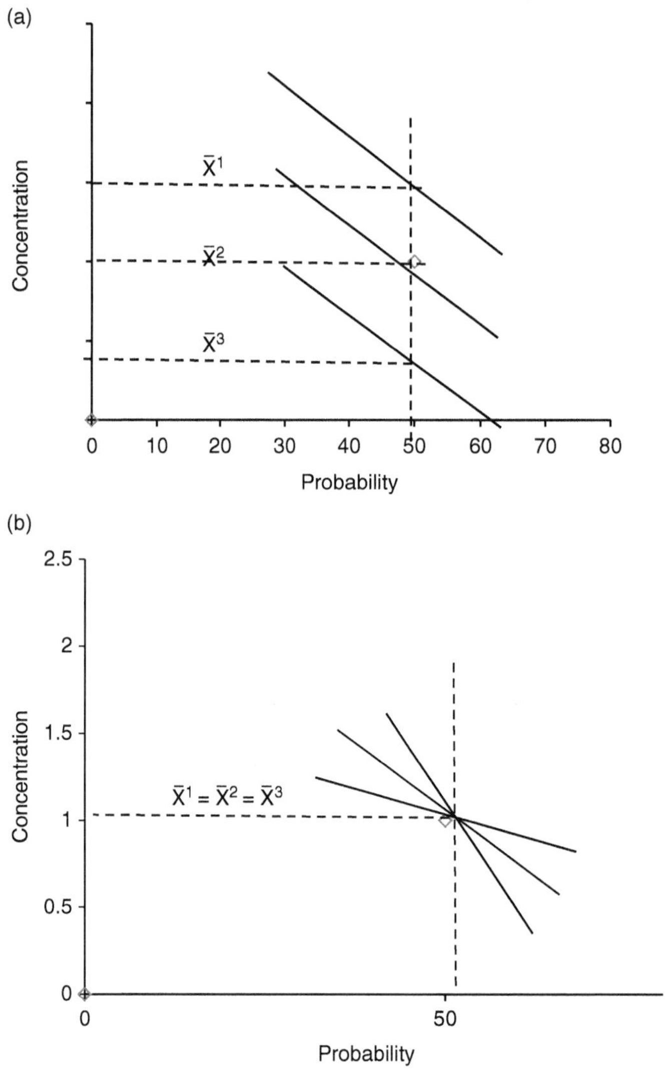

Figure 5.7 (a) Schematic of probability plots for constituents A, B, C, with different means but same standard deviation. (b) Schematic of probability plots for constituents A, B, C, with identical means but different standard deviation.

Assuming a straight line represents a reasonable characterization of the data, the values of the mean and SD may be read directly from the graph. The mean is simply the 50th percentile point. The SD of the distribution is determined from the slope of the fitted straight line. Because each of the populations is normal (a straight-line plot on probability paper) the values of the mean, ±1 SD can be estimated from the 16th and 84th percentiles.

The implications of different means but similar SD for different monitoring locations are depicted in Figure 5.7a. Alternatively, the implications of similar means but different SD are depicted in Figure 5.7b.

Example 5.6 Consider the chloride concentration data listed in Table 5.2. Use the Weibull plotting position formula to plot the data on normal probability paper.

Table 5.4 Calculation sequence for probability graphing of chloride data.

Number of samples	Chloride concentrations (mg/l)	Rank	Rank-ordered concentration data	Weibull plotting position[a]
1	25.25	1	27.50	0.048
2	13.32	2	25.25	0.095
3	15.78	3	22.63	0.143
4	22.63	4	22.27	0.190
5	20.36	5	21.43	0.238
6	21.43	6	20.47	0.286
7	11.74	7	20.36	0.333
8	22.27	8	18.71	0.381
9	18.11	9	18.12	0.429
10	27.50	10	18.11	0.476
11	18.12	11	16.66	0.524
12	16.16	12	16.39	0.571
13	16.39	13	16.16	0.619
14	16.66	14	16.09	0.667
15	18.71	15	15.78	0.714
16	14.90	16	15.64	0.762
17	14.09	17	14.90	0.810
18	15.64	18	14.09	0.857
19	16.09	19	13.32	0.905
20	20.47	20	11.74	0.952

[a] Computed from m/(n + 1) where m is the rank and n = 20 for the total number of samples.

The columns in Table 5.4 indicate the details of the calculation sequence. The resulting data are plotted in Figure 5.8. The mean value estimate can be read as the ordinate value corresponding to the 50th percentile or 18 mg/l. The standard deviation can be estimated fairly precisely by ordinate values that correspond to the 16th and 84th percentiles, respectively, viz.

$$S = C_{50} - C_{84} = 18.0 - 13.8 = 4.2 \text{mg} / 1$$

These results compare favorably with the values computed from data as summarized below:

	From the graph (mg/l)	From the data (mg/l)
$\overline{X}$	18.0	18.3
S	4.2	4.1

As apparent from Figure 5.8, the data points plot as a relatively straight line, thus confirming that the data are adequately described by the normal distribution.

The linearity, or lack thereof, of a set of sample data is used as a basis for determining whether the distribution of the underlying population is the same as that of the probability paper.

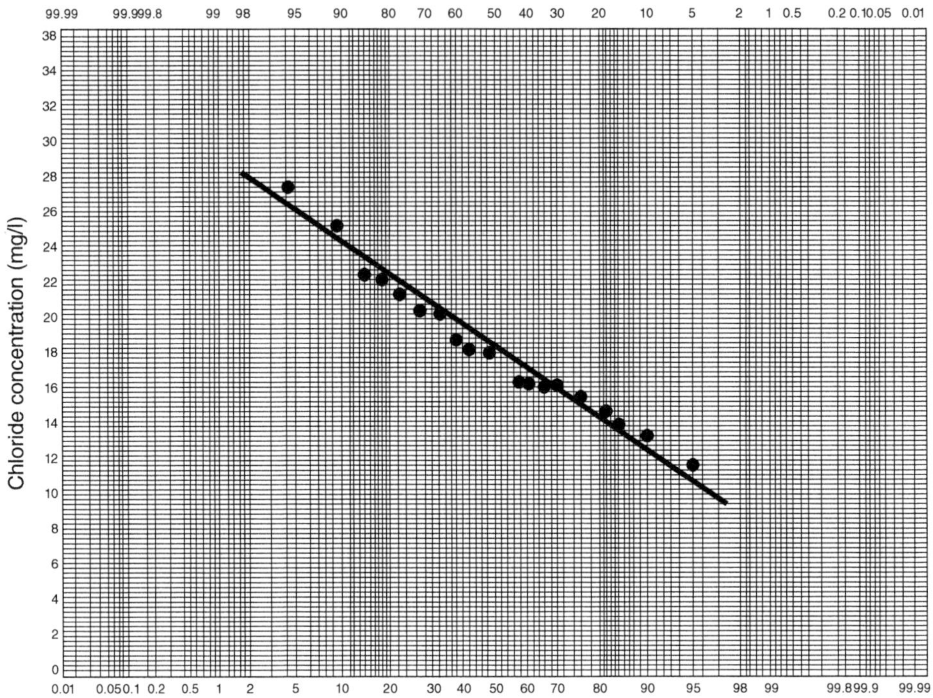

Figure 5.8 Normal probability graph for chloride concentrations.

The disadvantages of probability plots are that there is no commonly accepted rule for plotting the data and drawing the lines by eye and there is neither a simple nor a universally accepted way to judge how well the data points conform to the straight lines (Mage, 1982; McBean and Rovers, 1984).

Owing to sampling variations, there will be some departure from a straight line even when normally distributed data are plotted. We are concerned with the need to look for systematic deviations of the type indicating a long-tailed distribution or those indicating a skew distribution. A situation in which the normal distribution is a rather poor fit to the data is provided in Example 5.8.

Example 5.7 Consider the benzene concentration data listed in Table 5.2. Use the Weibull plotting position formula to plot the data on normal probability paper.

The columns in Table 5.5 summarize the details of the calculation sequence. The resulting data are plotted in Figure 5.9. As apparent from the plotted data that do not plot on a straight line, the normal distribution is not a good descriptor of the data.

A problem with this rather informal approach is that different people will draw different lines when independently analyzing the same probability plot. Hahn and Shapiro (1967) instruct the reader "If a straight line appears to fit the data, draw such a line on the graph by eye." A further limitation of this procedure occurs when the dataset is small to determine whether a straight-line fit is adequate.

Statistical goodness-of-fit tests exist for quantifying whether the dataset may be characterized by the normal distribution, as will be examined in Sections 5.3.4 and 5.3.5. However, a problem with statistical goodness-of-fit hypothesis testing is that few tests are useful with small sample sizes. The negative aspect of the plotting method utilizing the visual comparison of a line with data points is the subjectivity associated with the drawing of the lines.

Table 5.5 Calculation sequence for probability graphing of benzene data.

Number of samples	Benzene concentrations (mg/l)	Rank	Rank-ordered concentration data	Weibull plotting position[a]
1	1.1	1	23.5	0.048
2	1.4	2	10.15	0.095
3	1.8	3	7.40	0.143
4	2.05	4	5.35	0.190
5	2.11	5	5.12	0.238
6	2.15	6	4.80	0.286
7	2.23	7	3.50	0.333
8	2.54	8	3.15	0.381
9	2.58	9	2.94	0.429
10	2.61	10	2.85	0.476
11	2.85	11	2.61	0.524
12	3.15	12	2.58	0.571
13	3.50	13	2.54	0.619
14	4.80	14	2.23	0.667
15	5.12	15	2.15	0.714
16	7.40	16	2.11	0.762
17	10.15	17	2.05	0.810
18	5.35	18	1.8	0.857
19	2.94	19	1.4	0.905
20	23.5	20	1.1	0.952

[a] Computed from $m/(n + 1)$ where m is the rank and $n = 20$ for the total number of samples.

Probability plots are particularly useful for spotting irregularities within the data. It is easy to determine whether departures from normality are occurring more or less in the middle ranges of the data or in the extreme tails. Probability plots can also indicate the presence of possible outlier values that do not follow the basic pattern of the remainder of the data.

A second type of probability paper in common use is lognormal probability paper. Lognormal and normal probability papers have identical probability scales; the difference is that the lognormal probability paper has a logarithmic scale for the ordinate, whereas the latter has an arithmetic scale. Further discussion of lognormal paper is provided in Chapter 6.

A given dataset may be plotted directly, or, if the data do not plot as a straight line, a data transformation that produces a function that is normally distributed may be investigated. Alternative data transformations are infinite in number, but examples include square root, square, inverse, and others, as summarized in Table 5.6.

5.3.4 The Chi-Square Goodness-of-Fit Test

The chi-square (χ^2) test is one of the most versatile tests in statistical theory. The intent in this test is to evaluate whether the observed frequencies in a distribution differ significantly from the frequencies that might be expected according to some assumed hypothesis.

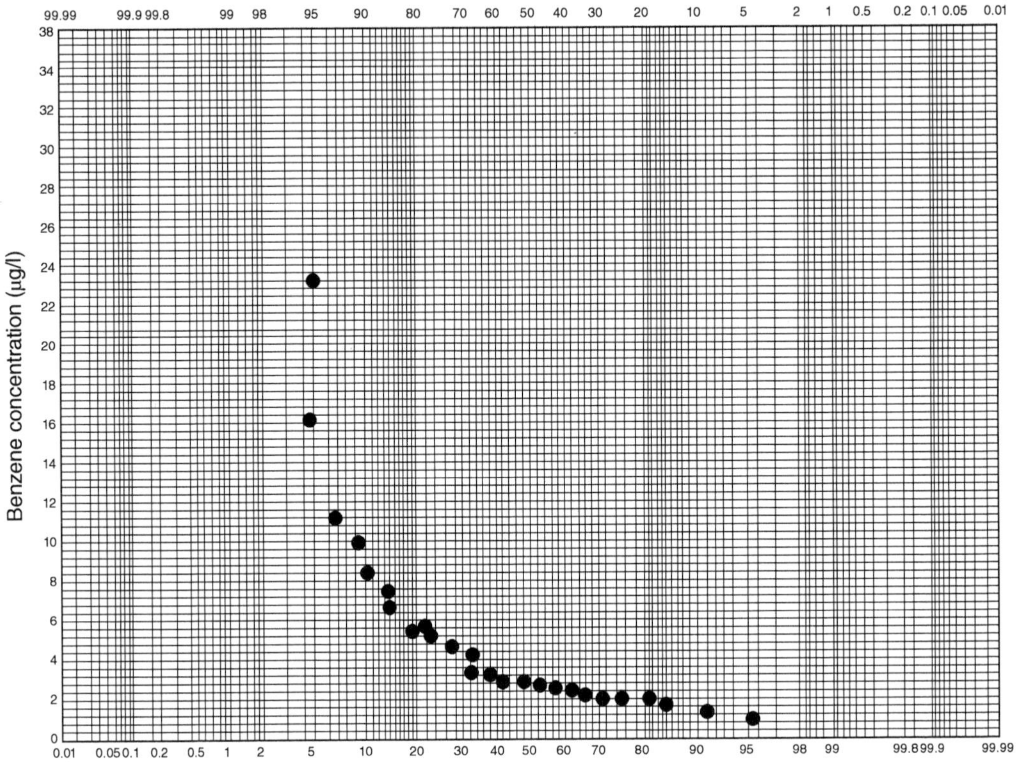

Figure 5.9 Normal probability plot of benzene concentrations showing that normal distribution is not a good characterization of the data.

Table 5.6 Common data transformations.

Data transformation	Calculation procedure
Arithmetic	$x' = x + C$
Reciprocals	$x' = \frac{1}{x}$
Arcsine	$x' = \arcsin\sqrt{x}$
Logarithmic (base 10)	$x' = \log x$
Logarithmic (base e)	$x' = \ln\ x$
Square root	$x' = \sqrt{x}$
Cubic root	$x' = \sqrt[3]{x}$

Thus the first step is to formulate a null hypothesis and an alternative hypothesis, namely:

H_0: The data distribution is normal, with mean $\bar{x}$ and standard deviation S (where $\bar{x}$ and S are specific values for a particular application).

H_A: The data distribution is not normal.

The chi-square goodness-of-fit test is always a one-tailed test because the hypotheses are unidirectional, since the random variable is either distributed as specified, or it is not.

Although an acceptable goodness-of-fit test, the chi-square test is neither the most sensitive nor the most powerful (Gan and Koehler, 1990).

The chi-square test for normality is accomplished by segmenting the range of the variable into a series of intervals or cells, representing distinct nonoverlapping ranges of data values. In each interval, an expected value is computed based on the number of data points that would be found in each of the intervals if the distribution being evaluated (in this case the normal distribution) compares to the observed frequencies of data within each of the intervals. The observed data are used to form a histogram, which shows the observed frequencies in a series of k cells. The chi-square cells are usually selected such that the cell width for each cell is the same (although this is not a requirement of the test). The test is demonstrated in Example 5.9.

Example 5.8 For the chloride concentration data from Example 5.5, use the chi-square test to determine whether the data follow the normal distribution at the 5% level of significance.

Assuming an interval of 2.50 mg/l, the chloride data are used to develop the observed frequencies (O_i). The expected frequencies (e_i) for each cell are computed using the probability distribution function associated with the intended population (in this case, the distribution being considered is the normal distribution). To compute the expected frequencies, the expected probability for each cell is determined and multiplied by the sample size n. The expected probability for cell i, p_i, is the area under the probability distribution function between the cell bounds for that cell. The sum of the expected frequencies then equals the total sample size n.

The specifics in relation to the application of the chi-square test are indicated in Table 5.7.

The χ^2 distribution works with k-1-p degrees of freedom, where k is the number of intervals or classes into which the observations are divided and p is the number of parameters that have been independently estimated from the data.

Table 5.7 Representative values of the t-distribution indicating the relationship to the normal distribution.

	Points of the t-distribution[a]		
Degrees of freedom	**0.05**	**0.01**	**0.001**
1	6.314	31.82	318.1
5	2.015	3.365	5.893
10	1.812	2.764	4.144
15	1.753	2.602	3.733
20	1.725	2.528	3.552
25	1.708	2.485	3.450
40	1.684	2.423	3.307
60	1.671	2.390	3.232
100	1.658	2.358	3.160
∞	1.645	2.326	3.090

[a] Note: This table gives the values of t that are exceeded with probability P.

A large value of χ^2 indicates there is a large difference between the observed and expected frequencies. The occurrence of large χ^2 would indicate the null hypothesis can be rejected. To do this, one must check to see whether the calculated value of χ^2 is greater or less than the critical value, χ_c^2 given in the appropriate table (see Table A.12). To do this, one must know the degrees of freedom. For this example, the number of degrees of freedom is 8 – 3 = 5. One degree of freedom was lost because the total number of observations is used to compute the expected frequencies, while two were lost for the mean and standard deviation, which were estimated from the sample of 20 observations.

For these data, the degrees of freedom, df, is

$$\text{df} = 8 - 3 = 5 \text{ at } \alpha = 0.05 \left(\text{i.e. } 5\% \text{ significance level}\right) \chi_c^2 = 11.07 > 3.32$$

The null hypothesis is not rejected and the data may be assumed to be normally distributed at a 5% level of significance.

It should be noted that differences from normality tend to show up more clearly in the tails of the distribution (this is as expected since the normal distribution is a central fitting distribution using only the mean and the SD); it is in the tails that the chi-square goodness-of-fit test does poorly because of the need to combine cells to avoid small frequencies. The result is that the chi-square goodness-of-fit test is not a very useful test of normality with small datasets (although a correction factor is needed if $n < 5$). The validity of the chi-square test is best if the number of observations in each group or interval is consistently greater than 5. Thus, the chi-square test requires a fairly large sample size.

If the chi-square test indicates the data are not normally distributed, it may not be clear what ranges of the data most violate the normality assumption. Most tests such as the ANOVA (see Chapter 11) are not overly influenced by deviations from the normal distribution in the central portion of the data. If the extreme tails are approximately normal in shape even if the middle part of the density is not, these parametric tests will still tend to produce valid results. For chi-square, departures from normality in the middle intervals are given nearly the same weight as departures from normality from the extreme tail intervals. As a result and as will be demonstrated later, the chi-square test is not as powerful for detecting departures from normality in the extreme tails of the data, the areas most crucial to the validity of parametric tests like the t-test or ANOVA (Miller, 1986).

5.3.5 The Kolmogorov–Smirnov Goodness-of-Fit Test

The Kolmogorov–Smirnov test is another test of the goodness-of-fit of data. The Kolmogorov–Smirnov (e.g. Lindgren, 1976) is used to test the hypothesis that data come from a specific (that is, completely specified) distribution. A minimum sample size of 50 is recommended for use of this test. The Kolmogorov–Smirnov test can take one of two forms. A one-sample test can be performed to examine how closely observed probabilities correspond to theoretical probabilities. To accomplish this the observed cumulative distribution of sample values are compared with a specified continuous distribution function (which, for purposes of this chapter, is the normal distribution). Alternatively, a two-sample test can be performed to test the agreement between two observed cumulative distributions. This application tests the null hypothesis, H_o, that two independent samples come from identical continuous distributions. It is sensitive to population differences with respect to location, dispersion, and skewness.

The Kolmogorov–Smirnov statistic is simply the maximum absolute difference between the theoretical and observed cumulative probability distribution, and this distribution is written as D.

Algebraically, the Kolmogorov–Smirnov test considers the deviations between a hypothesized cumulative distribution function such as the normal distribution function, F(x), and the observed cumulative histogram as

$$F^{*}\left(x^{(i)}\right)=\frac{i}{n} \tag{5.9}$$

in which $x^{(i)}$ is the ith largest observed value in the random sample of size n. The sample statistic, D, is calculated as

$$D=\max_{i=1,n}\left\{\left|F^{*}\left(x^{i}\right)-Fx\left(x^{(i)}\right)\right|\right\} \tag{5.10}$$

Of interest then is whether the difference D is larger than can reasonably be expected. The test involves consideration of the following hypotheses:

H_0: That x has a specified distribution (the null hypothesis for the Kolmogorov–Smirnov goodness-of-fit test is that the sample data that produced the observed probability distribution has been drawn from a population that possesses the specified theoretical distribution).
H_A: That the distribution of x is other than that specified.

If indeed the sample is from the underlying distribution (in this case the normal distribution), then the Kolmogorov–Smirnov statistic D has a distribution that is independent of the underlying distribution. Since the sampling distribution of D under the null hypothesis is known, values of the critical statistic, $d_{n,}\alpha$, as a function of sample size "n" and level of significance a are contained in many standard references (e.g. Benjamin and Cornell, 1970, p. 667). Tables of critical values of D are available (see Table A.13) and are a function of sample size n and level of significance α. Therefore, the hypothesis H_o is accepted if

$$D \le d_{n,\alpha} \tag{5.11}$$

If the calculated value of D is greater than the table of critical values at a specified significance level, the null hypothesis can be rejected at that level.

The degrees of freedom for the Kolmogorov–Smirnov goodness-of-fit test are the number of items in the observed frequency distribution. Harter (1980) provides modified asymptotic formulas for the statistic $d_{n,}\alpha$.

Lilliefors (1967) modified the Kolmogorov–Smirnov statistic tables to test whether a set of observations is from a normal population when the mean and variance are not specified but must be estimated.

The Kolmogorov–Smirnov test has the following advantages over the chi-square test (Benjamin and Cornell, 1970):

a) For continuous distributions, it does not lump data and compare discrete categories, but rather compares all data in an unaltered form.
b) The value of the test statistic, D, is usually computed more easily than the chi-square statistic.
c) Unlike the chi-square test, the Kolmogorov–Smirnov test is an exact test for all sample sizes. However, as a general statement, the departures from normality have to be large to be detected in sample sizes less than approximately 25.

The Kolmogorov–Smirnov test is generally more efficient than the chi-square test for goodness-of-fit for small samples, and it can be used for very small samples where the chi-square test does not apply. As well, it is to be noted that the test as presented above was applied to compare the data with the normal distribution; however, there is no need for a restriction of the test to the normal distribution since the theory applies to any continuous distribution.

5.3.6 The Shapiro–Wilk W Test

The Shapiro–Wilk W test (Shapiro and Wilk, 1965) is another statistical goodness-of-fit test that performs well on small sample sizes and tests the null hypothesis that the data values are random samples from a normal distribution against an unspecified alternative distribution. This test is considered one of the best numerical tests of normality and is particularly useful for detecting departures from normality in the tails of a sample distribution. It can be used in conjunction with a probability plot to measure how well the plotted quantiles are following a straight line (i.e. how well the sample values are correlated with normal quantiles).

The test can be performed on any sample size from 3 to 50. As would be expected, the power of the test increases as the sample size gets larger.

Steps in the calculation involve:

i) Order the sample data.
ii) Compute a weighted sum of the differences between the most extreme observations
iii) Divide the weighted sum by a multiple of the standard deviation and square the result to get the Shapiro–Wilk statistic W:

$$W = \left\{ \frac{b}{S\sqrt{n-1}} \right\}^2 \tag{5.12}$$

where the numerator is computed as

$$b = \sum_{i=1}^{k} a_{n-i+1}\left(x_{n-i+1} - x_{(i)}\right) = \sum_{i=1}^{k} b_i \tag{5.13}$$

where $x_{(i)}$ represents the smallest ordered value in the sample and coefficients a_i depend on the sample size n. The coefficients can be found for any sample size from 3 to 50 in Table A.7. The value of k can be found as the greatest integer less than or equal to n/2.

Normality of the data should be rejected if the Shapiro–Wilk statistic is too low when compared to the critical values provided in Table A.6. Otherwise, one can assume the data are approximately normal for purposes of further statistical analyses.

The Shapiro–Wilk test statistic (W) will tend to be large when a probability plot of the data indicates a nearly straight line. Only when the plotted data show significant bends or curves will the test statistic be small (Miller, 1986; Madansky, 1988).

Example 5.9 Use the chloride data of Example 5.1 to compute the Shapiro–Wilk test of normality.

Step 1: Order the data from smallest to largest, as per the table below. Also list the data in reverse order alongside the ordered data.

Step 2: Compute the difference $x_{(n-i+1)} - x_{(i)}$ in column IV of the table by subtracting column II from column III.

I	II	III	IV	V	VI
i	$x_{(i)}$	$x_{(n-i+1)}$	$x_{(n-i+1)} - x_{(i)}$	a_{n-i+1}	b_i
1	11.74	27.50	15.76	0.4734	7.461
2	13.32	25.25	11.93	0.3211	3.831
3	14.08	22.63	8.54	0.2565	2.191
4	14.90	22.27	7.37	0.2085	1.537
5	15.64	21.43	5.79	0.1686	0.976
6	15.78	20.47	4.69	0.1334	0.626
7	16.09	20.36	4.27	0.1013	0.433
8	16.16	18.71	2.55	0.0711	0.181
9	16.39	18.12	1.73	0.0422	0.073
10	16.66	18.11	1.45	0.0140	0.020
11	18.11	16.66	−1.45		b = 17.329
12	18.12	16.39	−1.73		
13	18.71	16.16	−2.55		
14	20.36	16.09	−4.27		
15	20.47	15.78	−4.69		
16	21.43	15.64	−5.79		
17	22.27	14.90	−7.37		
18	22.63	14.09	−8.54		
19	25.25	13.32	−11.93		
20	27.50	11.74	−15.76		

Step 3: Compute k as the greatest integer less than or equal to n/2. Since n = 20, k = 10.

Step 4: Look up the coefficients a_{n-i+1} from Table A.6 and place in column V. Multiply the differences in column IV by the coefficients in column V and add the first k products to get quantity b. In this case, b = 17.329.

Step 5: Compute the standard deviation of the sample, S = 4.05. Then

$$W = \left[\frac{17.329}{4.05\sqrt{19}}\right]^2 = 0.964.$$

Step 6: Compare the computed value of W = 0.964 to the 5% critical value for sample size 20 in Table A.7, namely, $W_{0.05,20} = 0.905$. Since W > 0.905 the chloride data do not show significant evidence of nonnormality.

Example 5.10 Use the benzene data of Example 5.2 to compute the Shapiro–Wilk test of normality. Step 1: Paralleling the analyses in Example 5.10, the tabular array becomes as follows.

I	II	III	IV	V	VI
i	$x_{(i)}$	$x_{(n-i+1)}$	$x_{(n-i+1)} - x_{(i)}$	$a_{(n-i+1)}$	b_i
1	1.1	23.5	22.4	0.4734	10.60
2	1.4	10.15	8.75	0.3211	2.81
3	1.8	7.40	5.60	0.2565	1.44
4	2.05	5.35	3.30	0.2085	0.69
5	2.11	5.12	3.01	0.1686	0.51
6	2.15	4.80	2.65	0.1334	0.35
7	2.23	3.50	1.27	0.1013	0.13
8	2.54	3.15	0.61	0.0711	0.04
9	2.58	2.94	0.36	0.0422	0.02
10	2.61	2.85	0.24	0.0140	0.00
11	2.85	2.61	−0.24		b = 16.59
12	2.94	2.58	−0.36		
13	3.15	2.54	−0.61		
14	3.50	2.23	1.27		
15	4.80	2.15	−2.65		
16	5.12	2.11	−3.01		
17	5.35	2.05	−3.30		
18	7.40	1.8	−5.60		
19	10.15	1.4	−8.75		
20	23.5	1.1	−22.4		

Step 2: For n = 20, k = 10.
Step 3: b = 16.59 and standard deviation = 4.99.

Step 4: $W = \left[\dfrac{16.59}{4.99\sqrt{19}}\right]^2 = 0.582$.

Step 5: Since $W_{0.05,20} = 0.905$ and $W < 0.905$, the data show significant evidence of nonnormality by the Shapiro–Wilk test.

As a result of the above, the data should be transformed using natural logs and rechecked using the Shapiro–Wilk test. This alternative is described in Example 6.4.

Example 5.11 Air quality measurements for mercury have been collected downwind of a factory, as detailed in the table below. What is the probability that the mercury concentrations will fall within the following ranges? X > 10, 4 < x <15 and x < 5

Date	Mercury concentration
5 September 2011	3
6 September 2011	6
7 September 2011	15
8 September 2011	21
9 September 2011	22
10 September 2011	15
11 September 2011	14
12 September 2011	18
13 September 2011	15
14 September 2011	2

From the data, we calculate

$$x_{mean} = 13.1$$
$$S = 7.09$$
$$C_s = -0.552 \text{ (moderately skewed)}$$
$$C_k = 0.049 \text{ (acceptable)}$$
$$COV = 0.542 < 1$$

We can perform the Shapiro–Wilk test to check for normality:

Rank	x_i	x_{n-i+1}	$x_{n-i+1} - x_i$	a_{n-i+1}	b_i
1	2	22	20	0.5739	11.478
2	3	21	18	0.3291	5.9238
3	6	18	12	0.2141	2.5692
4	14	15	1	0.1224	0.1224
5	15	15	0	0.0399	0
6	15	15			
7	15	14			
8	18	6			
9	21	3			
10	22	2			

$$b = \sum b_i = 20.0934$$

$$W = \left\{\frac{b}{S\sqrt{n-1}}\right\}^2 = \left\{\frac{20.0934}{7.09\sqrt{10-1}}\right\}^2 = 0.8924$$

Compare the computed value of W = 0.8924 to the 5% critical value for sample size n = 10 in Table A.7, namely, $W_{0.05,\,10} = 0.842$. Since W > 0.842 the data do not show significant evidence of nonnormality. Therefore, we can assume the sample values follow a normal distribution with a mean of 13.1 and a standard deviation of 7.09:

$$z = \frac{x - 13.1}{7.09}$$

$$\begin{aligned} &P(x > 10) \\ &z = \frac{(10 - 13.1)}{7.09} = -0.44 \\ &P(z > -0.44) = 0.5 + 0.17 = 0.67 \\ &\therefore P(x > 10) = 67\% \end{aligned}$$

$$\begin{aligned} &P(4 < x < 15) = P(x < 15) - P(x < 4) \\ &P(x < 15) \rightarrow z = \frac{(15 - 13.1)}{7.09} = 0.27 \\ &P(x < 4) \rightarrow z = \frac{(4 - 13.1)}{7.09} = -1.28 \\ &P(z < 0.27) - P(z < -1.28) = (0.5 + 0.1064) - (0.5 - 0.3997) = 0.5061 \\ &\therefore P(4 < x < 15) = 50.61\% \end{aligned}$$

$$\begin{aligned} &P(x < 5) \\ &z = \frac{(5 - 13.1)}{7.09} = -1.14 \\ &P(z < -1.14) = (0.5 - 0.3729) = 0.1271 \\ &P(x < 5) = 12.71\% \end{aligned}$$

5.3.7 The Shapiro–Francia Test

When there are more than 50 observations, the Shapiro–Francia test involves a slight modification of the Shapiro–Wilk test. This test has the same advantages as the Shapiro–Wilk test, and like the Shapiro–Wilk test (Shapiro and Francia, 1972), the test statistic of Shapiro–Francia (W′) will tend to be large when the data follow the normal distribution. This test statistic is defined as

$$W' = \frac{(\Sigma m_i x_i)^2}{\Sigma (x_i - \bar{x})^2}$$

Steps in the calculation involve:

i) Order the sample data.
ii) Compute a weighted sum of the observations:

$$W' = \Sigma m_i x_i \tag{5.14}$$

The values for m_i can be approximately computed as

$$m_i = \phi^{-1}\left[\frac{i}{(n+1)}\right] \tag{5.15}$$

where ϕ^{-1} denotes the inverse of the standard normal distribution with zero mean and unit variance.

Normality of the data should be rejected if the Shapiro–Francia statistic is too low in comparison with the critical values provided in Table A.8. Otherwise one can assume the data are approximately distributed for purposes of further statistical analyses.

iii) Divide the square by the weighted sum by a multiple of the standard deviation to obtain the Shapiro–Francia Statistic W′:

$$W' = \frac{(\Sigma m_i x_i)^2}{(n-1)S^2 \Sigma m_i^2}.$$

iv) Normality of the data is rejected if W′ is too low when compared to the tabulated critical value.

5.3.8 Data Transformations

The preceding seven subsections have all been directed toward alternative tests for assessing whether a dataset is adequately described by the normal distribution. Obviously any and/or all of the tests may reject the hypothesis that the data follow a normal distribution. In this circumstance, a possible approach is to apply a data transformation so that the normal distribution is appropriate. In fact, there always exists a transformation (or a series of transformations) such that the resulting distribution is normal (although in many applications it may be difficult to determine the one or more transformations that are required). Some of the more commonly used transformations are indicated in Table 5.7.

In some circumstances, the situation exists in which the data contain seasonal variability, in addition to random sampling and analytical errors. For this situation, if the data record is sufficiently lengthy, a sinusoidally varying term may be fit to the data and this component thus removed, leaving only the residuals. Such removal of seasonal variability is extensively done in surface water hydrology. Nevertheless, for many environmental quality problems, the ability to "fit" a sinusoidally varying term to a brief record is highly suspect.

A negative aspect of these data transformation approaches is that they have a degree of arbitrariness with respect to the transformation to be employed for a specific application. However, without the data transformation, the characteristics of the data may preclude the utilization of specific tests. The lognormal distribution is a good alternative distribution and will be the focus of analysis in Chapter 6.

5.3.9 Summary of Goodness-of-Fit Tests

With large quantities of data, we are in the position of developing an accurate characterization of the parameters of the population (e.g. the kurtosis and skewness) and from these parameters determine if the population is normal (with a specified level of confidence). Alternatively, we can use a chi-square goodness-of-fit test for testing for normality. However, when this dataset is small (e.g. $n < 25$), these measures or tests (kurtosis, skewness, and chi-square goodness-of-fit) are not accurate indicators of normality.

Unfortunately, all of the available tests for normality do, at best, a fair job of rejecting non-normal data when the sample size is small (e.g. $n < 20–30$). For example, small samples of untransformed lognormal data can be accepted by a test of normality even though the skewness of the data may lead to poor statistical conclusions later. Most environmental datasets are like this, and therefore utilization of the normal or lognormal distribution as adequate is likely. Plotting tests and COV tests give a reasonable basis for judgment with respect to applicability of specific distributions.

5.4 The t-Distribution

All of the preceding sections in this chapter have been specifically focused on the normal distribution. However, the normal distribution is actually a special case or subset of the t-distribution. The t-distribution is an unbounded distribution having a mean of zero and a variance that depends on the scale parameter, df, the degrees of freedom. As the degrees of freedom increase toward infinity, the variance approaches unity, and the t-distribution becomes the standard normal distribution. In general, the standard normal distribution can be used instead of the t-distribution for sample sizes greater than 30.

As a way of introducing the t-distribution, it is useful to compare the structure of the t-table with the table of the standard normal distribution. Recall that the standard normal distribution is not a function of the degrees of freedom; as the number of degrees of freedom increases, the t-distribution approaches the normal distribution. The t-table in Table A.2 provides t values when the probability specified at the top of the table is only in one tail or direction of the distribution.

5.5 Extent of Use of the Normal Distribution

As indicated in Section 5.1, the normal distribution is widely applicable to many applications (and far beyond those concerns relevant to environmental quality data). In later chapters in this book, the normal distribution will be seen to be fundamentally important to many statistical techniques such as ANOVA (to be discussed in Chapter 9), and the testing of various hypotheses relies on the assumption of the normal distribution, subjects that will be addressed in Chapters 10 through 12.

It is also noteworthy that while many statistical methods invoke the assumption of normality, the statistical methods remain approximately valid even when there are moderate departures from normality. As such, these procedures are considered robust (not impacted by moderate departures from normality). Reference will be made to this feature (i.e. deviations from normality are acceptable) at various points in the chapters to follow. Similar features to those noted above in terms of the importance of the central limit theorem will also be shown to be the situation for the lognormal distribution in Chapter 6.

5.6 Summary Comments

The term "normal distribution" has arisen because of the frequency with which it is appropriate to describe phenomena. As subsequent chapters will demonstrate, many of the subsequent tests are premised on the data being (at least approximately) normally distributed. Some tests reviewed need large samples and some can function with very few. Some tests are specific to test the appropriateness of the normal distribution, while some are goodness-of-fit tests applicable to any distribution (e.g. Kolmogorov–Smirnov).

5.7 References

Benjamin, J. and Cornell, A. (1970). *Probability, Statistics and Decisions for Civil Engineers.* New York: McGraw-Hill.

Bloom, G. (1958). *Statistical Estimates and Transformed Beta Variables*, 68–75. New York: John Wiley & Sons, Inc. 143–146.
Bulmer, M.G. (1965). *Principles of Statistics*. Cambridge, MA: The MIT Press.
Gan, F.F. and Koehler, K.J. (1990). Goodness-of-fit tests based on P-P probability plots. *Technometrics* 32 (3): 289–303.
Hahn, G. and Shapiro, S. (1967). *Statistical Models in Engineering*. New York: Wiley, Interscience.
Harter, H.L. (1980). Modified asymptotic formula critical values of the Kolmogorov test statistic. *The American Statistician* 34: 110–111.
Hazen, A. (1914). Storage to be provided in impounding reservoirs for municipal water supply. *Proceedings of the ASCE* 39: 1943–2044.
Kimball, B.F. (1960). On the choice of plotting positions on probability paper. *Journal of American Statistical Association* 55: 546–560.
Lilliefors, H.W. (1967). On the Kolmogorov-Smirnov normality with mean and variance unknown. *Journal of the American Statistical Association* 62: 399–404.
Lindgren, B.W. (1976). *Statistical Theory*, 3e. New York: MacMillan Press.
Madansky, A. (1988). *Prescriptions for Working Statisticians*. New York: Springer-Verlag.
Mage, D.T. (1982). An objective graphical method for testing normal distributional assumptions using probability plots. *The American Statistician* 36 (2): 116–120.
McBean, E. and Rovers, F. (1984). Alternatives for handling detection limit data in impact assessments. *Ground Water Monitoring Review* Spring: 42–44.
Miller, R.G. Jr. (1986). *Beyond ANOVA Basics of Applied Statistics*. New York: John Wiley & Sons, Inc.
Shapiro, S.S. and Francia, R.S. (1972). An approximate analysis of variance test for normality. *Journal of American Statistical Association* 67 (337): 215–216.
Shapiro, S.S. and Wilk, M.B. (1965). An analysis of variance test for normality (complete samples). *Biometrika* 52: 591–611.
Viessman, W., Lewis, G.L., and Knapp, J.W. (1989). *Introduction to Hydrology*, 3e. New York: Harper and Row Publishers.

5.8 Problems

5.1 We are frequently concerned whether a set of concentration data are normally disturbed. Two of the tests to check whether normality is acceptable as a descriptor of data involve calculation of the skewness and kurtosis, and yet, when we have small datasets, these two tests are not very reliable. Why is that?

5.2 As long as the concentration of chemical XYZ in fish fillets in general is less than 5.2 mg/l, then we can assume that the potential for contamination in fish is acceptable (not sufficient to cause illness in people who might consume the fish). We have found sufficient data on the chemical analyses of XYZ in fish fillets (the portions of fish that humans will consume) in river A to have a mean of 12 mg/l and a standard deviation of 9, and the data are able to be described by a Gaussian distribution.

a) What is the probability that a fish fillet exceeds the concentration of XYZ that endangers human health?
b) You have just located some additional data on the concentrations of XYZ that you were previously not aware of, which involved a concentration for two samples of XYZ in the fish fillet that are 37 and 3. The high value looks suspect – how would you test to determine if it correct data (describe the test, and do not do the calculations)?

5.3 A random variable has a mean of 0.8 and a standard deviation of 0.2. Assuming the data can be described by a normal distribution, find the probability that the value of a sample point is

- Less than 0.62
- Greater than 0.9
- Between 0.65 and 0.88

5.4 A random variable has a mean of 25 mg/l and a standard deviation of 5 mg/l. Assuming the data can be described by a normal distribution, estimate the value of the concentration such that:

a) 25% of the area lies above it
b) 7% of the area lies below it

5.5 Air quality measurements for mercury have been collected downwind of a factory, as detailed in the table below. What is the probability that the mercury concentrations will fall within the following ranges? X > 10, 4 < x < 15 and x < 5

Date	Mercury concentration
5 September 2011	3
5 September 2011	6
7 September 2011	15
8 September 2011	21
9 September 2011	22
10 September 2011	15
11 September 2011	14
12 September 2011	18
13 September 2011	15
14 September 2011	2

5.6 Calculate the mean, variance, skewness, and kurtosis for the data below.

Date	Mercury concentration
5 September 2011	3
6 September 2011	6
8 September 2011	15
8 September 2011	21
9 September 2011	22
10 September 2011	15
11 September 2011	14
12 September 2011	18
13 September 2011	15
14 September 2011	2

5.7 You have water quality information as detailed in the table below for PCB concentrations. Using the Shapiro–Wilks test, determine if the normal distribution is a reasonable descriptor for the dataset. As well, compute the skewness and kurtosis of the data and assess whether the data are reasonably characterized by those statistical measures. Comment on the differences in your findings.

Date	PCB concentrations (μg/l)
January 2008	112.0
February 2008	54.9
March 2008	73.8
April 2008	67.1
May 2008	85.8
June 2008	89.7
July 2008	100.8

5.8 Rank from best (most effective) to worst (least effective) of the tests for determining whether a particular set of data can be characterized by a Gaussian distribution.

1) Shapiro–Wilk test
2) Kolmogorov–Smirnov
3) Skewness
4) Kurtosis
5) Coefficient of variation

5.9 The data listed in table indicate the highest flows in river HYD in the indicated year, as reported for each of the years listed. Assuming that the normal distribution is an appropriate distribution to describe the flow data, estimate the 1000-year flood risk for the community that is adjacent to river HYD. List the assumption you are making in your determination.

Flow data for river HYD:

Year	Peak flow (m^3/s)
2001	28.0
2002	32.9
2003	15.4
2004	22.3
2005	28.9
2006	39.7
2007	14.5

5.10 You are preparing for a massive snowball fight that will take place in a nearby field. If the snowball is too small (less than 3 cm), then the snowball may freeze and injure someone if your throw is sufficiently accurate. If a snowball is too large (greater than 7.5 cm), you may not be able to throw it with sufficient accuracy as to be able to reach the target.

You have access to a newly released device from a robotics lab that automatically makes snowballs (it grabs basically a handful of snow and scrunches it into the form of a snowball). The snowball maker is known to create a snowball with a mean of 5 cm and a standard deviation of 3 cm, and the sizes can be described by the normal distribution.

What is the probability/risk that the snowball you are about throw is going to be adequate to toss at the other team?

5.11 For the following water quality data, estimate the mean and standard deviation using the probability plot procedure

8.2,5.3,< 5,< 5,10.1,9.3,7.6,< 5

5.12 You have determined from an analysis of the data that the statistics describing bacteria levels on a strawberry can be described by a t-distribution and that you have a mean of 4 and a standard deviation of 1.2. What would be your estimate of the probability that the number of bacteria would exceed 4?

5.13 What is the purpose of the Shapiro–Wilk test? And why is this test much better than the other alternatives?

5.14 Is there a difference between the "Gaussian distribution," the "normal distribution," and the "t-distribution"? Does the central limit theorem apply to both distributions? If not, why not?

5.15 Typically, we do not use the skewness coefficient and the kurtosis coefficient to decide that a dataset with 10 numbers can be adequately described by the normal distribution. Why is this? What is likely the best procedure for testing whether the data at a location can be adequately described by as a normal distribution?

5.16 You have been told that the data that describes the probabilities of illness from ingestion of contaminated water is normally distributed. You know the average concentration is 7 ug/l. From examining records illness you know that 5% of the people became ill when the concentrations exceeded 17.5 ug/l. What the probability of people becoming ill when the concentrations exceed 12 ug/l?

5.17 Is it true that the t-distribution is a subset of the normal distribution? Explain your answer.

5.18 The following data have been collected: 0.049, 0.09, 0.10, 0.10, 0.14, 0.19, 0.198, 0.29, 0.32, and 2.00 mg/l. Do not consider the last value as an outlier

a) Using the lognormal distribution, predict the 95% exceedance limit.
b) Estimate the 95% exceedance limit using the log Pearson type III distribution.
c) Comment on how the two values differ and why they differ.

6

The Lognormal Distribution

6.1 Introduction

Discussion in Chapter 5 focused in detail on the features of the normal distribution, a very general and important distribution. However, two attributes of the normal distribution limit its applicability to environmental quality concerns. Specifically,

i) The normal distribution is symmetric about the mean and yet many distributions of environmental quality data are not symmetric (for example, the skewed water quality distribution in Figure 6.1).
ii) The tails of the normal distribution extend to infinity in both the negative and positive directions. Since negative concentrations are infeasible, the normal distribution may be a poor characterization of environmental quality data.

As a result of (i) and (ii), situations arise where it is inappropriate to characterize the distribution of an environmental quality constituent as Gaussian or normal. An important alternative that is applicable to a wide array of environmental concerns is the lognormal distribution. Notably, the lognormal distribution precludes values less than zero and the distribution has no upper bound. Consequently, the lognormal distribution has a distributional shape that has considerable potential to describe environmental quality data (McBean and Rovers, 1992).

6.2 Important Features of the Lognormal Distribution

6.2.1 The Central Limit Theorem

To better understand why the lognormal distribution is often appropriate to describe environmental quality data, it is useful to establish the parallel to the normal distribution. To do this, consider a variable that is the product of a large number, say, "m," of independent factors (none of which dominates the resulting product), i.e.

$$y_i = x_{i1} \cdot x_{i2} \cdot x_{i3} \cdot x_{i4} \cdots x_{im} \tag{6.1}$$

where "m" is a large number. In other words, the variable of interest "y_i" is expressed as the product of a large number of subvariables, each of which is, in itself, difficult to study and describe. If we take the logarithm of both sides, (6.1) becomes

$$zi = \ln\ y_i = \ln\ x_{i1} + \ln x_{i2} + \ln\ x_{i3} + \cdots + \ln\ x_{im} \tag{6.2}$$

Risk Assessment: Procedures and Protocols, First Edition. Edward A. McBean.

Companion website: www.wiley.com/go/McBean/RiskAssessProcedure_1e

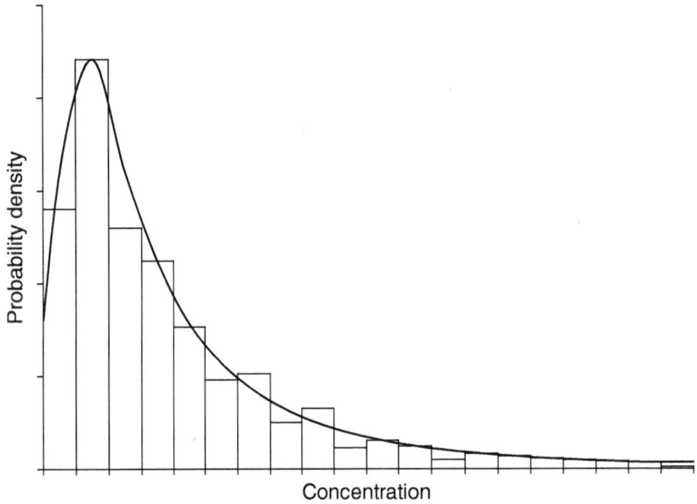

Figure 6.1 Example of the concentration data skewed to the right.

which is of the form of the relationship described in Section 5.1; the variable z_i is the sum of a large number of "m" variables. Thus, z_i is distributed as per the Gaussian or normal distribution. Therefore, all of the arguments that were made in Chapter 5 that related to the central limit theorem and the normal distribution as appropriate to support the concept of the normal distribution can, in a similar way, be used to support the concept of the lognormal distribution to describe data. In its simplest conceptual form then, the lognormal distribution is viewed as a normal distribution of the logarithms of a set of data. Thus, the lognormal distribution gets its name because the logarithms of random values are normally distributed.

For simple problems such as the summation of dice rolls, it is possible to support the idea that the sum of the dice rolls may be characterized using the central limit theorem as a normal distribution. In reality, however, it is difficult (if not impossible) to identify the large number of contributing variables that contribute in an additive and/or multiplicative way in environmental quality problems. The water quality in a stream is the consequence of many features – these include the runoff from farmers' fields, drainage from urban streets, wastewater discharges, and scour and deposition from the stream bottom, to name only a few. It is difficult to explicitly envision that the combinations from these numerous individual sources are either additive or multiplicative – it is simply necessary to recognize that the concept of the central limit theorem basically supports the idea that environmental phenomena and the consequences of innumerable individual features that combine in such a way that their combinatorial effect is not dominated by a single feature but a combination. The results may be described by a normal and/or lognormal distribution.

6.2.2 The Mathematics of the Lognormal Distribution

If the variables $y_1, y_2, y_3,\ldots, y_n$ of the original data, are lognormally distributed, then the transformed variables $z_1, z_2, z_3, \ldots, z_n$ where $z_i = \ln y_i$, for $i = 1, \ldots, n$, are normally distributed. The z_i, $i = 1, \ldots, n$ may then be described by the normal distribution equation (Equation [5.1]) as per the mathematical details presented in Chapter 5.

The population parameters μ_z and σ_z^2 may be estimated by $\overline{z}$ and S_z^2 in the manner described in Chapter 5. This is done first by transforming all of the y_i's to z_i's by taking logarithms ($z_i = \ln y_i$). The $\overline{z}$ and S_z^2 are then calculated as

$$\overline{z} = \sum_{i=1}^{n} \frac{z_i}{n} \tag{6.3}$$

and

$$S_z^2 = \frac{\sum_{i=1}^{n} (z_i - \overline{z})^2}{n - 1} \tag{6.4}$$

Example 6.1 The data listed below represent measurements of the hydraulic conductivity of the soils utilized in developing a compacted clay liner at a landfill.

Sample	Hydraulic conductivity (cm/s)
1	1.08×10^{-8}
2	8.76×10^{-9}
3	1.01×10^{8}
4	8.07×10^{-9}
5	5.63×10^{-9}
6	4.60×10^{-9}
7	7.42×10^{-9}
8	1.49×10^{-8}
9	3.48×10^{-9}
10	6.40×10^{-8}

a) Calculate the arithmetic mean and the geometric mean. Which of the two measures of central tendency, the arithmetic mean or the geometric mean, is a better indication of the central tendency of the data?

   Arithmetic mean = 1.38×10^{-8} cm/s
   Geometric mean = 9.31×10^{-9} cm/s

   A histogram of the data is illustrated in Figure 6.2. The geometric mean is more representative of the "bulk" of the data.

b) Is the logarithm of the (arithmetic) mean the same as the mean of the logarithm-transformed data?

   Logarithm of the mean = $\ln(1.38 \times 10^{-8}) = -18.1$
   Mean of the logarithms = -18.49

   Note that the mean of the original data is not the same as the logarithm of the mean of the original data. With the original data, the mean is very much related only to the high values.

Chow (1964), in a comprehensive investigation of the lognormal distribution, presents the following relationships for calculating $\overline{z}$ and S_z^2 without taking the logarithms of all of the data, as

$$\overline{z} = \frac{1}{2} \ln\left[\frac{\overline{y}}{\{C_v^2 + 1\}}\right] \tag{6.5}$$

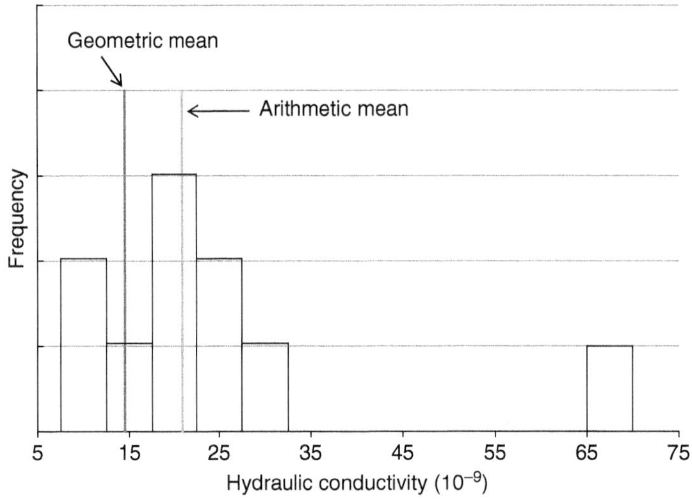

Figure 6.2 Histogram of hydraulic conductivity.

and

$$S_z^2 = \ln\left[C_v^2 + 1\right] \tag{6.6}$$

where C_v is the coefficient of variation of the original data ($C_v = S_y / \bar{y}$).

The mean and variance of the lognormal distribution are

$$\mu_y = E[y] = \exp\left(\mu_z + \frac{\sigma_z^2}{2}\right) \tag{6.7}$$

and

$$\sigma_y^2 = Var[y] = \mu_y^2\left[\exp\left(\frac{\sigma_z^2}{2}\right) - 1\right] \tag{6.8}$$

where "E[ ]" is the expectation operator (numerous reference sources discuss the expectation operator; see, for example, Chatfield (1978)).

The coefficient of variation of the y's is

$$C_v = \left[\exp\left(\frac{\sigma^2}{2} - 1\right)\right]^{0.5} \tag{6.9}$$

The coefficient of skew of the y's is

$$y = 3C_v + C_v^3 \tag{6.10}$$

With the widespread availability of spreadsheet calculation programs, the need for Equations (6.4–6.9) is considerably less than when the equations were first developed. As well, the above-noted equations are premised on the values being the population parameters (as opposed to sample statistics).

There are several very important features associated with the lognormal distribution. First, the lognormal distribution is a two-parameter distribution, which means to completely characterize the distribution, only two parameters need to be quantified. Therefore, as with the normal distribution, the lognormal distribution is a "central fitting" distribution, not a fitting to the "tails" of the distribution. As a result, the lognormal distribution may not provide acceptable accuracy in describing the extreme values of the data.

Because there are just the two fitting parameters implicit in the description of the lognormal distribution and yet the distribution shape is not symmetrical, there is a "built-in" skewness associated with the distribution. Once the mean and variance are specified, the value of the skewness is fixed. If it is desired to change the skewness (and therefore to better reflect the tails of the data distribution), one needs to utilize more sophisticated distributions, the subject of Chapter 7.

Transformation to the logarithmic scale is not done to make "large numbers look smaller." Performing a logarithmic or other monotonic transformation preserves the basic ordering within a dataset so that the data are rescaled with a different set of units. The changes occur because the logarithms of lognormally distributed data are more nearly normal or Gaussian in character, thus satisfying a key assumption of many statistical procedures.

6.2.3 Probability Paper

The primary discussion on probability plotting is contained in Section 5.3.3 and should be the first point of reference; discussion herein simply follows along from the discussion in Section 5.3.3. The same probability plot technique outlined in Chapter 5 may be used to investigate whether a set of data or residuals follows the lognormal distribution. The procedure is the same except that each observation is replaced by its logarithm. Hence, probability paper for the lognormal distribution is identical to normal distribution paper except that the ordinate is a logarithmic scale. Thus, lognormal probability paper is similar to the normal probability paper in terms of its effectiveness at providing the opportunity to visually characterize whether the lognormal distribution is an appropriate fit to a dataset.

Example 6.2 Use the Weibull plotting position formula and probability plotting paper to indicate the appropriateness of the following data to be characterized by the lognormal distribution:

Mercury concentration (μg/l): 2.2, 1.5, 1.0, 2.2, 3.8, 0.15, 0.45, 1.3, 6.4, 0.85

We can demonstrate the data are skewed from the coefficient of variation test:

$$\overline{x} = 3.30$$

$$S_x = 3.70$$

Hence, the coefficient of variation is 3.70/3.30 = 1.12 and greater than unity. The data distribution must be skewed.

It is useful as well to plot the data on probability paper since deviations of the plotted data from a straight line indicate the data are not well characterized by the normal distribution. The results using normal probability paper are shown in Figure 6.3a and for lognormal paper in Figure 6.3b.

Note that the ordinate on the probability paper may be equally expressed on the logarithmic axis or, equivalently, in arithmetic terms.

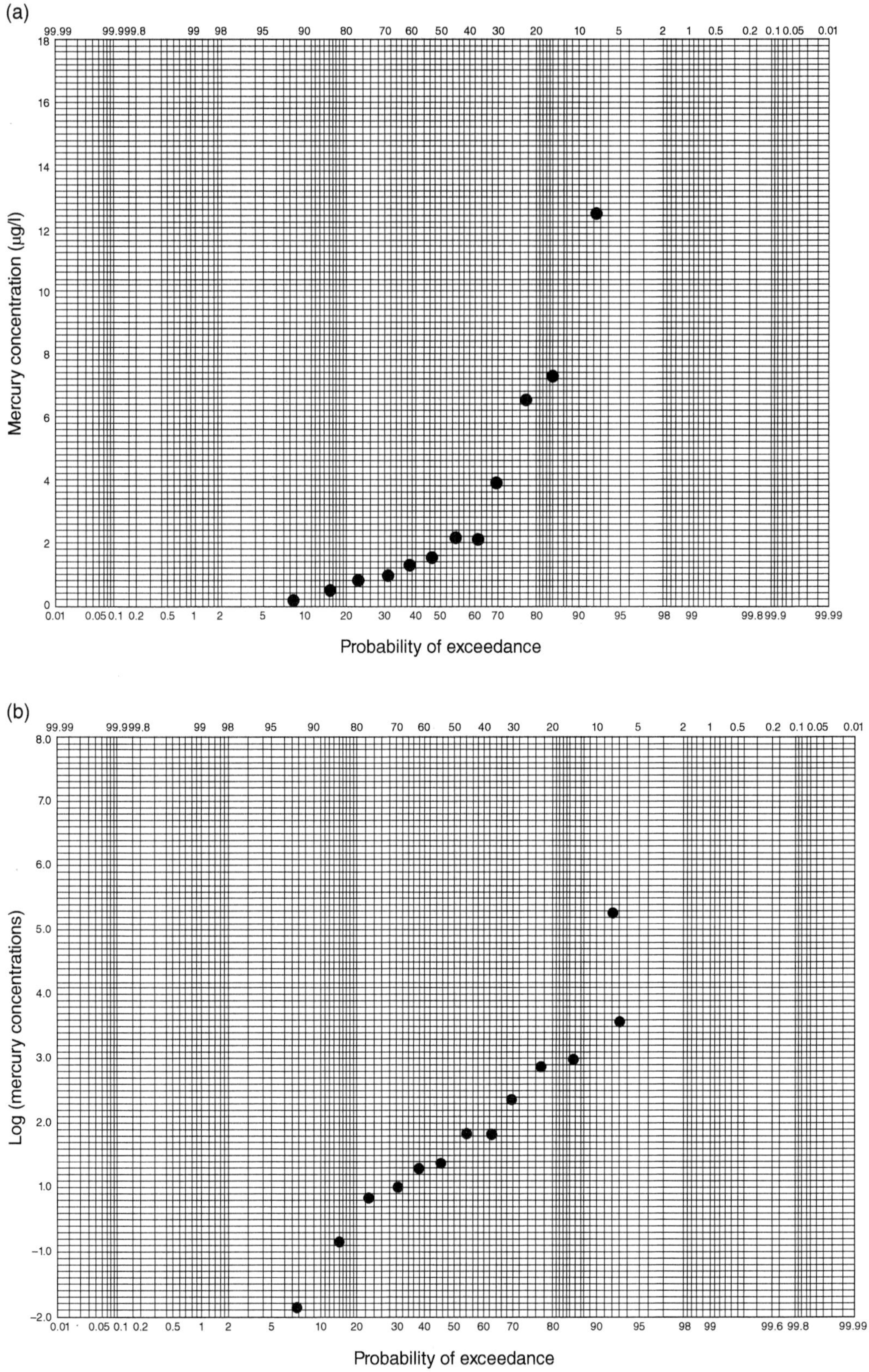

Figure 6.3 (a) Probability plot of mercury data, normal probability paper. (b) Probability plot of mercury data, lognormal probability paper.

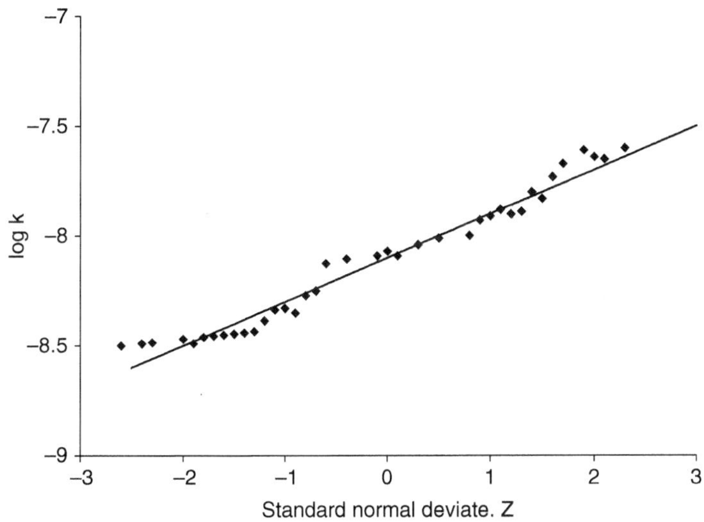

Figure 6.4 Probability plot of hydraulic conductivity of an engineered natural soil liner.

In a manner comparable to the discussion in Section 5.3.3, both the mean and the standard deviation may be estimated from the plotted lines on the probability paper.

Example 6.3 Of all soil physical properties, Warrick and Nielsen (1980) illustrate that hydraulic conductivity exhibits the greatest degree of variability. This implies that samples collected from a field site will not all yield the same hydraulic conductivity but rather will be discrete values from an underlying statistical distribution. In addition, given that the behavior of a porous medium is not necessarily best described by its mean value, the mean value of the hydraulic conductivity's statistical distribution does not represent the overall effective or bulk conductivity. Donald and McBean (1994) showed a plot of the logarithms of hydraulic conductivity for samples from an engineered natural soil liner for Metropolitan Toronto's Keele Valley landfill versus the standard normal deviate. The results are illustrated in Figure 6.4 and include a straight-line estimate of the lognormally distributed conductivity measurements. A reasonable characterization of the hydraulic conductivity data by the lognormal distribution is apparent since the data are reasonably characterized by the straight line. Note that Z indicates the number of standard deviations from the mean.

6.3 Tests for Lognormality

To determine whether the lognormal distribution is an adequate descriptor of a dataset, similar mathematical procedures to those described in Sections 5.3.1 through 5.3.7 may be applied by applying the mathematical procedures to the logarithms of the data. For this reason, the tests of "goodness of fit" to the distribution will not be reexamined.

Example 6.4 Determine if the benzene concentration data of Table 5.2 are adequately described by the lognormal distribution. Use the Shapiro–Wilk test (see Section 5.3.6 for a discussion of the Shapiro–Wilk test). This example represents an extension from Example 5.11.

Step 1: Order the data from smallest to largest and take the natural logarithms as listed in the following table.
Step 2: Compute the differences.

i	X_i	$X_{,(i)}$	X_{n-i+1}	$X_{n-i+1} - X_i$	a_{n-1+i}	b_i
1	11.74	2.463	3.314	0.851	0.4734	0.403
2	13.32	2.589	3.229	0.640	0.3211	0.206
3	14.09	2.645	3.119	0.474	0.2565	0.122
4	14.90	2.701	3.103	0.402	0.2085	0.084
5	15.64	2.750	3.065	0.315	0.1686	0.053
6	15.78	2.759	3.019	0.260	0.1334	0.035
7	16.09	2.778	3.014	0.236	0.1013	0.024
8	16.16	2.783	2.929	0.146	0.0711	0.010
9	16.39	2.797	2.897	0.100	0.0422	0.004
10	16.66	2.813	2.912	0.099	0.0140	0.001
11	18.11	2.912	2.813			b = 16.59
12	18.12	2.897	2.797			
13	18.71	2.929	2.783			
14	20.36	3.014	2.778			
15	20.47	3.019	2.759			
16	21.43	3.065	2.750			
17	22.27	3.103	2.701			
18	22.63	3.119	2.645			
19	25.25	3.229	2.589			
20	27.50	3.314	2.463			

Step 3: For n = 20, k = 10.
Step 4: Coefficients a_{n-i+1} from Table A.6; b = 0.942.
Step 5: Compute the standard deviation of the log-transformed data as S = 0.218:

$$W = \left[\frac{0.942}{0.218\sqrt{19}}\right]^2 = 0.983$$

Step 6: $W_{0.05,20} = 0.905$ from Table A.6. Since W > 0.905, the benzene data do not show significant evidence of nonnormality in the log-transformed sense.

6.4 Generation of Lognormal Concentration Data

Numerous computer programs provide a simple procedure to generate a sequence of uniformly distributed random numbers and normally distributed random numbers. However, there may be circumstances in which a set of lognormally distributed random numbers is of interest. This can be accomplished as follows.

If z = ln y, where "x" is the actual concentration of a chemical, then the probability distribution of y is normal with a mean μ_y and variance μ_y. The mean and variance of z are readily computed from the mean and variance of y (Aitchison and Brown, 1957):

$$\mu_z = \ln(\mu_y) = 0.5\left[\ln\left\{\frac{\sigma_y}{\mu_y}\right\}^2 + 1\right] \tag{6.11}$$

$$\sigma_z = \left[\ln\left(\frac{\sigma_y}{\mu_y}\right)^2 + 1\right]^{0.5} \tag{6.12}$$

For generating datasets for Monte Carlo experiments, the population mean, μ_y, and the coefficient of variation, σ_y/μ_y, can be specified and used to calculate μ_y and σ_y. Then a dataset of true concentrations, y_i, of any number of values, n, can be generated by

$$y_i = \exp[\mu_z + s_z\varepsilon_i] \tag{6.13}$$

where ε_i is a randomly chosen value from a normal distribution with a mean of zero and variance of one.

A linear trend may be added by causing the specified population μ_y to increase at a constant rate, m, with time t, as

$$\mu_z = \ln(\mu_y + mt) - 0.5\left[\ln\left[\left(\frac{\sigma_y}{m_y}\right)^2 + 1\right]\right] \tag{6.14}$$

6.5 References

Aitchison, J. and Brown, J.A. (1957). *The Lognormal Distribution*. Cambridge: Cambridge University Press.

Chatfield, C. (1978). *Statistics for Technology*. New York: John Wiley & Sons.

Chow, V. (1964). *Handbook of Applied Hydrology*. New York: McGraw-Hill.

Dagan, G. (1979). Models of groundwater flow in statistically homogeneous porous formations. *Water Resources Research* 15 (1): 47–63.

Donald, S.B. and McBean, E.A. (1994). Statistical analyses of compacted clay landfill liners. Part 1: model development. *Canadian Journal of Civil Engineering* 21: 872–882.

McBean, E.A. and Rovers, F.A. (1992). Estimation of the probability of exceedance of contamination concentration risk problems. *Ground Water Monitoring Review* Winter: 115–119.

Smith, L. and Freeze, R.A. (1979). Stochastic analysis of steady state groundwater flow in a bounded domain. Two-dimensional simulations. *Water Resources Research* 15 (6): 1543–1559.

Warrick, A.W. and Nielsen, D.R. (1980). Spatial variability of soil physical properties in the field. In: *Applications of Soil Physics* (ed. D. Hillel), 319–344. New York: Academic Press, Inc.

6.6 Problems

6.1 You have identified seven data values of arsenic in air (all in μg/m^3) in a community near a gold mine as tabulated.
1.43, 19.3, 4.13, 37.6, 1.77, 1.01, 5.10
Arithmetic mean = 10.0
Standard deviation of raw data = 13.7
Geometric mean = 1.51
Standard deviation of log-transformed (natural logarithmic) data = 1.36

a) Assuming the arsenic concentrations are lognormally distributed, what is the probability that the arsenic concentration is larger than 45 ug/m^3?
b) Plot the data on the probability paper provided, assuming the data are characterized by the lognormal distribution. You will see that the plotted values on the probability paper do not all lie along a straight line. Why is this?
c) Using the probability paper, and assuming you were to sample the air quality 100 times, what would your best estimate be of the highest value from the results?

6.2 Log-transform the data from the previous question, and re-do steps in Question 6.1 to determine if the lognormal distribution is a reasonable descriptor for the dataset.

6.3 Add 5 to each of the values listed in the table below.

Date/location	MW1	MW2	MW3	MW4	MW5	MW6	MW7	MW8	MW9	MW10
2 April 1995	5.74	2.36	1.83	1.84	8.65	2.50	3.33	2.61	1.84	1.94
9 April 1995	2.06	2.21	3.39	3.53	5.99	1.39	2.08	3.15	4.50	6.21
16 April 1995	1.67	2.89	4.14	1.48	2.23	6.81	2.14	2.84	3.94	2.47
23 April 1995	4.92	3.26	3.29	3.29	6.22	3.40	1.29	3.74	1.75	2.55
30 April 1995	1.94	1.30	1.21	2.17	4.39	1.66	4.40	2.81	6.55	3.78
7 May 1995	1.81	4.47	2.29	2.65	3.76	2.32	2.53	3.00	2.96	2.44
14 May 1995	1.57	1.85	6.35	4.66	2.61	1.33	1.43	3.13	2.02	3.56
21 May 1995	4.30	3.49	1.17	5.39	4.26	2.05	2.12	1.32	2.19	2.21
28 May 1995	3.28	1.77	4.84	2.99	2.17	6.25	3.95	1.11	1.85	1.18
4 June 1995	2.53	2.43	1.83	3.30	6.01	1.99	2.12	1.73	2.98	2.92

Assuming the lognormal distribution is a good descriptor of the PCB data, calculate the 95% exceedance concentration.

6.4 What is the basis by which we assume the normal distribution is generally appropriate to characterize concentration data that we might utilize in a risk assessment? If not the normal distribution, what is an alternative distribution that we frequently use, and why is that a very frequent option that is utilized?

6.5 What is the basis by which we assume the normal distribution is generally appropriate to characterize concentration data that we might utilize in a risk assessment? If not the normal distribution, what is an alternative distribution that we frequently use, and why is that a very frequent option that is utilized?Plotted below is a probability of recurrence plot of chemical concentration using normal probability paper. Relevant to the figure (note – do not do a lot of calculations in the following – just indicate approximate values), determine:

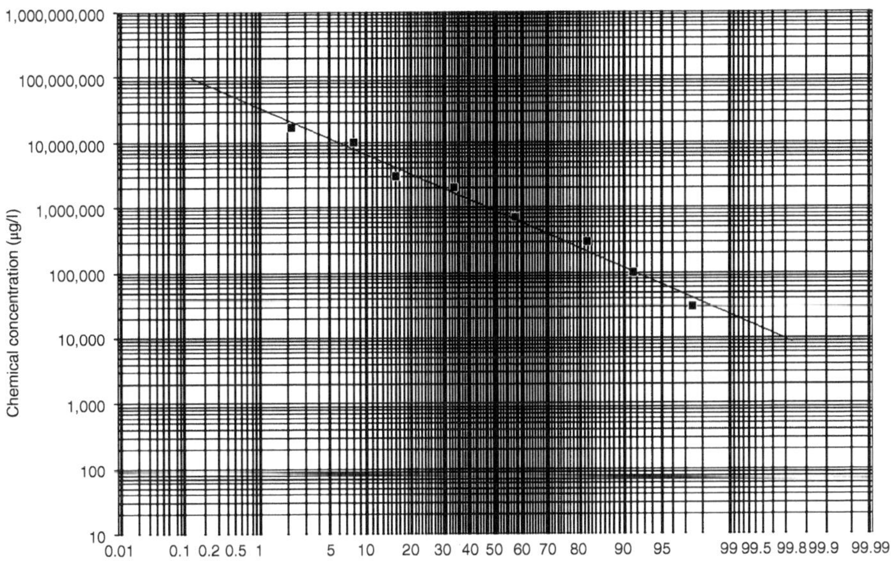

a) If you were to describe these concentration data as to the type of distribution, what would you indicate?
b) What is your estimate of the geometric mean concentration of the chemical at the site?
c) What is your best estimate of the concentration for the mode of the chemical at the site?
d) For purposes of your exposure risk assessment, you are concerned with the exceedance probability of the concentration such that there is only a 5% change of exceedance. What would be your specification of this concentration?
e) The plotted points do not lie on the best fit solid line in the figure. Why is that, and how important is that in your statements in (a)–(d) above?

6.6 You have identified seven data values of arsenic in air (all in ug/m^3) in a community near a mine as tabulated below.

Sample no.	Arsenic concentration (µg/m^3)	Ln-transform of arsenic
1	1.43	0.358
2	12.1	2.493
3	4.8	1.569
4	39.2	3.669
5	1.77	0.571
6	1.01	0.010
7	5.1	1.629

Mean	9.34	1.471
Std. dev.	13.71	1.297
Skew	2.262	0.690

a) Assuming the arsenic concentration data are lognormally distributed, what is the probability that the arsenic concentration is larger than 45 ug/m^3 in terms of exposure risk to the villagers?
b) Plot the data on the probability paper provided, assuming the data recharacterized by the lognormal distribution. You will see that the plotted values on the probability paper do not all lie along a straight line, and yet, you are assuming the lognormal distribution is a useful indicator of the air quality. Why do the plotted points not all fall along a straight line?
c) Using the probability paper, and assuming you were to sample the air quality 100 times, what would your best estimate be of the highest value from the results?

7

Other Distributions Useful for Characterizing Environmental Quality Data

When you find yourself in a hole, stop digging.

—Will Rogers

7.1 Introduction

Numerous other probability distributions exist (beyond the normal and lognormal distributions) that may be appropriate for characterizing environmental quality data. Examples of distributions include the binomial, Poisson, Gumbel, and log Pearson III; plus many more exist. By selecting a particular distribution, we do not know of course what the next value might be, but, by the distribution that we select, the probability associated with the various values that might occur can be assigned. As will be seen, circumstances sometimes require distributional characteristics different from those of the normal and lognormal distributions described in Chapters 5 and 6. These other distributions have been presented here on the basis of their ability to "fit" the data due to their greater flexibility. The intent of this chapter is to examine aspects of the important features of some of the other probability distributions as they pertain to their usefulness in application to environmental quality problems.

Statistical distributions in the technical literature are frequently demonstrated in applications in which data lengths number in the thousands. Large datasets such as these seldom exist in environmental quality situations. For example, some of these additional distributions require characterization of the skewness coefficient, which, as indicated in Chapter 4, may be considerably in error when the dataset is small. The intent of the various sections in this chapter is to identify some of the both positive and negative attributes of these other distributions as they pertain to environmental quality data.

7.2 The Poisson Distribution

In 1837, Poisson published the derivation of a distribution that bears his name. Poisson's approach was to derive a distribution for a series of independent events in which the number of trials was large, the probability of occurrence of the outcome was small, and the probability remained constant over the trials. To utilize the Poisson distribution, the random variable takes on only discrete, usually integer values; if this is not the case, the analyses should be carried out using a continuously distributed random variable. Thus, the Poisson distribution describes the number of events or occurrences.

Risk Assessment: Procedures and Protocols, First Edition. Edward A. McBean.

Companion website: www.wiley.com/go/McBean/RiskAssessProcedure_1e

The Poisson distribution is governed by the average rate of occurrence, μ, which can be estimated by summing the Poisson events of all samples in the background pool of data and dividing by the number of samples in the pool. Once the average rate of occurrences has been estimated, the formula for the Poisson distribution is given by

$$P(x;\mu) = \frac{e^{-\mu}\mu x}{x!} \quad \text{for } x = 0, 1, 2, 3, \ldots \tag{7.1}$$

where x represents the Poisson count and μ represents the average rate of occurrence.

The cumulative distribution is

$$F(x;\mu) = \sum_{i=0}^{x} P(i;\mu) \tag{7.2}$$

In other words, the cumulative probability F(1;3) means the probability of obtaining x = 0 or 1, where the mean of the distribution is 3, is

$$F(1;3) = P(0;3) + P(1;3) \tag{7.3}$$

Using (7.1) then Equation (7.3) is calculated as

$$\begin{aligned} F(1;3) = \frac{e^{-3}3^0}{0!} + \frac{e^{-3}3^1}{1!} &= 0.0497 + 0.1494 \\ &= 0.1991 \\ &\cong 0.20 \end{aligned}$$

In words, if the mean of the Poisson distribution is 3, there is a 20% probability of having a zero or one-integer response.

Examples of environmental engineering applications of the Poisson distribution include the counts of the number of microorganisms of a certain type, as quantified by a lab test, or the number of α-particles emitted per unit of time during radioactive decay.

The probabilities P(x) are nonnegative and sum to one:

$$\begin{aligned} \sum_{x=0}^{\infty} P(x) &= \sum_{x=0}^{\infty} \frac{e^{-m} m^x}{x!} \\ &= e^{-\mu} \frac{u^x}{x!} \\ &= e^{-\mu} e^{+\mu} \\ &= 1 \end{aligned} \tag{7.4}$$

The function of the factor $e^{-\mu}$ that does not vary with x is to ensure that the probabilities add up to unity.

The parameter μ equals the mean of the probability function; the variance is also equal to μ, or

$$\text{Mean} = \mu$$

and

$$\text{Variance} = \sigma^2 = \mu$$

Both the mean and the variance are equal to μ, while the skewness is $1/\sqrt{\mu}$ so that the distribution is always skew to the right. However, the degree of skewness decreases as μ increases.

The Poisson distribution can thus be used to determine probabilities for discrete random variables that satisfy the following three assumptions:

1) The random variable is the number of times that an event occurs in a single trial, which is defined as a basic unit of time, space, or quantity. A trial may be defined as a unit of time (e.g. a millisecond, hour, or year), a unit of space (i.e. length, area, volume), or some other quantity. The number of events in a trial is usually an integer value from 0 to 1, 2, 3, ... The random variable is the number of events that occur per trial.
2) The average number of occurrences of the random variable in a single trial remains constant from trial to trial.
3) The discrete values are nonoverlapping and are independent of one another.

Given these three assumptions, it can be seen that the Poisson distribution is the limiting form of the binomial distribution where there are a large number of trials but only a small probability of success at each of them.

In the context of environmental quality monitoring, further applications of the Poisson distribution include the number of "hits" out of a large number of VOC measurements. This problem has been modeled by the Poisson distribution (e.g. Silver and Dunn, 1986). Alternatively, Gibbons (1987) considered the molecule as a unit of observation and postulated that the number of molecules of a particular compound out of a much larger number of molecules is the result of a Poisson process. Poisson's approach is justified in that the number of units (i.e. molecule) examined is large and the probability of the occurrence (i.e. a molecule being classified as benzene) is small and thus considered the occurrence of low-level hits of volatile organic priority pollutant compounds statistically modeled as a Poisson process.

Example 7.1 As a simple example of the Poisson distribution to assist with the explanation, the parts per billion in the sequence of nine monitoring periods (of air mass through a screen), is

$$3,1,10,2,4,6,8,2\,(\text{units})$$

At low levels, the values are only reported to specific magnitudes. What is the probability that in the next monitoring period, there will be one or less ppb?

The mean of the samples is 4.33.

The probability of 0 is $P(0;4.33) = \dfrac{e^{-4.33}4.33^0}{0!} = 0.013$

The probability of 1 is $P(1;4.33) = \dfrac{e^{-4.33}4.33^1}{1!} = 0.057$

The probability that in the next monitoring period there will be one or less ppb measured is 0.013 + 0.057 = 0.070.

Other examples of application of the Poisson distribution will be developed in later chapters as specific areas of application.

7.3 Extreme Value Distributions

There is a family of extreme value distributions that, as the name implies, have been utilized to characterize extreme events (e.g. what is the flood flow that is exceeded only once, on average, in 100 years?). The utilization of an extreme value distribution involves the fitting of a continuous distribution such as Pearson or log Pearson and/or an asymptotic distribution such as Gumbel's extreme value. Whereas the normal and lognormal distributions are referred to as central fitting distributions, the "tails of distribution" (i.e. the extreme values) are better described by distributions other than the normal or lognormal distributions.

Chemical concentration data are constrained on the low side by the detection limit of the analytical technique and essentially unconstrained on the high side. Because of this, the probability distribution is often considered as lognormally distributed (as per Chapter 6). On the negative side, the flexibility of the lognormal distribution to fit the data is limited because the skewness is fixed; only two parameters define the entire lognormal distribution. Therefore, for example, in risk assessment studies, in which much of the concern may be focused on a relatively extreme event, the reliance on only two statistics does not allow for much flexibility for the fitting of the distribution.

7.3.1 The Gumbel Distribution

Gumbel (1954, 1958) showed that the exponential function is useful in characterizing the distribution of annual flood peak discharges. Gumbel proposed a double exponential distribution for the largest value. Specifically, in its cumulative form, the Gumbel model is defined as

$$F(X) = \exp\left[-e^{-\alpha(x-\beta)}\right] \tag{7.5}$$

in which F(x) is the cumulative distribution of the large value x and α and β are fitting parameters. The probability distribution is

$$f(x) = \left[-\alpha(x-\beta)e^{-\alpha(x-\beta)}\right] \tag{7.6}$$

The parameters α and β are related to the mean and standard deviation of the data by the relationship

$$\mu = \beta + \frac{0.5772}{a} \tag{7.7}$$

and

$$\sigma = \frac{\pi}{\alpha\sqrt{6}} = \frac{1.282}{\alpha} \tag{7.8}$$

Alternatively, these same two equations may be rewritten as

$$\beta = \mu - 0.450\sigma \tag{7.9}$$

and

$$\alpha = \frac{1.282}{\sigma} \tag{7.10}$$

Several features about the Gumbel distribution are noteworthy: The distribution is a two-parameter distribution. The skewness coefficient is fixed at 1.1396, a positive and constant value. The kurtosis coefficient of the Gumbel distribution is also constant as 4.5. Since the kurtosis coefficient is greater than 3.0, the "peakedness" of the distribution is different from the normal distribution.

The data utilized in defining the fitting parameters of the distribution must be independent values and extremes. Thus, the Gumbel model is commonly adopted to describe hydrologic phenomena such as the maximum daily flow in a year or the annual peak hourly discharge during a flood (e.g. Chow, 1951, 1964). The rationale for use of this model is that a single value utilized in the case of maximum daily flow in a year represents the single highest value (i.e. 1 out of 365) of flow in the year.

Example 7.2 A history of record of the peak annual flows in a stream is analyzed to determine the estimates of the mean and variance as

$$\mu = 18\ \text{m}^3/\text{s}$$

$$\sigma^2 = 16 \quad \text{or} \quad \sigma = 4\ \text{m}^3/\text{s}$$

Estimate the probability that the flow will exceed $30\,\text{m}^3/\text{s}$.

Solving for the parameters α and β gives

$$\alpha = \frac{1.282}{4} = 0.321$$

and

$$\begin{aligned}\beta &= \mu - 0.45\sigma \\ &= 18 - 0.45(4) = 16.2\end{aligned}$$

The probability that the peak flow in a particular year will exceed $30\,\text{m}^3/\text{s}$ is

$$\begin{aligned}P(x \geq 30) &= 1 - F(30) \\ &= 1 - \exp\left[-3^{-0.321(30-16.2)}\right] \\ &= 1 - 0.988 = 0.012\end{aligned}$$

There is only a 1.2% chance that the flow will exceed $30\,\text{m}^3/\text{s}$ in a particular year.

Probability paper for the Gumbel distribution has been developed, as illustrated in Figure 7.1. All of the features described in Section 5.3.3 on plotting on probability paper apply to Gumbel probability paper, regarding the use of the Gumbel paper. The Gumbel probability paper contains an extreme value probability scale versus either an arithmetic or logarithmic scale to test whether the data approximate a Gumbel or log-Gumbel cumulative distribution function.

By a symmetrical argument to that in Equation (7.8), the distribution of the smallest of many independent variables with a common unlimited distribution with an exponential-like lower tail can be addressed. The cumulative distribution for this case is

$$F(z) = 1 - \exp(-e - \alpha(z - \beta)) \tag{7.11}$$

with the mean of

$$\mu_z = \beta - \frac{0.5772}{a} \tag{7.12}$$

and

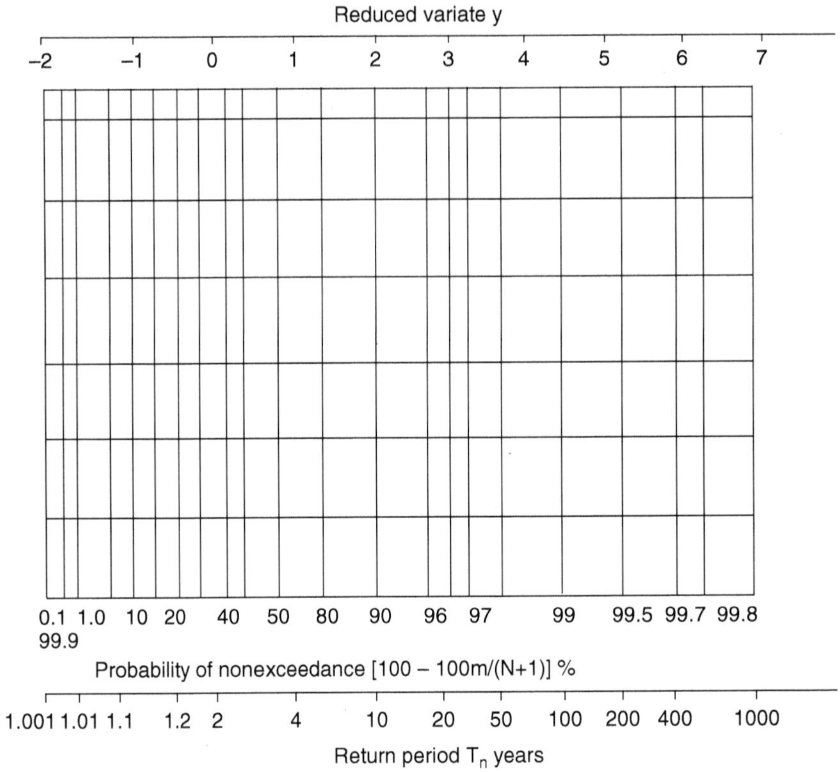

Figure 7.1 Gumbel probability paper.

$$\alpha_z = \frac{\pi}{\alpha\sqrt{6}} \tag{7.13}$$

and coefficient of skewness of −1.1396.

Two log-Gumbel distributions may be defined and known as the extreme value type II and III distributions; the former may be applied to floods (maxima), while the latter distribution (also known as the Weibull distribution) is applied to minima events. However, as noted in Benjamin and Cornell (1970), the extreme small value model noted immediately above is not always a good choice if the variable being characterized is necessarily positive as, for example, streamflows and/or concentrations must be. This occurs because the distribution is unbounded, extending to infinity in both a positive and a negative sense.

7.3.2 Log Pearson Type III Distribution

Chow (1951, 1964) demonstrated that for many types of frequency analyses, the extreme value Z_T (where Z_T is the value having the potential for being equaled or exceeded once on average in T values) can be written as

$$Z_T = \overline{x} + k_T S \tag{7.14}$$

where $\overline{x}$ is the mean of a dataset, k_T is a frequency factor, and S is the standard deviation.

Values for the frequency factor, k_T, are a function of the distribution being utilized in the analysis. Equation (7.17) is of a general form. For the case of log-transformed data,

$$Y_T = \overline{y} + k_T' S \tag{7.15}$$

where $\overline{y}$ is the mean of the log-transformed data, k_T' is the frequency factor for the normal distribution, and S is the standard deviation of the log-transformed data.

If the magnitude of the frequency factor is selected in accordance with the log Pearson type III distribution, then Equation (7.15) becomes the equation for estimating the return frequency distribution assuming the log Pearson type III distribution. Thus, the log Pearson distribution represents one particular form of (7.14), and others exist.

An important merit of the log Pearson distribution is that k_T' is a function of the frequency factor and the skewness. In other words, the log Pearson III distribution is a three-parameter distribution and provides the flexibility to fit to the tail of the data distribution. The flexibility to provide this extra fitting arises because of the existence of the third parameter.

Values of k_T for the Pearson type III distribution are included in Table A.14.

If the skew coefficient falls between −1.0 and 1.0, approximate values of the frequency factor for the Pearson type III can be obtained from

$$k_T = \frac{2}{C_S}\left[\left(z - \frac{C_S}{6}\right)\frac{C_S}{6} + 1\right]^3 - 1 \tag{7.16}$$

where z is the standard normal deviate for the selected recurrence interval T and C_s is

$$C_S = \frac{a}{S^3} = \frac{n}{(n-1)(n-2)}\sum_{i=1}^{n}(x_i - \overline{x})^3 \tag{7.17}$$

Further discussion on the log Pearson III distribution is presented in Viessman et al. (1989).

Example 7.3 As a demonstration of the log Pearson distribution (after McBean and Rovers, 1992), consider the monitoring data of 0.049, 0.07, 0.07, 0.07, 0.07, 0.14, 0.14, 0.158, 0.29, 0.32, and 12 μg/l. These concentrations represent monthly groundwater sample concentration data from a monitoring well in close proximity to a stream. The data have been plotted on lognormal probability paper on Figure 7.2 using the Weibull plotting position. The failure of the data points to plot as a reasonably straight line indicates the lognormal distribution is not a good characterization of the data (after assurance is obtained that the high value is not an error in sampling).

Figure 7.3 illustrates a plot of the data on log Pearson type III paper. Note how the lower ordinate of the log Pearson paper differs from the lognormal probability paper.

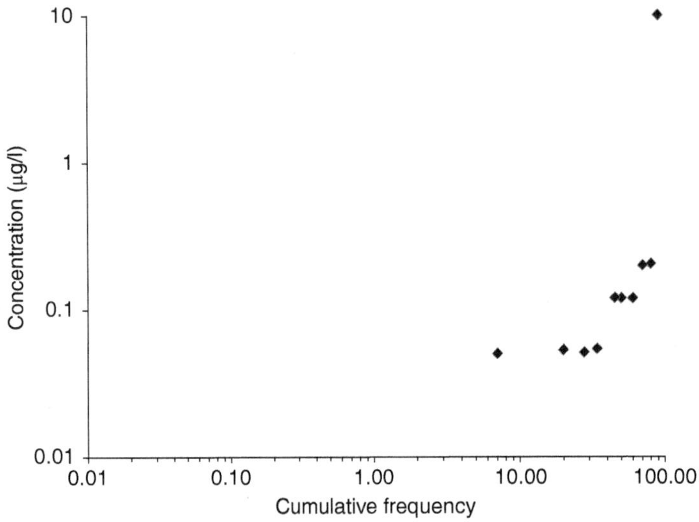

Figure 7.2 Lognormal distribution plot of monitoring data.

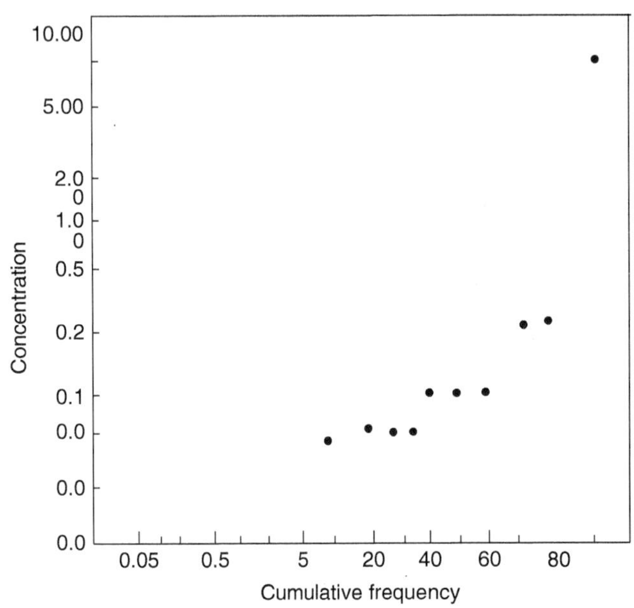

Figure 7.3 Log Pearson type III distribution plot of monitoring data.

7.4 References

Benjamin, J.R. and Cornell, C.A. (1970). *Probability, Statistics, and Decisions for Civil Engineers*. McGraw-Hill.

Chow, V.T. (1951). A general formula for hydrologic frequency analysis. *Transactions American Geophysical Union* 32 (2): 231–237.

Chow, V.T. ed. (1964). *Handbook of Applied Hydrology*. New York: McGraw-Hill Inc.

Gibbons, R.D. (1987). Statistical models for the analysis of volatile organic compounds in waste disposal sites. *Ground Water* 25 (5): 572–580.

Gumbel, E.J. (1954). Statistical theory of droughts. *Proceedings of the ASCE* 80 (439): 1–19.

Gumbel, E.J. (1958). *Statistics of Extremes*. New York: Columbia University Press.
McBean, E. and Rovers, F. (1992). Estimation of the probability of exceedance of contaminant concentrations. *Ground Water Monitoring Review* Winter: 115–119.
Silver, C.A. and Dunn, D. (1986). Statistical Analysis of Rare Events in Groundwater, report prepared for the Environmental Institute for Waste Management Studies, The University of Alabama.
Viessman, W., Lewis, G., and Knapp, J. (1989). *Introduction to Hydrology*, 175. New York: Harper and Row Publishers Inc.

7.5 Problem

7.1 Plotted below are the probabilities of exceedance values for three locations of measurement. Location A is upgradient from the site, and locations B and C are downgradient (the arrow depicts the direction of groundwater flow).
Is there evidence that the gasoline station is impacting the groundwater?
Ensure you provide an explanation of the assumptions and procedural checks involved in your answer.

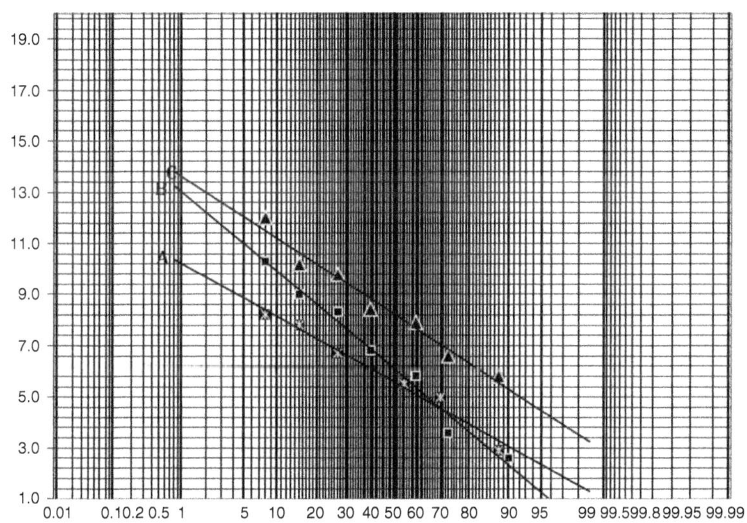

Part III

Hypothesis Testing of Environmental Quality

8

Identification of System Changes and Outliers Using Control Charts

8.1 Introduction

When examining a time series of data, of interest may be if there are changes occurring over time. For example, perhaps in a particular environmental monitoring record, of interest may be whether there has been a change in concentration(s), e.g. has a wastewater treatment system encountered upset conditions such that the quality of the treated effluent has deteriorated? Alternatively, there may be one or two sampling results contained within the record that are suspiciously different from the remainder – has an error occurred in, for example, either the reading of an instrument or a transcription error in the recording of the data? For these types of situations, we are interested in statistical procedures capable of identifying if there is a significant reason to believe that changes to the system have occurred and/or whether data outliers are likely present in the data record.

This chapter describes several analytical procedures appropriate to addressing these types of questions. It may well be that subsequent, more detailed assessments are needed beyond what is presented in this chapter. More detailed procedures will be examined in subsequent chapters, with the purpose at this point being to describe only the capabilities of a series of relatively simple procedures.

Of interest initially is the development of three statistical intervals for detecting system changes:

1) *Tolerance intervals* (see Section 8.2): Tolerance intervals are designed to contain a designated proportion of the population (e.g. 95% of all possible sample measurements).
2) *Confidence intervals* (see Section 8.3): Confidence intervals are designed to contain a specified population parameter with a designated level of confidence or probability, denoted as $1 - \alpha$. The interval will fail to include the true parameter in approximately α percent of the cases where such intervals are constructed.
3) *Prediction intervals* (see Section 8.4): Prediction intervals are constructed to contain the next sample value(s) from a population or distribution with a specified probability. For example, after sampling a monitoring location for some time, the data may be used to construct an interval that, with a specified probability, will contain the next monitoring result(s) (assuming the distribution has not changed and/or to detect significant changes if they have occurred).

In summary, a tolerance interval usually contains a specified proportion of the population, a confidence interval is created to contain a statistical parameter, and a prediction interval contains one or more future observations. These intervals are all based on sample data (as opposed to populations) and thus have uncertainty associated with them.

Risk Assessment: Procedures and Protocols, First Edition. Edward A. McBean.

Companion website: www.wiley.com/go/McBean/RiskAssessProcedure_1e

The final portion of this chapter describes a series of different procedures useful for identifying statistical outliers. Data outliers are extreme (high or low) values that are widely divergent from the majority of the data. Of interest is whether the values are legitimate members of the dataset and therefore contain important information or represent, for example, errors in measurement. Detection of data outliers is addressed in Section 8.5.

8.2 Tolerance Intervals

A tolerance interval establishes a concentration range that is constructed to contain a specified proportion (P percent) of the population with a specified confidence, such that a specified large proportion of monitoring observations should fall within this interval. Tolerance intervals are usually constructed assuming the data (or the transformed data) are symmetrically distributed (specifically, the t-distribution and, as the number of samples becomes large, the normal distribution), but they can also be constructed assuming other distributions.

The steps in developing a tolerance interval are as follows:

Procedure

Step 1: Compute the mean, $\overline{x}$, and the standard deviation, S_x, of the data.
Step 2: Construct the one-sided upper tolerance limit, TL, as

$$\mathrm{TL} = \overline{x} \pm kS_x \tag{8.1}$$

where k is the one-sided normal tolerance factor assuming the data are adequately described by the normal distribution. The magnitude of "k" is selected from tables in accordance with the desired tolerance limit. For example, a tolerance limit with 95% coverage gives an upper and lower bound between which 95% of the individual observations of the distributions should fall. The magnitude of "k" is determined from Table A.3 for different tolerance limit probabilities and degrees of freedom. It is useful here to emphasize that the t-distribution approaches the normal distribution as the number of sample points becomes large and becomes identical as the number of samples become infinite. For further discussion of this feature, see Section 5.4.

Example 8.1 Development of Tolerance Limits

Determine the 95% tolerance limits for the following data: 13.55, 6.39, 13.81, 11.20, and 13.88. Thus, n = 5, $\overline{x} = 11.77$, and S_x = 3.20. The value of k is determined from Table A.4 for (5 − 1) = 4 degrees of freedom and for a two-sided tolerance limit (i.e. k = 2.776). Hence, the 95% tolerance limit on the individual values is

TL = 11.77 ± 2.776 (3.20) or a tolerance interval from 2.89 to 20.65.

In words, 95% of measurements should be within the range of 2.89–20.65. Figure 8.1a indicates the tolerance limits and plotted monitoring values.

Assume that an additional five samples are now collected. Construct the resulting 95% tolerance limits, where the additional data are 5.88, 10.22, 6.86, 12.43, and 7.94. Now n = 10, $\overline{x} = 10.22$, and S_x = 3.22. Incorporating this additional information, the 95% tolerance limit on the individual values becomes

TL = 10.22 ± 2.262 (3.22) or a tolerance interval from 2.94 to 17.50.

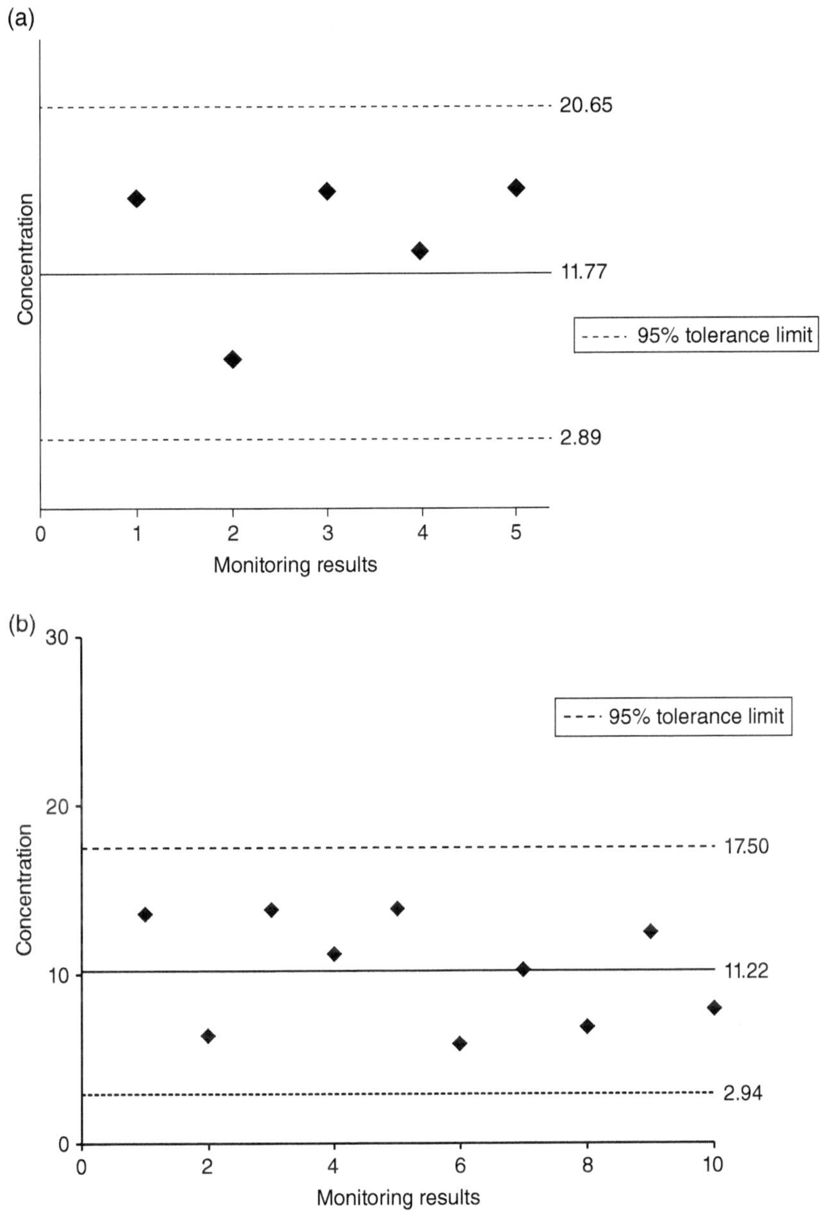

Figure 8.1 (a) 95% tolerance limits derived from five sampling results. (b) 95% tolerance limits derived from 10 sampling results.

The width of the 95% tolerance limits has decreased as a result of the additional sampling results. Figure 8.1b indicates the new tolerance limits and plotted monitoring values.

If the data are not normally distributed, a possible approach involves transforming the data using, for example, a logarithmic transformation, prior to the determination of the tolerance interval. In this situation, the resulting tolerance intervals will not be equidistant from the mean when transformed back to the original domain as demonstrated in Example 8.2.

Example 8.2 Development of Tolerance Limit for Non-Gaussian Data
The data as follows were obtained for air quality for sulfur oxides, namely, 10.2, 147.6, 8.1, 12.2, 8.6, 54.3, 12.4, 15.6, and 34.6.

a) Determine the 95% tolerance interval without using any data transformation.
The mean of the data is 33.73, the standard deviation is 45, n = 9, k = 2.31 from Table A.4, and the 95% tolerance interval is

$$TL = 33.73 \pm 2.31(45) = -70.2 \text{ to } 137.6.$$

The results are plotted on Figure 8.2a. The inappropriateness of the lower tolerance limit is apparent since air quality concentrations cannot assume negative values.

b) Determine the 95% tolerance interval following a logarithmic transformation (i.e. using the logarithms of the data, $\bar{y} = 2.98$ and $S_y = 0.988$).

$$TL = 2.98 \pm 2.31(0.988) = 0.70 \text{ to } 5.26$$

The results are shown in Figure 8.2b when the ordinate is plotted as a logarithmic scale:

$$\text{Mean} = e^{2.98} = 19.7$$

The tolerance limit extends from $e^{0.70}$ to $e^{5.26}$ or 2.0 to 192.

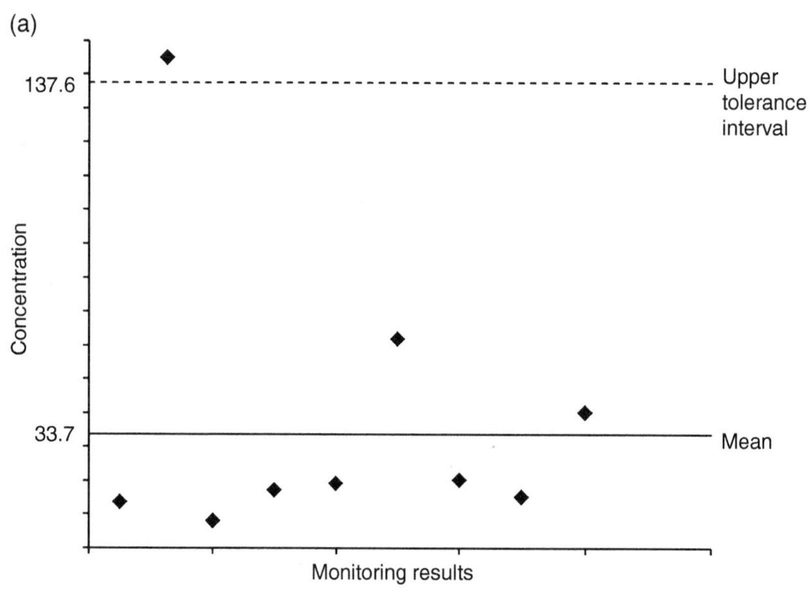

Figure 8.2 (a) Data and tolerance interval with no data transformation, (b) data and tolerance interval with logarithmic transformation, and (c) data and tolerance interval calculated after logarithmic transformation and back-transformed onto arithmetic scale.

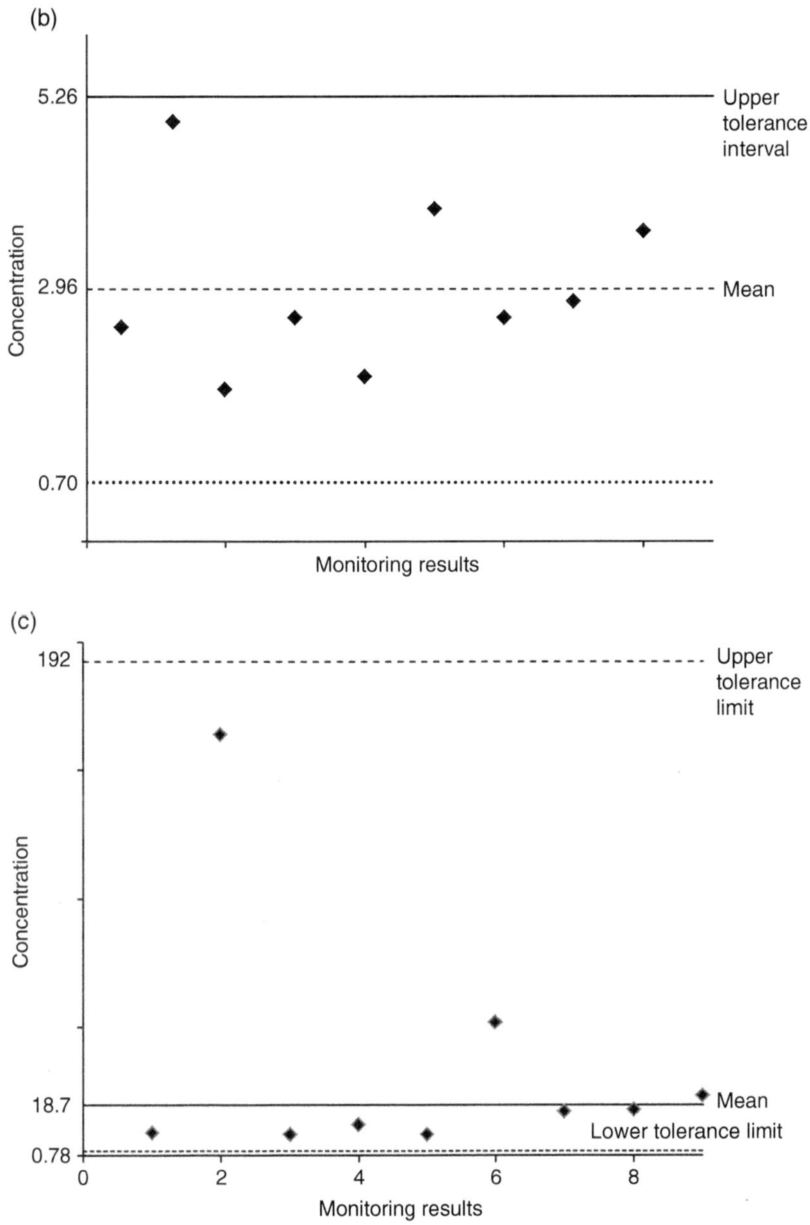

Figure 8.2 (Continued)

The results are shown in Figure 8.2c when the ordinate is plotted on an arithmetic scale. This figure demonstrates how the tolerance limits are asymmetric when transformed back to the arithmetic scale.

Example 8.3 Defining an Upper Tolerance Limit

The concentrations of trichloroethylene in soil at a former industrial site are as listed below. Define the 95% upper tolerance limit on the data.

Weibull			
Number of sample	Concentration (μg/kg)	Plotting position	ln of concentration
1	730	0.031	5.677
2	800	0.063	5.769
3	810	0.094	5.781
4	1000	0.125	5.992
5	1700	0.156	6.523
6	1900	0.188	6.634
7	2000	0.219	6.685
8	2400	0.250	6.868
9	2800	0.281	7.022
10	5600	0.313	7.715
11	6100	0.344	7.801
12	6700	0.375	7.894
13	8700	0.406	8.156
14	10000	0.438	8.294
15	12000	0.469	8.477
16	12000	0.500	8.477
17	16000	0.531	8.765
18	17000	0.563	8.826
19	25000	0.594	9.211
20	31000	0.625	9.426
21	41000	0.656	9.706
22	46000	0.688	9.821
23	63000	0.719	10.316
24	100000	0.750	10.598
25	240000	0.781	10.780
26	290000	0.813	10.969
27	290000	0.844	10.969
28	360000	0.875	11.186
29	960000	0.906	12.167
30	1000000	0.938	12.207
31	6100000	0.969	14.016
Mean	309920		9.804
Standard deviation	1103425		2.37

a) Is the normal distribution a reasonable means of characterizing the data?
A check of the coefficient of variation, namely,

$$\frac{1\ 103\ 425}{309\ 920} = 3.6$$

indicates the value is much greater than unity and the normal distribution is not a good description of the data. Similarly, plotting the data on normal probability paper, as depicted in Figure 8.3a, clearly shows that the data do not plot as a straight line.
Conclusion: The normal distribution is not a good descriptor of the data.

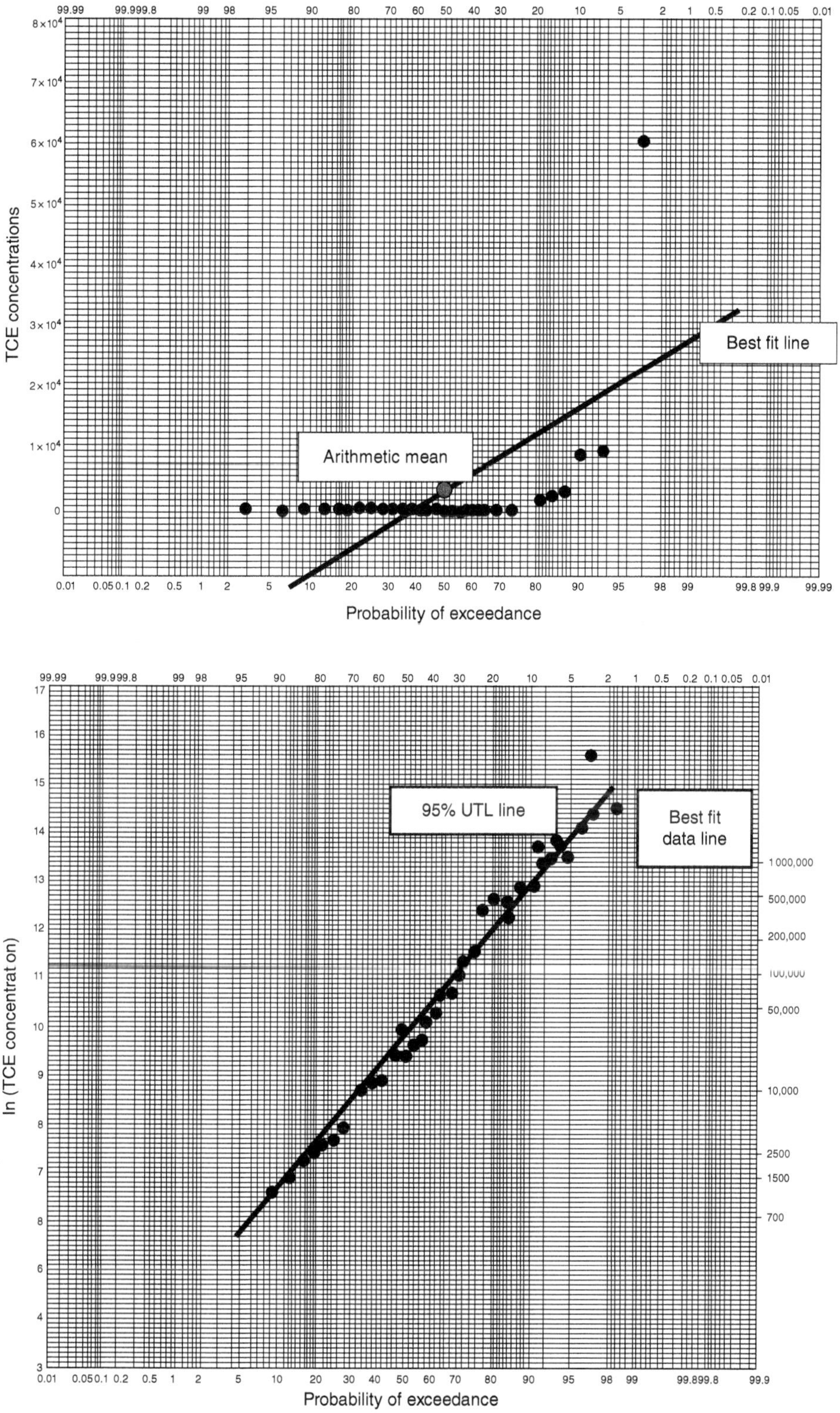

Figure 8.3 (a) Normal probability plot of TCE data from Example 8.3. (b) Lognormal probability plot of TCE data from Example 8.3.

b) Is the lognormal distribution a reasonable means of characterizing the data?

Plotting the data as indicated in Figure 8.3b as a lognormal distribution indicates the data plot as a straight line and the consequently lognormal distribution is an appropriate description of the data. As a result, the 95% UTL can determined as

$$UTL_{95} = \overline{x} + kS$$

where $\overline{x}$ is the mean of the log-transformed data (9.804); k, as obtained from Table A.4, is 1.697 corresponding to the t value for 5% (one-sided) and the number of samples (n = 31) and thus degrees of freedom (df = 30); and S is the standard deviation of the log-transformed data (2.37)

or

$$UTL_{95} = 9.804 + 1.697\ (2.37)$$
$$= 13.826$$

or

$$UTL_{95} = e^{13.826} = 1\,010\,433\ \mu g/kg$$

In words, we would expect to see only approximately 5% of the individual values exceeding 1 010 433 µg/kg. These results appear reasonable, since only one of the values out of 31 total is in excess of the 95% UTL.

An extension of the tolerance interval procedure involves construction of an interval that contains a specified fraction of the concentration measurement with a specified degree of confidence. For example, we may want to have the tolerance interval contain 95% of the measurements of concentration with a confidence of at least 95%.

The procedure to quantify tolerance interval (or tolerance limit) is similar to that indicated in Equation (8.1) except that the value of "k" is selected to reflect the degree of confidence. To construct a one-sided upper tolerance interval with an average coverage of (1 – β) percent, the k multiplier can be computed directly from the t-distribution table as

$$k = t_{n-1,1-\beta}\sqrt{1 + \frac{1}{n}}$$

where the t-value represents the (1 – β)th upper percentile of the t-distribution with (n – 1) degrees of freedom.

Tabulated values of k for different lengths of record of data are listed in Table A.3.

Example 8.4 Calculate the upper limit of the tolerance interval that contains 95% of the measurements of concentration with confidence at least 95% for the data from Example 8.3.

From Example 8.3, $\overline{x} = 9.804$, S = 2.37 and from Table A.3, k = 2.19.

The upper tolerance limit with 95% confidence is calculated as

$$= 9.804 + 2.19\ (2.37)$$
$$= 15.0$$

or, in arithmetic terms,

$$e^{15.0} = 3269000\ \mu g / kg$$

This value is larger than the 95% UTL indicated in Example 8.3.

This occurs because in this case, we now have the tolerance interval that contains 95% of the measurements of concentration with confidence of at least 95%.

8.3 Confidence Intervals

Confidence intervals use historical information for the construction of one or more lines or limits to characterize the uncertainty associated with a particular parameter, such as the mean. They are constructed from sample data and thus are random quantities. In other words, each set of sample data will generate a different confidence interval. The limits on the confidence limits may be updated periodically as additional information becomes available.

Construction of confidence intervals for the population mean in general requires the assumption that the observations are drawn from a statistically convenient distribution, usually the normal distribution.

8.3.1 Confidence Limits Using the Normal Distribution (and the t-Distribution)

Statistical confidence limits define an interval such that there is a specified probability, $(1 - \alpha)$, that a parameter will be contained in the interval (e.g. 95%). The probability that the confidence interval contains a specified population parameter is written as

$$\Pr\left\{x_{(1)} \le x \le x_{(u)}\right\} = 1 - \alpha \tag{8.2}$$

where $100(1 - \alpha)$ is the level of confidence, and for integers l and u, a confidence interval $\{x_{(l)}, x_{(u)}\}$ specifies the confidence interval.

If x is normally distributed, then the precision with which the sample mean $\bar{x}$ approximates μ_x can be assessed with the t-statistic as

$$t = \frac{\bar{x} - \mu_x}{\sqrt{S_x^2 / n}} \tag{8.3}$$

If the individual members of the set of x's are independent and normally distributed with the same mean and variance, then t from (8.3) follows the t-distribution with $n - 1$ degrees of freedom. Knowing the distribution of t, one can construct a confidence interval for μ_x. Rearranging (8.3) yields

$$\bar{x} - \frac{tS_x}{\sqrt{n}} \le \mu_x \le \bar{x} + \frac{tS_x}{\sqrt{n}} \tag{8.4}$$

If the set of x's is not normally distributed, the exact distribution of t in (8.4) generally is not known. However, the central limit theorem can be used to prove that if x has a distribution with finite mean and variance, then for large n, the distribution of t approaches that of the normal distribution with zero and a variance of unity.

As a result, in mathematical terms, when the degrees of freedom are infinite, the t-distribution becomes the normal distribution:

$$\text{Probability}\left(-1.96 < \frac{\overline{x}-\mu}{\sigma/\sqrt{n}} < 1.96\right) = 0.95 \tag{8.5}$$

or, rewriting,

$$\text{Probability}\left(\overline{x} - \frac{1.96\sigma}{\sqrt{n}} < \mu \leq \overline{x} + \frac{1.96\sigma}{\sqrt{n}}\right) = 0.95 \tag{8.6}$$

In practice, the true standard deviation, σ, is unknown. Then, if a random sample of size n is drawn, an estimate of the standard error of the mean $\overline{x}$ is given by $S_x/\sqrt{n}$. For small samples (e.g. $n < 30$), the confidence limits are increased (since S_x is only an estimate of σ) by use of the t-distribution.

For environmental quality phenomena, the population parameters are rarely known, and thus, Equation (8.4) is utilized to specify the confidence interval.

The values of t thus reflect the level of significance, α, and the number of samples, n. Values for 95% confidence interval for the value of t are as listed in Table A.2 for a two-sided characterization (e.g. for a 95% confidence limit, 2.5% in the one tail of the distribution and 2.5% in the other tail of the distribution). As apparent from the values listed in Table A.2, as n increases, the t value decreases. The values for a one-sided characterization are listed in Table A.2.

Example 8.5 Confidence Limits on the Mean

Using the data from Example 8.1, calculate the 95% confidence limits on the mean.

For the first five samples,

$$\text{Confidence limit} = 11.77 \pm \frac{2.76(3.20)}{\sqrt{5}} = 11.77 \pm 3.97 \text{ or } 7.80 - 15.74.$$

For the 10 samples,

$$\text{Confidence limit} = 10.22 \pm \frac{2.26(3.22)}{\sqrt{10}} = 10.22 \pm 2.31 \text{ or } 7.91 - 12.53.$$

The width of the 95% confidence limits on the mean have decreased as a result of the additional sampling. As would be expected, the confidence limits on the mean are much narrower than the 95% tolerance intervals on the individual values as summarized below.

Sample size	Tolerance interval	Confidence limit
5	2.89–20.65	7.80–15.74
10	2.94–17.50	7.91–12.53

For small sample sizes and/or for data with large spread (large standard deviations), the confidence intervals will be wide. As a result, when the data have a large spread, a substantial shift in, for example, water quality levels will be required for the subsequently calculated mean to fall outside the confidence limits.

As more data become available, the confidence interval will likely become narrower as noted in Example 8.2, and the test will become more sensitive to smaller shifts in water quality.

It is noteworthy that there is no particular reason that a 95% confidence interval is used. The level of confidence required is a function of the degree of certainty required to make a detection on the result of the statistical test. For example, a 99% confidence limit is considerably larger than the 95% confidence limit (for the 99% confidence limit, there is only 1% likelihood that the mean would be expected to lie outside the confidence limits). Suppose one wishes to be more "certain" of any conclusions that a change has occurred by reducing the chance of a calculated mean falling outside the confidence interval when no change had actually occurred. In other words, we have reduced the false positive or type I error to $1 - \alpha = 0.99$.

Example 8.6 Using the data in Part (c) above, develop the 95% and 99% confidence limits for the mean.

For n = five samples from Example 8.1,

- The 95% confidence limit is $11.77 \pm \frac{2.776(3.20)}{\sqrt{5}}$ or 7.80–15.74.
- The 99% confidence limit is $11.77 \pm \frac{4.60(3.20)}{\sqrt{5}}$ or 5.15–18.35.

For n = 10 samples from Example 8.1,

- The 95% confidence limit is $10.22 \pm \frac{2.262(3.20)}{\sqrt{10}}$ or 7.91–12.53.
- The 99% confidence limit is $10.22 \pm \frac{3.25(3.22)}{\sqrt{10}}$ or 6.91–13.53.

The false positive error, the risk of the mean being outside the confidence interval, is less for the 99% lines but at the expense of a larger range.

The 99% confidence limit will be less sensitive in detecting changes than the 95% confidence limit. This illustrates the penalty or loss of sensitivity and power, which is associated with increased confidence in conclusions that changes have occurred.

8.3.2 Confidence Limits for Lognormally Distributed Data

If data are lognormally distributed, as determined by the procedures described in Chapter 6, confidence limits for the mean can be completed using the log-transformed data. In other words, following a logarithmic transform, the calculation procedure described in Section 8.3.1 is followed.

It is noteworthy that the distribution of the estimate of the mean can usually be described by the normal distribution, almost irrespective of the distribution of the data.

8.3.3 Distribution-Free or Nonparametric Confidence Limits

Distribution-free confidence limits may be developed, which do not depend on the assumption of normality or lognormality. Distribution-free tolerance intervals are, however, wider and therefore less sensitive to changes in environmental quality than their normal counterparts. The interested reader is referred to Conover (1980) for further consideration of distribution-free confidence limits.

8.4 Prediction Interval Characterizations

Prediction intervals are intended to describe an interval within which future samples should lie, assuming the distribution or system has not changed.

8.4.1 The t-Distribution Prediction Intervals

The t-distribution may be utilized as the basis for developing prediction intervals. The prediction interval can be over time or over space. The spatial prediction interval might involve checking whether results at one monitoring location are representative of those at another monitoring location.

Example 8.7 Evidence of contamination is indicated if subsequent data indicate deterioration in quality by being above the upper prediction interval.

The data listed below are nickel concentrations measured at a groundwater monitoring well. Calculate the prediction interval and determine whether there is evidence of contamination as demonstrated by the monitoring results determined in months 4 and 5.

The initial data (for the first 3 months) from which the upper prediction interval is determined is as follows:

Sample date	Nickel concentration (μg/l)	
July	10.7	mean = 15.3
	11.4	
	20.6	
	18.4	
August	13.1	mean = 16.3
	22.6	
	11.4	
	18.1	
September	17.1	mean = 18.6
	26.4	
	12.3	
n = 11		
$\overline{x} = 16.55$		
S = 5.23		

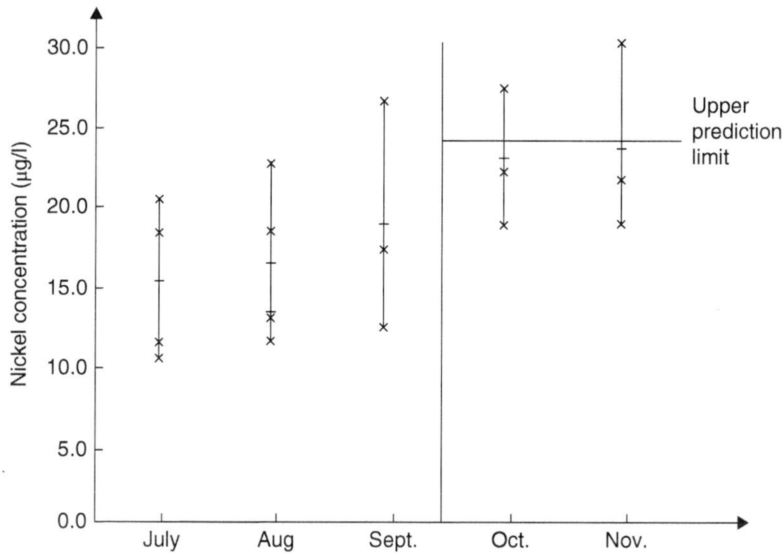

Figure 8.4 Plot of nickel concentrations over time for data from Example 8.7.

Subsequent monitoring results for the next 2 months are:

October	22.1	
	18.6	
	27.1	n = 3
		$\bar{x}$ = 22.60
		S = 4.27
November	21.2	
	18.6	
	29.2	n = 3
		$\bar{x}$ = 23.0
		S = 29.2

The individual data points are plotted on Figure 8.4.

Step 1: Test the first 3 months of "background" data for approximate normality. Only the data up to the end of the first 3 months are included since these are used to construct the upper prediction interval.

The Shapiro–Wilk test is utilized to test for normality of the data, as described below.

Shapiro–Wilk test

i	$x_{(i)}$	$X_{(n-i+1)}$	a_{n-i+1}	b_i
1	10.7	26.4	0.5601	14.79
2	11.4	22.6	0.3315	7.49
3	11.4	20.6	0.2260	4.66

i	$x_{(i)}$	$x_{(n-i+1)}$	a_{n-i+1}	b_i
4	12.3	18.4	0.1429	2.63
5	13.1	18.1	0.0695	1.26
6	17.1	17.1		b = 30.83
7	18.1	13.1		
8	18.4	12.3		
9	20.6	11.4		
10	22.6	11.4		
11	26.4	10.7		

$$W = \left[\frac{b}{S\sqrt{n-1}}\right]^2 = \left[\frac{30.83}{5.23\sqrt{10}}\right]^2 = 1.86$$

The critical value at the 5% level for the Shapiro–Wilk test on 11 observations is 0.850. Since the calculated W = 1.86 is well above the critical value, there is not sufficient evidence to reject the assumption of normality.

Step 2: Compute the prediction interval using the background data. Since there are 2 future months of data to be compared to the upper prediction limit, the number of future sampling periods is k = 2. At each sampling period, a mean of three independent samples will be computed, so m = 3 in the upper prediction interval formula. The Bonferroni t-statistic is $t_{10,2,0.95}$ with k = 2 and 11 degrees of freedom (df) and is equivalent to the t-statistic at 0.975 level with 10 df. In other words, $t_{10,0.975} = 2.23$

Compute the upper one-sided upper prediction limit using the formula

$$UL = \bar{x} + t_{n-1,k,0.95} S\sqrt{\frac{1}{m} + \frac{1}{n}}$$

Then the upper prediction limit is given by

$$UL = 16.55 + (2.23)5.23\sqrt{\frac{1}{3} + \frac{1}{11}} = 24.1$$

Now comparing the upper prediction limit to the compliance data, the means of the observations for months 4 and 5 are 22.60 and 23.0 μg/l, respectively.

Since the mean concentrations for months 4 and 5 are below the upper prediction limit, there is not sufficient evidence to indicate that a change in concentration at the monitoring facility has occurred.

8.5 Detection of Data Outliers

Data outliers are extreme (high or low) values that are widely divergent from the main body of a dataset. The presence of one or more outliers within a dataset may greatly influence any calculated statistics resulting in biased results. By increasing the variability within a sample, outliers will decrease the sensitivity of subsequent statistical tests.

However, there is also the possibility that the outlier is a legitimate member of the dataset. The data point could be the result of a particular cause that is the reason for the data collection effort, the very "effect" that is the purpose of an environmental monitoring program. Therefore, the purpose of outlier detection tests is to determine whether there is sufficient statistical evidence to conclude that an observation appears extreme and does not fit the distribution of the rest of the data or, alternatively, that the observation should remain in the dataset.

Before concluding that data outliers kept in the dataset necessarily indicate environmental change, and hence an additional dimension of risk, the results may have arisen from a faulty instrument – poor calibration of a water quality probe that measures dissolved oxygen, for example. Sometimes, the values in a dataset include errors in the transcription of data too. Other sources of mistakes include such features as unobserved technician or laboratory error, contaminated sampling equipment, the misreading of instruments, blunders in experimental procedures, an intentional cover-up because no value was taken, inconsistent sampling of analytical chemistry methodology, and so on. As well, data outliers may arise due to the vagaries of nature (the tendency for some features to frustrate the analyst), and/or they may be indicative of the "real" value such as the monitoring result could be a consequence of a catastrophic unnatural occurrence such as a spill. The question is to estimate, to the extent the dataset will allow, whether the observation should be included or excluded in a particular circumstance.

If one of the error-producing processes indicated above can be identified with an outlier, the value may be safely deleted from the dataset. Therefore, the first endeavor is to try and trace the conditions and experimental procedure of the aberrant observation through lab notebooks. Obviously the suspect value will be excluded if a rational basis for an exclusion can be identified. The subsequent data analysis can then be performed without the value.

A number of procedures have been developed as alternative methods for detecting outliers. As will be seen, many of the procedures premise the exclusion of a suspect data point on the basis of an assumed underlying normal distribution. As well, when the dataset is small, the outlier must be substantially different to be identified as an outlier. The various procedures for identifying an outlier include:

i) *Probability plotting approach for outlier detection*: This outlier identification procedure involves the preparation of a probability plot of the data followed by a visual inspection. This approach is difficult because it is difficult to assign specific rules that would ensure replicability of approach. As a result, this procedure is more commonly utilized to identify the need for more detailed analyses by one of the more formal procedures outlined below. Regardless, the data illustrated in Figure 8.5 certainly imply that the single very high value is a data outlier worthy of closer examination.
ii) *Scattergram analysis test for outlier detection*: The use of a "scattergram" or figure indicating interparameter correlation between two constituents involves development of a diagram such as depicted in Figure 8.6. For example, if there tends to be correlative behavior between two constituents, a single monitoring result with inconsistent behavior such as identified by the point lying a significant distance from the general trend of the data is indicative that the individual value is potentially an outlier.
iii) *Standard deviation test for outlier detection*: Another procedure for detecting outliers involves examining the changes in magnitude of a parameter estimating the dispersion of the data. A typical parameter utilized for this test is the standard deviation, or variance. Inflation or increase in the variance can result from including outliers. It is noteworthy that

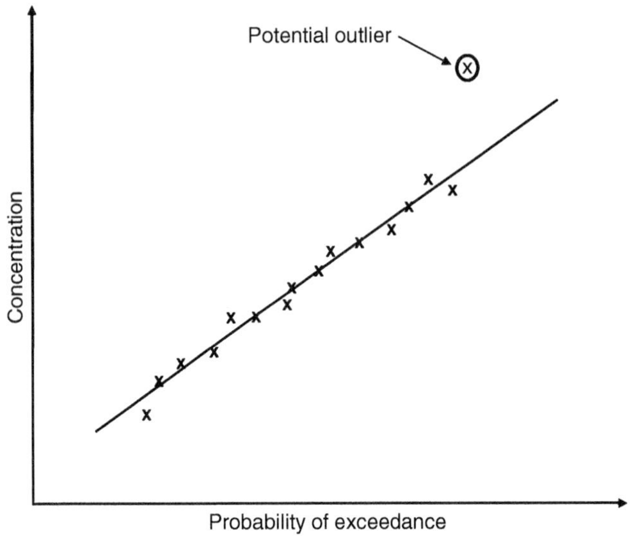

Figure 8.5 Use of probability plot as a means of identifying possible data outliers.

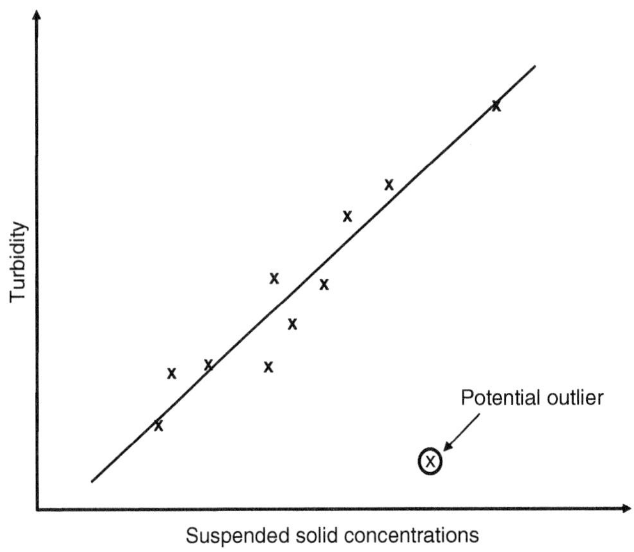

Figure 8.6 Scatter plot for purpose of identifying possible data outliers.

this test is useful only when the dataset is not large since a single value will not likely create a significant change in the standard deviation when the dataset is greater than 20–30 in length. The steps for this procedure involve the following:

a) Rank-order the data.
b) Calculate the cumulative mean and standard deviation with and without the observations being tested as a possible outlier.

Example 8.8 For the aluminum concentration data listed below, utilize the standard deviation test for outlier detection to determine if there is likely an outlier among the data. The aluminum concentrations (μg/l) are 56, 27, 85, 87, and 362.

Rank-ordered concentration	Cumulative means	Cumulative standard deviations
27	27	—
56	41.5	20.5
81	54.7	27.0
86	62.3	26.8
362	122.2	136.0

Conclusion: The single large concentration data point has a considerable influence on the calculated standard deviation. The results suggest the individual value is likely an outlier.

iv) *Standard normal test of outlier detection*: A more formalized procedure for detecting outliers than the procedure outlined in (iii) is a stepwise procedure using the following steps:

a) Rank-order the data.
b) Calculate the mean and the standard deviation of all the data.
c) Calculate T_n = (largest value – mean)/standard deviation.
d) Determine if T_n is "really large," utilizing the standard normal distribution tables. Really large is typically interpreted as being three standard deviations (i.e. if the distribution is normal), then a single value that is three standard deviations larger than the mean implies that it is indeed an unusual value.

Example 8.9 For the aluminum concentration data listed in Example 8.8, use the standard normal test to determine if the value of 362 is likely an outlier.

From the data, the following statistics result:

Mean = 122.2
Standard deviation = 136.0

$$t_c = \frac{362 - 122.2}{136.0} = 1.76$$

From the normal distribution table (see Table A.1) the probability of a value exceeding 1.76 standard deviation (the area under the curve) is 1 – (0.50 + 0.46) = 0.04. The single value does not lie more than three standard deviations from the mean and is thus not considered an outlier by this procedure.

v) *Dixon's test for outlier detection* (after Dixon and deMassey, 1969): In Dixon's test for outlier detection, the set of observations is first ordered according to the individual magnitudes. The ratio of the difference of an extreme value from one of its nearest neighbor values in the range of values in the sample is then calculated using a formula that varies with sample size. This ratio is then compared to a tabulated value and, if found to be equal to or greater, is considered to be an outlier at the $p \leq 0.05$ level or $p \leq 0.01$ level. The formula for the ratio varies with sample size and according to whether it is the smallest or largest value that is suspected as the outlier.

The tables associated with Dixon's test are included as Table A.16. The test may be utilized on sample sizes less than 25 in number.

Example 8.10 Assume the data are as listed in the table below, where the second column of the table indicates the reported concentration of the ordered observations.

a) Utilize Dixon's test to determine whether the highest value, 57.0, is an outlier.

Ordered data point	Concentration (mg/l)	Ordered data point	Concentration (mg/l)
1	0.5	12	12.1
2	0.6	13	12.4
3	0.8	14	12.5
4	1.0	15	13.1
5	1.1	16	14.5
6	1.9	17	20.2
7	2.9	18	22.1
8	4.6	19	24.7
9	8.8	20	24.9
10	9.2	21	44.0
11	11.1	22	46.9
		23	57.0

For the highest value recorded, x_{23}, the Dixon's coefficient is calculated from

$$r = \frac{x_{23} - x_{21}}{x_{23} - x_3} = \frac{57 - 44}{57 - 0.8} = 0.23$$

The critical value for 95% is, from Table A.16, a magnitude of 0.42. The Dixon coefficient is less than the critical value; thus x_{23} is not considered an outlier at the 95% level of significance. Thus x_{23} is not considered an outlier.

Dixon's test also allows testing for two outliers (e.g. testing whether both x_{23} and x_{22} are outliers).

It is apparent from the discussion on Dixon's test as described above that the procedure uses the extremes of the data (i.e. both the highs and the lows). Thus, when portions of the dataset are censored (to be discussed in Chapter 12), Dixon's procedure cannot be utilized.

Since Dixon's test uses individual values (the lowest and the highest), it differs from the previously described tests that consider deviations from the mean.

Example 8.11 Concentration measurements for benzo(a)pyrene are as indicated below: 2.77, 2.80, 2.90, 2.92, 3.45, 3.95, 4.44, 4.61, 5.21, and 7.46.

Utilize Dixon's test to examine whether the highest value is an outlier:

$$r = \frac{x_{10} - x_9}{x_{10} - x_2} = \frac{7.46 - 5.22}{7.46 - 2.80} = 0.48$$

Since $r = 0.48$ is larger than the critical value of 0.477 at 5% (see Table A.16), then 7.46 is considered as an outlier (but not for 1%).

vi) *Maximum normal residual and the extreme studentized deviate tests for outlier detection*: The maximum normal residual procedure (after Stefansky, 1972) examines the dataset for the maximum deviation for a sample as

$$\text{Maximum normal residual}\,(MNR) = \frac{\max\left|x_i - \bar{x}\right|}{\sqrt{\Sigma\left(x_i - \bar{x}\right)^2}} \tag{8.7}$$

Apparent from Equation (8.7) is that the MNR procedure, like the first procedures described in this section, requires estimation of the mean. If the dataset includes numerous censored values, such as nondetect measurements, the ability to quantify the mean is influenced.

Example 8.12 Use the extreme studentized deviate test to determine if there is an outlier in the arsenic concentrations listed below.

Observation	Concentration	Deviation	Studentized deviate
1	12.0	−19.1	−1.11
2	22.1	−9.0	−0.52
3	18.6	−29.5	−1.72
4	75.0	43.9	2.55[a]
5	36.1	5.0	0.29
6	19.2	−11.9	−0.69
7	20.4	−10.7	−0.62
8	31.7	0.6	0.03
9	52.4	21.3	1.24
10	41.0	9.9	0.58
11	17.9	−13.2	−0.77
12	27.0	−4.1	−0.24

[a] The maximum deviate.

The mean and standard deviations of the concentration data are as follows:

$$\overline{x} = 31.1$$
$$S_x = 17.97$$

n = number of observations = 12.
The deviation is calculated as

$$\text{Deviation} = \text{concentration} - \text{mean concentration}$$

$$\text{Studentized deviation} = \text{studentized residual}$$

$$= \frac{\text{Deviation}}{S_x\left(1 - \frac{1}{n}\right)^{0.5}} \tag{8.8}$$

where Equation (8.8) has been derived by manipulating Equation (8.7) into a simpler form.

Consider the outlier ($Z_{97.5} = 1.96$) where Z is the standard normal deviate. Now, test whether the highest value, namely, 75.0, is an outlier. Since $Z_{97.5} = 1.96$ (from Table A.1) and 2.55 > 1.96 then, the value of 75.0 is considered as an outlier. It is noted that this procedure becomes identical to the standard normal test in (iv) above, if the same condition for rejection is selected.

vii) *Barnett and Lewis' outlier detection test*: An objective test for outliers that turns out to be the same as the maximum normal residual test described above is described by Barnett and Lewis (1984). If a set of data is ordered from low to high, x_{low}, x_2, …, x_{high} and the mean and standard deviation are calculated, then suspected high or low outliers can be tested by the following procedure:

$$t^* = \frac{x_H - \overline{x}}{S_x} \quad \text{for a high value} \tag{8.9}$$

or

$$t^* = \frac{\overline{x} - x_L}{S_x} \quad \text{for a } \textit{low} \text{ value} \tag{8.10}$$

Compare the value of t* for the 5% or 1% level of significance. If the calculated T is larger than the table value for the number of measurements, n, the x_H or x_L is an outlier at that level of significance.

Number of measurements	Critical value 5%	Critical value 1%	Number of measurements	Critical value 5%	Critical value 1%
3	1.15	1.15	16	2.44	2.75
4	1.46	1.49	18	2.50	2.82
5	1.67	1.75	20	2.56	2.88
6	1.82	1.94	30	2.74	3.10
7	1.94	2.10	40	2.87	3.24
8	2.03	2.22	50	2.96	3.34
9	2.11	2.32	60	3.03	3.41
10	2.18	2.41	100	3.21	3.60
12	2.29	2.55	120	3.27	3.66
14	2.37	2.66			
15	2.41	2.71			

Example 8.13 Utilize the Barnett and Lewis outlier detection test on the following data to determine if the highest value is an outlier:

$$0.5, 0.6, 0.8, 1.0, 1.1, 1.9$$

From the data,

Mean $\overline{x} = 0.98$
Standard deviation $S_x = 0.50$

From Equation (8.9), $t^* = \frac{1.9 - 0.98}{0.50} = 1.84$. Since n = 6, $t_c = 1.82$ for 5% level of significance. Therefore,

$$t^* > t_c$$

$$1.84 > 1.82$$

and the high value is determined to be an outlier. It is to be noted that if we adopted a 1% level of significance, $t_c = 1.94$ and thus

$$t^* < t_c$$

$$1.84 < 1.94$$

which does not identify the high value as an outlier.

viii) *Chauvenet's criterion for outlier detection*: For data that are distributed in accord with the normal distribution, with a single extreme value, a simple method such as Chauvenet's criterion may be employed (Meyer, 1975). Chauvenet's criterion states that if the probability of a value deviating from the mean is greater than 0.5 n, expressed as a percentage, there are adequate grounds for rejection.

Example 8.14 For the monitoring values listed below,

1,6,7,8,8,9,9,9,10,10,10,10,10,11,11,11,12,12,13,14

consider whether the "1" is an outlier.

The mean of the data is 9.55 and the standard deviation is 2.80, and n, the number of samples is 20. One approach involves rejecting the value of "1" if its probability of occurrence were less than 0.5 (20) = 10%. From the table of Z scores in Table A.1, we see that 10% of the values in a normal distribution are beyond ±1.645 standard deviations of the mean or, in other words, (1.645) (2.80) = 4.61.

This means we would reject values beyond this range from the mean, i.e. 9.55 − 4.61 = 4.94 or greater than 9.55 +4.61 = 14.16. We therefore would reject the value of "1" but accept the remainder of the data.

It is to be noted that as the sample size gets larger, the rejection zone for Chauvenet's criterion will also increase. A value of "n" of approximately 20 is essentially a maximum for use of this procedure.

ix) *Winsorization of data as an approach to deal with outliers*: If an outlier is rejected, it can be completely removed from the dataset. Alternatively, in some situations, it is reasonable to replace the outlier with the next largest (or smallest) observation. This latter technique called winsorization (see Dixon [1960] for further details) is often used to calculate the sample mean when one observation is significantly higher or lower than the rest.

Another variation on this approach is to replace the highest and lowest values in a set of data.

Example 8.15 Assume a grouping of data consists of the values 75, 56, 47, 42, 39, 29, and 11. We would replace the high value 75 with a second 56 and the lowest value, 11, with a replicate 29. This would give a modified dataset as 56, 56, 47, 42, 39, 29, and 29, which is then treated as if it was the original dataset.

Winsorizing should not be performed if the extreme values constitute more than a small minority of the entire dataset.

x) *Regression modeling to identify outliers*: The approach utilizing regression modeling to identify outliers involves first, the fitting of a particular regression model to the data (this is a subject that is developed in Chapter 9). The degree to which the individual data points deviate from the assumed regression model is then considered. For example, one approach is to reject individual data points whose residual is more than four times the residual standard deviation. However, further discussion of this regression-based procedure is delayed to Chapter 9.

8.6 Summary of Approaches for Identifying Data Outliers

Outliers, by increasing the variability within a sample, decrease the sensitivity of any subsequent statistical tests. Outliers are observations that appear to be inconsistent with the rest of the data. The magnitude of the reported value may be in error or it may be genuine and indicate a valuable data point, one of the reasons we are collecting the monitoring data.

The principal safeguards against errors in data are vigilance in carrying out the operations, the measurements, and the recording of the data. As apparent from the preceding sections, a number of tests exist for the detection of outliers. Different pros and cons exist for the various tests for outlier detection, and the selection of the appropriate test to utilize for one circumstance will differ for alternative circumstances. It is to be noted that while many of the tests require that data be characterized by the normal distribution, a data transformation may be feasible, which still allows for tests to be utilized. The sample size and presence/absence of censored data are probably the most important features that will determine which outlier detection test is the most appropriate for use in an application.

Additional techniques to identify outliers are described in Grubbs (1969) and Beckman and Cook (1983).

8.7 References

Barnett, V. and Lewis, T. (1984). *Outliers in Statistical Data*, 2e. New York: John Wiley & Sons.

Beckman, R.J. and Cook, R.D. (1983). Outliers. *Technometrics* 25: 119–163.

Conover, W.J. (1980). *Practical Nonparametric Statistics*, 2e. New York: John Wiley & Sons Inc.

Dixon, W.J. and deMassey, F.J. (1969). *Introduction to Statistical Analyses*, 3e. New York: McGraw-Hill.

Dixon, W.J. (1960). Simplified estimation from censored normal samples. *Annals of Mathematical Statistics* 31: 385–391.

Grubbs, F.E. (1969). Procedures for detecting outlying observations in samples. *Technometrics* 11: 1–21.

Meyer, S.L. (1975). *Data Analysis for Scientists and Engineers*, 17–18. New York: John Wiley.

Stefansky, W. (1972). Rejecting outliers in factorial designs. *Technometrics* 14: 469–479.

8.8 Problems

8.1 Comment on why the mean changes so much when the number of waste sorts in attempts to characterize solid waste is so small?

8.2 You have sample data as per the following list of vehicular accidents in Toronto, reported during individual days.

1	37.20
2	24.60
3	36.10
4	29.20
5	30.00
6	34.90
7	31.10

8	27.90	
9	50.60	
10	53.30	
11	20.80	
12	25.50	
13	32.70	
14	63.00	
15	45.10	
16	19.00	
17	31.70	
18	36.10	
19	44.20	
20	26.80	
21	29.70	
22	48.60	
23	33.40	
24	33.90	
25	46.50	
	36.08	Mean
	10.56	Standard deviation

a) What is the probability that the number of accidents is between 40 and 50?
b) The mayor asks for the 90% range of your estimate of the average number of accidents that are occurring, after your 25 days of sampling (i.e. you are 90% confident the number of accidents per days is within this range).

8.3 The background measurements for a groundwater monitoring well are 0.8, 3.1, 1.7, 0.6, 1.1, 2.8, 1.8, and 0.9 for magnesium. Calculate the 95% limits on the mean of the background concentrations using these data.

8.4 What is the relationship between the t-distribution and the normal distribution?

8.5 In your assembly of data, you have identified two locations at you groundwater data as follows:

- At location 1, you have found that benzene concentrations in the groundwater are best described by a normal distribution:
  Data: 5.3, 7.7, 2.7, 8.6
- At location 2, you have found that benzene concentrations in the groundwater are best described by a lognormal distribution:
  Data: 1.1, 1.4, 5.3, 2.2, 12.9

a) Which of these two monitoring locations has the highest upper tolerance limit (95% of the values are less than the specified level)?
b) Estimate the lower confidence limit for the estimate of the mean at the two locations (95% confident that the mean is great than the specified value).

8.6 You have collected samples at two locations relevant to the physical configuration of the site as indicated in the figure, where you suspect there is a hidden industrial discharge to the river. You are interested in whether the data suggest there is a statistically significant difference as a result of the incoming discharge from the industrial plant. At both locations A and B, you have assembled 20 monitoring results.

a) What are the assumptions that you are going to make to allow utilization of the t-test to determine if the industrial outfall is truly impacting the water quality in the river?
b) Prepare a flowchart as to steps in the calculations that you would undertake. Be certain to indicate the options that you would utilize to assess whether the industrial discharge truly impacts the river.
c) You have the potential to collect additional data, but it involves additional effort. What would be a means by which you could make an assessment whether the collection of additional data would make a difference?

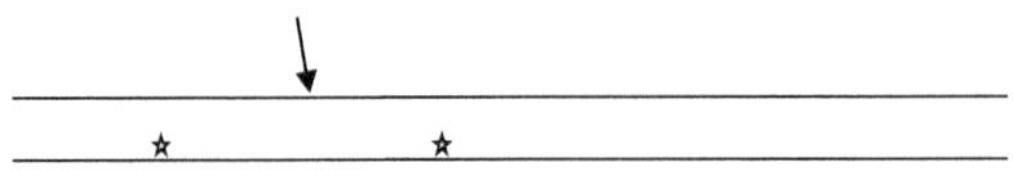

8.7 It is frequently stated that there is an underlying assumption that in the determination of the confidence limits on the mean diameter of a manufacturing process that produces widgets, that the "normal" or "Gaussian" distribution appropriately describes this mean. Briefly mean

a) Why is this true and yet we employ the t-distribution, not the Gaussian distribution in our analyses?
b) Is it really necessary that the distribution of the diameter of the widgets produced by the manufacturing process is actually normally disturbed?

8.8 Depicted in the figure is a factory smokestack and a downwind plot of land (it is an incineration stack used for air emissions from a mass burn facility for solid wastes as a means of disposal). Downwind, as schematically depicted in the figure, are a series of locations where there were soil samples obtained and analyzed in the laboratory for dioxins (a carcinogen). The results of the lab analyses are indicated in the table below. Develop a detailed flowchart that you would provide to a person to do the calculation to estimate the 95% exceedance level of dioxin concentration at any location on the tract of land. When describing the test you outline in the flowchart, include the values from the relevant statistical tables in the text that you would want utilized, although do not do the calculations.

Sample no.	Dioxin concentrations (ug/m^2)
1	28
2	31
3	17
4	84
5	32

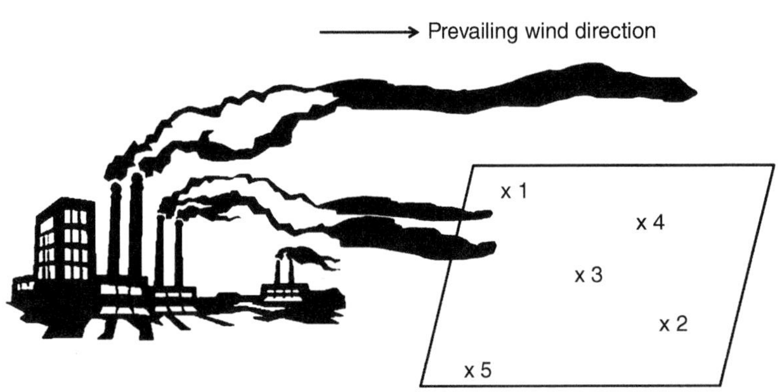

9

Hypothesis Testing: Testing Statistical Significance of Differences Between Data for Single Constituents

9.1 Introduction

Consider the following situation in which you flip a coin a total of 10 times and get seven heads and three tails. What are your chances of tossing the coin and getting a head on the next throw? If you consider the sample measurements at your disposal, you would estimate there to be a 70% chance of getting a head. Alternatively, if you believe the coin to be fair, and consider the total population of possible tosses, you would estimate there is a 50% chance of getting a head on the next throw.

The two alternative scenarios described above illustrate the basic difference between probability and statistical inference. Specifically,

1) With probability, which is based on deductive reasoning, you begin knowing something about the population (e.g. that the coin is fair) and then reason from the population to the sample (e.g. how likely will the next toss of the coin be a head?).
2) On the other hand, statistical inference is based on inductive reasoning. In this situation, you begin knowing only about the sample results (seven heads out of 10 tosses of the coin) and not the population. You then reason from the sample to the population (e.g. is the coin fair, based on the sample results?).

For environmental quality problems, we do not know the population, but only about the sample monitoring results we collect. Thus, we are forced to employ statistical inferences in order to estimate the behavior of the population from the sample results. As the coin-tossing example illustrates, even if the average behavior of the coin is to land 50% of the tosses as heads, specific sample results will vary from experiment to experiment. To estimate the coin's true behavior, we must account for between-sample variability. This is, then, precisely the problem we face in risk assessments. We have only sample measurements and we utilize statistical inference to gain an understanding of the population.

Carrying these ideas further, a relevant question in environmental quality data assessment is whether there has been a significant change in concentration for an individual chemical constituent. Examples of questions that may be of interest include the following.

Example 9.1 Are there significant differences between sampling records collected at different locations? For example, the objective of a groundwater monitoring program at a manufacturing facility may be to determine if a facility is impacting the groundwater. To determine whether contamination is occurring, monitoring wells are placed upgradient and downgradient of the facility, as depicted in Figure 9.1a. Samples are analyzed at regular intervals. The data illustrated in Figure 9.1b indicate the time history of water quality data as reported at the two locations. Do the data indicate that the facility has detrimentally impacted the groundwater quality?

Risk Assessment: Procedures and Protocols, First Edition. Edward A. McBean.

Companion website: www.wiley.com/go/McBean/RiskAssessProcedure_1e

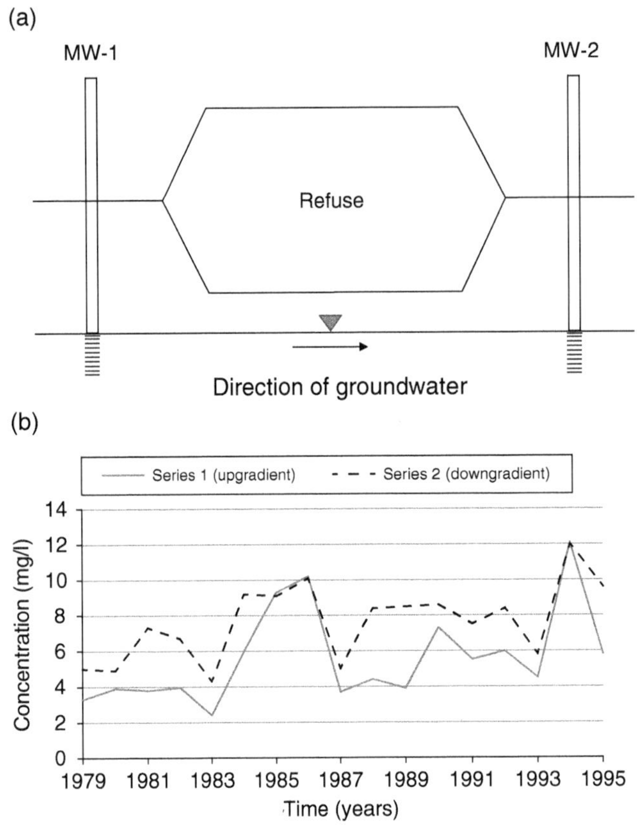

Figure 9.1 (a) Schematic of upgradient and downgradient monitoring of groundwater. (b) Arsenic concentration versus time.

Example 9.2 Are there significant changes in chemical concentrations collected over time at a particular location? Environmental quality data prior to and following the construction of a manufacturing facility in 1987 are available. Do the monitoring data indicate that the facility has detrimentally impacted the groundwater quality? By segmenting the data into two groups, one preconstruction and the other postconstruction, an assessment can be made to determine if a significant change in groundwater quality has occurred.

In the type of assessment being addressed in this chapter, a series of alternative procedures useful in identifying whether a significant change in concentration for a particular constituent has occurred will be examined. Later chapters address additional issues associated with problems that entail monitoring results with numerous constituents and situations where the datasets are censored (e.g. reported as "less than" some concentration; information is available about the value, but not the value itself). As the concerns change, the statistical technique(s) appropriate for application may also change as will become apparent in the later chapters; the focus in this chapter is on the simplest problems with later chapters dealing with more complex issues.

The mathematical analyses to determine whether there has been a change in concentration must reflect a number of considerations, including the following:

1) The available data are samples, not populations.
2) Variabilities in concentrations may exist due to such influences as seasonality.
3) Uncertainties exist in the sampling and laboratory analysis, resulting in "noise" in the dataset.

A statistical inference procedure must be able to incorporate the types of influences indicated above, influences that complicate the problem of detecting differences over time and space.

The most frequent approach to examine whether a difference exists in an environmental monitoring record is to determine if there is a difference in the mean concentration over space or time. In words, the analyst develops a hypothesis and then tests to determine whether the hypothesis is reasonable. A hypothesis test may be, therefore, a way of mathematically examining whether a population of measurements of a random variable, A (as represented by a sample from the population), overlaps another random variable, B, to a sufficiently small degree (or not at all), such that A and B must be considered statistically different or, alternatively, that A simply represents another sample beyond B from the same population. An important element of the hypothesis testing is that these procedures represent a class of statistical analysis techniques that are designed to extrapolate information from samples of data in order to make inferences about populations.

9.2 Details of Hypothesis Testing

Consider the following as an example of a hypothesis.

Example 9.3 A landfill is located immediately adjacent to a river as depicted on Figure 9.2, and is suspected of discharging arsenic to the river. The formulation of the hypothesis is the following: The hypothesis is "there is no statistically significant difference between the instream arsenic concentrations as measured upstream and downstream of the site."

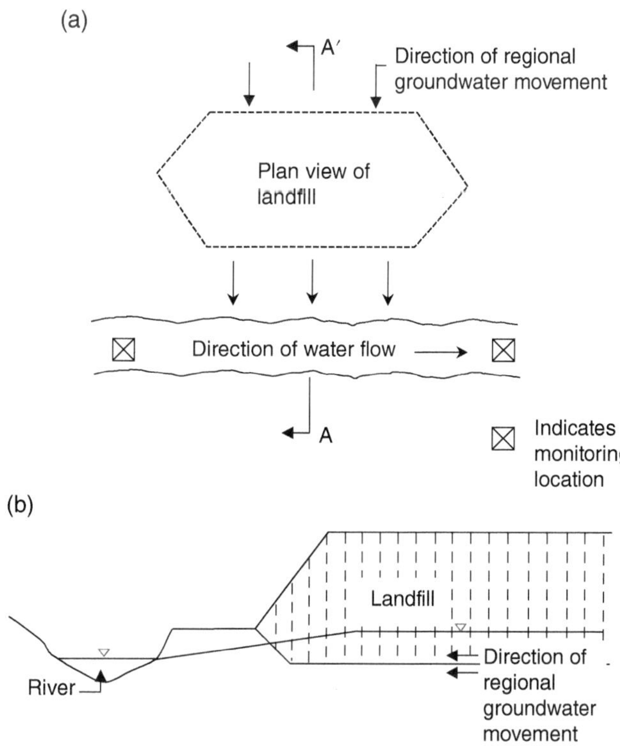

Figure 9.2 Plan view of a schematic of a situation where a landfill may be contaminating a stream: (a) plan view and (b) cross section A–A′.

The testing of a statistical hypothesis is then the application of a methodology for deciding whether:

1) To accept a particular hypothesis (termed the null hypothesis) that is equivalent to concluding that there is **not sufficient evidence** to say there is a difference in the concentrations as measured at the two locations or
2) To reject the null hypothesis. This is equivalent to concluding that there **is sufficient evidence** to say there is a difference in the concentrations as measured at the two locations. Specific to this hypothesis is that the results indicate the landfill is increasing the arsenic concentrations in the river.

Accepting the hypothesis means there is insufficient evidence to show the landfill is impacting the stream; rejecting the hypothesis means the landfill is impacting the arsenic levels in the stream.

Statistical tests thus, being with,

H_o, the null hypothesis, which is a statement such as *there is no difference*, or no impact of the landfill on the stream
H_1 or H_a, the **alternative hypothesis**, that *there is a difference*

If the objective is to compare two or more specific parameters such as the means of two populations, the hypothesis will be statements formulated to indicate the absence or presence of differences in the means.

In words, if there is a significant difference between the mean concentrations, is it due to (i) the inherent randomness associated with estimating a mean from two sets or samples of data taken from the same population (i.e. the same population is being sampled twice) or (ii) an actual difference exists between the two sample means as a result of contaminants being discharged by the landfill to the stream?

Assume we are interested in determining if a change in the mean of the instream concentration is a consequence of the landfill. This is accomplished by conducting a hypothesis test that compares the differences between the means of the random variables at the upstream and downstream locations to the standard error of the mean (the standard error of the mean (the standard deviation divided by the square root of the number of samples)). We accomplish this by calculating a ratio of difference between the means divided by the standard error of the mean; if the ratio is smaller than a tabulated critical value, then the null hypothesis cannot be rejected and the conclusion is there is no statistically significant difference. Alternatively, if the calculated ratio is larger than the tabulated value, then the null hypothesis is rejected and the conclusion is that there is a statistically significant difference.

Nevertheless, none of these determinations can be stated with 100% confidence. Remember we only have a sample from the population, and thus, we can only state our conclusion with a probability statement. When a result is obtained that is very unlikely to have arisen by chance, the result is said to be statistically significant. In other words, we are stating that it is difficult to ascribe the findings to chance, and the difference must, in all common sense, be accepted as a real difference. The observation was probably not just luck although there is still the possibility that the result happened by chance. Since the judgment is based on probability, it falls short of absolute certainty; the degree of certainty is α and we have control over the magnitude of α. The investigator decides what probability represents the limit beyond which he/she is unwilling to believe a random event has occurred (typically these values are chosen as 0.05 and 0.01, or, equivalently, 5% and 1%). Any event with a calculated probability of occurrence less than that limit is called statistically significant (although it is not an impossible event, it is considered sufficiently unlikely that the hypothesis must be doubted and there is presumed to be a real difference). For example, the difference between the mean of a water quality constituent is presumed to be a real difference. Thus, α is referred to as the level of significance.

If a significance test results in the acceptance of the null hypothesis, it does not follow that we have grounds for supposing this hypothesis to be true but merely that we have no grounds for supposing it to be false.

9.3 Steps for Significance Testing

It was demonstrated in earlier chapters that the normal distribution is a subset of the t-distribution. The t-distribution is the basis for the significant difference testing examined in this chapter.

The procedure to carry out significant difference testing consists of the following steps:

Step	Example of the step
(i) Formulate a suitable (in terms of the problem context) null hypothesis, H_o	Assume there is no difference between the upgradient and downgradient groundwater monitoring results
(ii) Calculate an appropriate test statistic t*	t-value: $t^* = \dfrac{\text{Difference between the means}}{\text{Standard error of the mean}}$
(iii) Define an acceptance/ rejection region	Define a critical statistic, t_c
(iv) Draw a conclusion based on whether the calculated test statistic is in the acceptance or rejection region	Reject the null hypothesis if $t^* > t_C$. In words, if $t^* > t_c$ there is evidence to indicate the hypothesis is incorrect or that there is a significant difference between the upgradient and downgradient monitoring results

Note that if the result proves nonsignificant ($t^* < t_C$), it is equivalent to a verdict of "not proven"; there is still the opportunity to consider further evidence that may become available.

In response to the significance testing problem, a number of alternative procedures have been proposed. The selection of the best procedure involves careful scrutiny of the characteristics of the problem at hand and the assumptions implicit in the particular discrimination test being considered. This involves not just the choice of significance level but also the choice of a statistic test and the requirements of the number of samples. The following sections develop each of these features specific to the various tests.

9.4 Student's t-Test

9.4.1 Development of the Equations

9.4.1.1 Comparing One Sample with the Population Mean

The original derivation of the Student t-test was presented by W. S. Gosset in 1908, who published his results in the scientific literature under the pseudonym of "Student." Gosset did this to prevent his findings from being used by his competitors in the brewery industry. His formulation revolutionized the statistics of small samples.

The Student t-test (henceforth referred to as the t-test) expresses

$$t^* = \frac{|\bar{x} - \mu|}{S/\sqrt{n}} \tag{9.1}$$

where t^* is the deviation of the estimated mean from the population mean, measured in terms of the unit $S/\sqrt{n}$. The quantity $S/\sqrt{n}$ is the sample estimate of $\sigma/\sqrt{n}$. Note that the current analysis is one of comparing a sample as a deviation from the population mean. Subsequent analyses will compare one sample relative to another, which is of greater interest in the environmental field. However, the derivations in this section are kept to the restricted situation of deviation of sample mean from the population mean.

A low value of t^* from Equation (9.1) indicates little difference between the means, and a high value of t^* is indicative of a large difference; the latter provides more justification for rejecting the null hypothesis.

Associated with the determination of the critical statistic, t_c, against which t^* is compared is the number of "degrees of freedom." Then,

If $t^* > t_c$, we reject the null hypothesis and conclude there is a difference.
If $t^* < t_c$, we cannot reject the null hypothesis. The evidence suggests the sample has not deviated from the population.

The degree of freedom is a value that represents the size of the sample involved in the test. When t_c is calculated from a single random variable of size n, the degrees of freedom are (n − 1). Note also that the magnitude of t_c is a function of the level of significance, α. A table of the values of the critical statistic, t_c, is in Table 9.1 for the 5% level of significance. The value for t_c is selected from the table associated with the level of significance and the sample size (as quantified by the number of degrees of freedom). When the degrees of freedom are infinite, $t_c = 1.645$ when $\alpha = 5\%$ for a one-sided test of significance (i.e. the 5% is all located in the one tail of the distribution as noted in Figure 9.3a). (An alternative consideration may involve a two-sided test of significance as indicated in Figure 9.3b). Further discussion of the differences between one- and two-sided tests will be presented in Section 9.4.1.2 to follow. With 30 degrees of freedom, t_c has increased to 1.697; as the same size decreases, the differences between the means of the sample and the population must be larger before the hypothesis is rejected.

Example 9.4 A lengthy history of monitoring record from a wastewater treatment facility has established that the long-term population mean for BOD_5 is 12 mg/l. Current operational characteristics over the last 4 weeks of sampling have resulted in the following BOD_5 concentrations in the effluent stream (14.6, 12.8, 13.7, and 15.4). Do the current operating characteristics suggest that there is an operational problem in the wastewater treatment facility?

The mean of the sample concentrations is $\overline{x} = 14.1$ mg/l and standard deviation S = 1.12 mg/l. The t-statistic from Equation (9.1) is then calculated as

$$t^* = \frac{|\overline{x} - \mu|}{S/\sqrt{n}}$$

$$t^* = \frac{14.1 - 12}{1.12/\sqrt{4}} = 3.75$$

The four samples indicate 4 − 1 = 3 degrees of freedom and for a 5% level of significance (from Table 9.1), $t_c = 2.353$ and $\alpha = 5\%$. The result is that $t^* > t_c$. In words, this means that in all likelihood, the results indicate a difference in the mean operating characteristics of the treatment facility has occurred.

Note that Gosset's original derivation tested only whether a sample differed from its population mean, μ, or alternatively, allowed the specification of confidence limits on the estimate of the mean.

Table 9.1 Values of critical statistic t_c for 5% level of significance.

Number of degrees of freedom	Value of t_c	
	One-sided test	Two-sided test
1	6.314	12.706
2	2.920	4.303
3	2.353	3.182
4	2.132	2.776
5	2.015	2.571
6	1.943	2.447
7	1.895	2.365
8	1.860	2.306
9	1.833	2.262
10	1.812	2.228
11	1.796	2.201
12	1.782	2.179
13	1.771	2.160
14	1.761	2.145
15	1.753	2.131
16	1.746	2.120
17	1.740	2.110
18	1.734	2.101
19	1.729	2.093
20	1.725	2.086
21	1.721	2.080
22	1.717	2.074
23	1.714	2.069
24	1.711	2.064
25	1.708	2.060
30	1.697	2.042
Infinity	1.645	1.960

Note: The values noted in the table correspond to those associated with α of 5% in a one-sided test. A more comprehensive table is included in Table A.2.

Example 9.5 The background measurements for a groundwater monitoring well are 0.8, 3.1, 1.7, 0.6, 1.1, 2.8, 1.8, and 0.9 for magnesium. Calculate the 95% limits on the mean of the background concentrations using these data.

The t-test allows calculations of the confidence levels of the mean, knowing the estimate of S of the standard deviation but not the population standard deviation σ. If σ is known and we assume a Gaussian distribution for the data, 95% confidence limits for μ are given by the relation

$$\overline{x}-\frac{1.96\sigma}{\sqrt{n}} \leq \mu \leq \overline{x}+\frac{1.96\sigma}{\sqrt{n}}$$

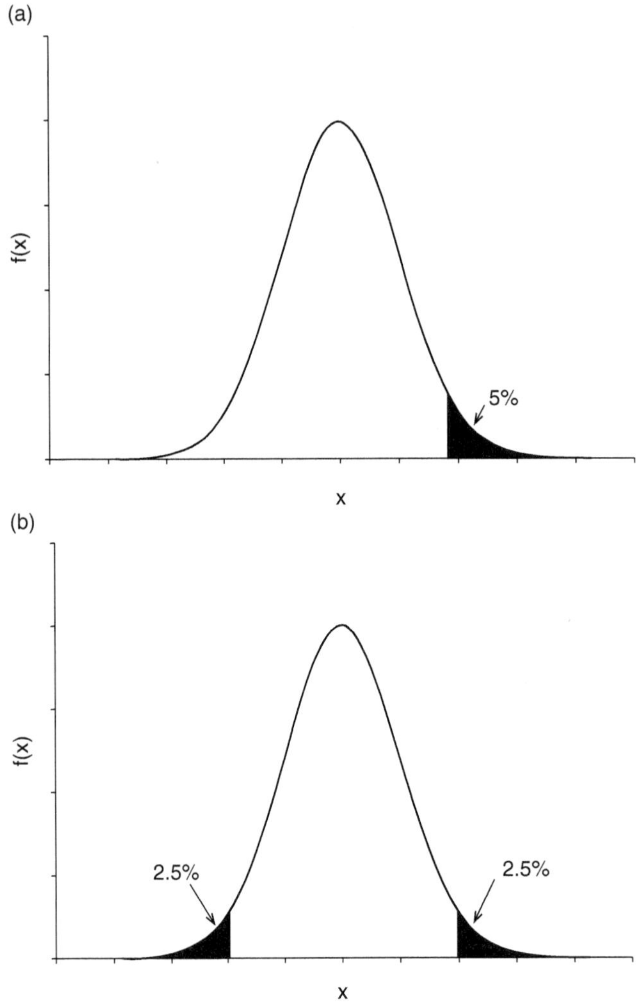

Figure 9.3 (a) A one-sided test of significance. (b) A two-sided test of significance.

When σ is replaced by S, the only change needed is to replace the number 1.96 in the equation by a quantity represented as $t_{\alpha/2}$ (where $\frac{\alpha}{2}$ is used to indicate a two-sided test, above and below the mean as further described below). Thus, our equation becomes

$$\bar{x} - \frac{t_{\alpha/2}S}{\sqrt{n}} \leq \mu \leq \bar{x} + \frac{t_{\alpha/2}S}{\sqrt{n}}$$

So from the data provided we have $\bar{x} = 1.60$ and $S = 0.94$.

For n = 8, the degrees of freedom are 8 − 1 = 7 and $t_{\alpha/2} = 2.365$ for $\alpha = 5\%$ (from Table A.2 go to 7 degrees of freedom, and since the values in the table are for a one-sided test, we need to go to the 0.975 column). Alternatively, we could have used Table A.4 (0.050 two-tailed test and found the same 2.365 value). The resulting confidence bounds are

95% confidence limit :

$$\mu = 1.60 \pm \frac{2.365(0.94)}{\sqrt{8}} = 1.60 \pm 0.79 \quad \text{or between 0.81 and 2.39.}$$

Now consider an alternative aspect of the t-distribution. The t-test allows calculations of the confidence levels for the mean, knowing the estimate, S, of the standard deviation but not the population standard deviation, σ. If σ is known and assuming a Gaussian distribution describes the data, 95% confidence limits for μ are given by the relation

$$\overline{x} - \frac{1.96\sigma}{\sqrt{n}} \leq \mu \leq \overline{x} + \frac{1.96\sigma}{\sqrt{n}} \tag{9.2}$$

When σ is replaced by S, the only change needed is to replace the number 1.96 in Equation (9.2) by a quantity represented as $t_{\alpha/2}$ (where $\alpha/2$ is used to indicate the test is a two-sided test, or above and below the mean as further described below). Thus, Equation (9.2) becomes

$$\overline{x} - \frac{t_{\alpha/2}S}{\sqrt{n}} \leq \mu \leq \overline{x} + \frac{t_{\alpha/2}S}{\sqrt{n}} \tag{9.3}$$

The t-distribution is symmetrical about the mean (as is also the situation for the normal distribution).

In other words, the t-distribution is used to establish confidence intervals and test hypotheses when the population variance is not known and the sample size is small (in fact, anything less than an infinite record is "small," but as apparent from Tables 9.1 and A.2, the t-distribution approaches the normal distribution for 30 or more samples).

When the degrees of freedom are infinite, $t_{0.05}$ for a two-sided test is 1.96. The two-sided test for $\alpha = 5\%$ means there is 2.5% chance of exceeding the upper bound and 2.5% chance of being lower than the lower bound. The t-statistic is usually written as $t_{\alpha/2}$ for a two-sided test, indicating $\alpha/2$ for each side. Alternatively, for a one-sided test, when the degrees of freedom are infinite, $t_{0.05} = 1.65$. As the degrees of freedom become large, the t-distribution becomes the normal or Gaussian distribution.

Example 9.6 Given chloride measurements from a river as 24.1, 27.9, 36.2, 26.3, 22.1, and 29.2, estimate the 95% and 99% confidence limits for the mean chloride concentration. From the data provided, $\overline{x} = 27.6$ and S = 4.92 mg/l. For n = 6 measurements, the degrees of freedom are 6 – 1 = 5 and $t_{\alpha/2} = 2.571$ for $\alpha = 5\%$ and $t_{\alpha/2} = 4.032$ for $\alpha = 1\%$ (from Table A.2). The resulting confidence bounds are:

a) For the 95% confidence limits

$$\mu = 27.6 \pm \frac{2.571(4.92)}{\sqrt{6}}$$
$$= 27.6 \pm 5.16 \quad \text{or between 22.4 and 32.8}$$

b) For the 99% confidence limits

$$\mu = 27.6 \pm \frac{4.032(4.92)}{\sqrt{6}}$$
$$= 27.6 \pm 8.10 \text{ or between 19.5 and 35.7}$$

When the number of degrees of freedom are infinite, $t_{0.05/2}$ = 1.96 (i.e. the t-distribution becomes the normal distribution). With 40 degrees of freedom, $t_{0.05/2}$ has increased to 2.021 (i.e. the confidence intervals on μ are larger using the t-test, because the sample size is only 40); with 20 degrees of freedom, it has become 2.086, and it continues to increase steadily as the degrees of freedom decrease. The t_c value increases with a smaller dataset, and the confidence interval on the mean becomes larger, reflecting the greater uncertainty in the parameter estimates.

Equation (9.3), when manipulated, becomes

$$t^* = \frac{|\bar{x} - \mu|}{S/\sqrt{n}} \tag{9.4}$$

where the absolute value sign, ||, is used to incorporate the +/− signs. Thus, t* is the deviation of the estimated mean from the population mean, measured in units of $S/\sqrt{n}$, the standard error of the estimate of the mean, and (9.1) has been obtained.

Example 9.7 Given the chloride measurements for a river as 24.1, 27.9, 36.2, 26.3, and 29.2, estimate the 95% confidence limits for the mean chloride concentration. From the data provided, $\bar{x} = 28.7$, S = 4.58, and n = 5. For n = 5, the degrees of freedom are 5 − 1 = 4, and $t_{\alpha/2}$ = 2.776 for α = 5%. The resulting confidence bounds are

$$\mu = 28.7 \pm \frac{2.776(4.58)}{\sqrt{5}}$$
$$= 28.7 \pm 5.7\ \text{mg/l}$$

Now, when additional data are collected, namely, 27.3, 29.1, 28.7, and 32.3, estimate the new confidence bounds on the mean concentration. From the entire dataset, $\bar{x} = 29.0$, S = 3.50, and n = 9. The value of $t_{\alpha/2}$ =2.262 for α = 5%. The resulting confidence bounds are

$$\mu = 29.0 \pm \frac{2.262(3.50)}{\sqrt{9}}$$
$$= 29.0 \pm 2.6$$

9.4.1.2 One-Sided Versus Two-Sided Tests

In one-sided or one-tailed tests of a mean or the difference between two means, the investigator pays attention only to deviations from the null hypothesis in one direction, ignoring the deviations in other directions. This happens primarily in situations where the investigator knows enough about the circumstances of the test to be certain that if μ_2 is not equal to μ_1, then μ_2 is greater than μ_1. For example, for the situation of monitoring of trichloroethylene upgradient and downgradient of a landfill, the only concern is whether the concentration of trichloroethylene is higher at the downgradient location than at the upgradient location. For this circumstance, the one-tailed test is appropriate. Conversely, if one is monitoring pH levels, the downgradient location might have higher or lower pH levels as a result of environmental contamination arising from the landfill. For this situation, a two-tailed test is appropriate.

9.4.1.3 Comparing Two Samples for Significance of Difference

The development of the t-test revolutionized the statistics of small samples. However, it was Fisher (1925) who realized that an extension of the test was possible to allow use of the test for identifying whether there is a significant difference between two sets of data.

Example 9.8 Assume that a random sample of size n_1 is collected from a population having unknown mean μ_1, the groundwater quality at location MW1 in Figure 9.1a. Further, a second sample of size n_2 is taken from a population having an unknown mean μ_2 at location MW2. The two sample means are $\overline{x}_1$ and $\overline{x}_2$. Does $\overline{x}_1$ differ significantly from $\overline{x}_2$? In other words, the problem is to test the hypothesis H_0: $\mu_1 = \mu_2$. The alternative hypothesis may be of the form H_1: $\mu_1 \neq \mu_2$, H_1: $\mu_1 > \mu_2$ or H_1: $\mu_1 < \mu_2$. A two-tailed test is appropriate in the first case, and a one-tailed test in the second and third.

If the variances are (approximately) equal, the t-statistic, t^*, is written as

$$t^* = \frac{|\overline{x}_2 - \overline{x}_1|}{SE_{Diff}}$$

where SE_{Diff} is the standard error of the difference.

One now compares the calculated t^* with tabulated one-sided $t_c = t_{df,\alpha}$ where the degrees of freedom df are defined as $df = n_1 + n_2 - 2$ and the level of significance is α. This is a more complex determination of the number of degrees of freedom from the previous analysis, since now two parameters are being estimated (as opposed to the single parameter in Section 9.4.1.1).

Example 9.9 The monitoring record for BOD_5 (all values in mg/l) in the outfall for a wastewater treatment facility lists the following results:

October 1994	12, 11, 14, 7
November 1994	8, 9, 6, 13

Are concentrations in October larger (statistically significant) than average concentrations for November, at 5% level of significance?

The statistics for the records are:

	October	November
Mean $\overline{x}$	11	9
Standard deviation, S	2.94	2.94

$$t^* = \frac{11-9}{2.94/\sqrt{4}} = 1.36$$

Degrees of freedom $df = 4 + 4 - 2 = 6$

t-statistic is $t_c = 1.94$

$t^* < t_c$

The test indicates there is insufficient evidence to indicate a statistically significant difference between the means for October and November.

9.4.1.4 Assumptions Implicit in the t-Test

The t-test has some built-in assumptions, some of which have not been acknowledged in the preceding example to avoid unnecessarily complicating the issues. If the assumptions are violated, then the t-test cannot be utilized (at least directly). The assumptions include:

a) Independence between observations.
b) Equal variances in both groups.
c) The assumption that the population(s) is normally distributed.

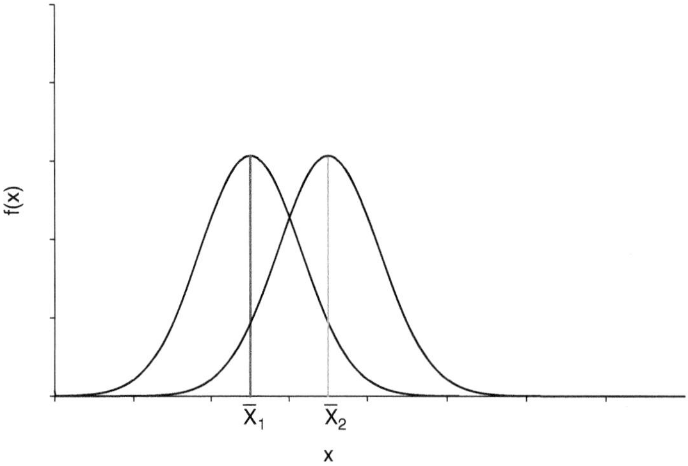

Figure 9.4 Schematic of probability distribution with some overlap.

The question of independence between samples is a difficult and subtle one when dealing with environmental quality measurements. Independence is probably seldom achieved, but, for example, as reported in McBean and Rovers (1985), a single groundwater sample split into subsamples or aliquots, with each aliquot then analyzed had a considerably smaller variance than analyses of successive groundwater samples. Clearly, aliquots are not independent observations.

Additional discussions of the assumptions implicit in the t-test are in Draper and Smith (1981), Milliken and Johnson (1984), Kerlinger and Pedhazur (1973), Koch and Link (1970), Krumbein and Graybill (1965), and Kufs (1992).

In summary, the t-test under the conditions for which the test holds is a powerful test. Nevertheless, a significant t-value is evidenced only in relation to a specified level of significance. It does not mean that every member of one population is larger than every member of a second population. In other words, some overlap of the distributions is likely, as indicated in Figure 9.4.

Common sense suggests that the larger the difference between $\bar{x}_1$ and $\bar{x}_2$, the less likely it is that the null hypothesis (that the means are the same or not different), H_o, is true.

Take the situation of two independent samples with means $\bar{x}_1$ and $\bar{x}_2$, which are estimates of their respective means, μ_1 and μ_2. The t-test becomes

$$t^* = \frac{(\bar{x}_1 - \bar{x}_2) - (\mu_1 - \mu_2)}{S_{\bar{x}_1 - \bar{x}_2}} \tag{9.5}$$

The denominator in (9.5) is a sample estimate of the standard error of the difference between the means, $(\bar{x}_1 - \bar{x}_2)$, as

$$\sigma^2_{\bar{x}_1 - \bar{x}_2} = \sigma^2_{\bar{x}_1} + \sigma^2_{\bar{x}_2} \tag{9.6}$$

Note that the variance of a difference is determined from the sum of the variances. If the two means, $\bar{x}_1$ and $\bar{x}_2$, are drawn from populations with variance σ^2, then (9.6) becomes

$$\sigma^2_{\bar{x}_1 - \bar{x}_2} = \frac{\sigma^2}{n} + \frac{\sigma^2}{n} = \frac{2\sigma^2}{n} \tag{9.7}$$

In most applications, the value of σ^2 is not known. However, each sample furnishes an estimate of σ^2, namely, S_1^2 and S_2^2. With samples of the same size n, the best combined estimate is their pooled average:

$$S^2_{\bar{x}_1-\bar{x}_2} = \frac{S_1^2 + S_2^2}{2} \tag{9.8}$$

The statistical analysis for groups of unequal sizes follows almost exactly the same pattern as that for groups of equal sizes. We assume the variance is the same in both populations as

$$\begin{aligned}\sigma^2_{\bar{x}_1-\bar{x}_2} &= \frac{\sigma^2}{n_1} + \frac{\sigma^2}{n_2} = \frac{2\sigma^2}{n} \\ &= \sigma^2\left(\frac{1}{n_1} + \frac{1}{n_2}\right) \\ &= \sigma^2\left(\frac{n_1 + n_2}{n_1 n_2}\right)\end{aligned} \tag{9.9}$$

The above equations assume the standard deviation is known. In most situations in environmental engineering, the standard deviation can only be estimated from the data from the two sampling locations. Thus, using the data from the two locations, we estimate

$$S_1^2 = \frac{\sum_{i=1}^{n}\left(x_{1i} - \bar{x}_1\right)^2}{n_1 - 1} \tag{9.10}$$

and

$$S_2^2 = \frac{\sum_{i=1}^{n}\left(x_{2i} - \bar{x}_2\right)^2}{n_2 - 1} \tag{9.11}$$

The unbiased estimate of $S^2_{\bar{x}_1-\bar{x}_2}$ is then

$$S^2_{\bar{x}_1-\bar{x}_2} = \frac{(n_1 - 1)S_1^2 + (n_2 - 1)S_2^2}{n_1 + n_2 - 2} \tag{9.12}$$

If the null hypothesis is true, it can be shown that the t-statistic follows a distribution with $(n_1 + n_2 - 2)$ degrees of freedom.

9.4.2 Effect of Unequal Variances

As evidenced in the derivation of the t-test, there is a built-in assumption that stipulates the variances should be the same (at least approximately). As noted earlier, a key to the valid use of the t-test is the assumption that the distributions of the two random variables, x_1 and x_2, have the same variance. Therefore, before using the t-test to investigate the difference between the sample means, a prior test must be done. The intent of the prior test is to investigate whether the sample variances are sufficiently alike to warrant the assumption that they are independent estimates of the sample population variance.

Suppose we have a normal distribution with variance σ^2. Two random samples, sizes n_1 and n_2, are drawn from this population, and the two sample variances, S_1^2 and S_2^2, are calculated in the usual way. As S_1^2 and S_2^2 are both estimates of the same quantity, σ^2, we expect the ratio S_1^2 / S_2^2 to be "close" to unity.

Let S_1^2 and S_2^2 be independent estimates of σ_1^2 and σ_2^2, and assume the observations in the two samples are normally distributed. Then we are often interested in testing the hypothesis that S_1^2 and S_2^2 are both estimates of the same variance σ^2. In other words, we want to test the null hypothesis, H_o, as

$$H_o : \sigma^2 = \sigma_2^2 = \sigma^2$$

For alternative hypotheses of the form

$$H_1 : \sigma_1^2 > \sigma_2^2 \quad \text{or} \quad H_1 : \sigma_1^2 < \sigma_2^2$$

a one-tailed test would be appropriate. If the ratio S_1^2 / S_2^2 (or S_2^2 / S_1^2) is much greater than unity, we would be inclined to reject H_o.

The significance of the difference between sample variances can be tested using the variance ratio test. The F distribution is the sampling distribution of the ratio of two independent unbiased estimates of the variance of a normal distribution and has widespread application in the analysis of variance (to be considered in Chapter 10). The variance ratio is defined as

$$F = \frac{\text{Greater estimate of the variance of the population}}{\text{Lesser estimate of the variance of the population}} \tag{9.13}$$

It follows that the larger the value of F, the less likely it is that the null hypothesis (that the two samples are drawn from the same population) is valid. However, with small sample sizes (n_1 and n_2 small), F may be different from unity. Large samples, if the null hypothesis is true, should give a value of F, which differs little from unity. Tables 9.2 and 9.3 show the value of F, which will be exceeded with a given degree of probability for various sample sizes of 5% and 1%, respectively. Note that the number of degrees of freedom for a sample of n items is equal to (n − 1).

The results of the analyses can be of two forms:

i) Pass the F-test.
ii) Fail the F-test. A failure of the F-test means that in the present form, the t-test cannot be used since the assumption of equal variance within the test is violated.

Table 9.2 5% level of variance ratio.

		Number of degree of freedom in the greater variance estimate							
		1	2	3	4	5	10	20	∞
Number of degrees of freedom in lesser variance estimate	1	161	200	216	225	230	242	248	254
	2	18.5	19.0	19.2	19.2	19.3	19.4	19.4	19.5
	3	10.1	9.6	9.3	9.1	9.0	8.8	8.7	8.5
	4	7.7	6.9	6.6	6.4	6.3	6.0	5.8	5.6
	5	6.6	5.8	5.4	5.2	5.0	4.7	4.6	4.4
	10	5.0	4.1	3.7	3.5	3.3	3.0	2.8	2.5
	20	4.3	3.5	3.1	2.9	2.7	2.3	2.1	1.8
	∞	3.8	3.0	2.6	2.4	2.2	1.8	1.6	1.0

Source: From Fisher and Yates (1953).

Table 9.3 1% level of variance ratio.

		Number of degree of freedom in the greater variance estimate							
		1	2	3	4	5	10	20	∞
Number of degrees of freedom in lesser variance estimate	1	4100	5000	5400	5600	5800	6000	6200	6400
	2	98	99	99	99	99	99	99	99
	3	34	31	29	29	28	27	24	26
	4	21.0	18.0	17.0	16.0	16.0	15.0	14.0	13.0
	5	16.0	13.0	12.0	11.0	11.0	10.0	9.6	9.0
	10	10.0	7.6	6.6	6.0	5.6	4.8	4.4	3.9
	20	8.1	5.8	4.9	4.4	4.1	3.4	2.9	2.4
	∞	6.6	4.6	3.8	3.3	3.0	2.3	1.9	1.0

Source: From Fisher and Yates (1953).

In situations without prior reason to anticipate inequality of variances, the alternative to the null hypothesis is a two-sided test ($\sigma_1 \neq \sigma_2$). On occasion, where one variance, if different, is known to be larger, a one-tailed t-test is warranted.

Example 9.10 Benzene concentrations were measured at two monitoring well locations (MW1 and MW2) as listed in the table below. Are the variances of the measurements at the two monitoring locations significantly different at 5% level of significance?

	MW1	MW2
	1.6	1.8
	1.9	1.9
	2.3	2.5
	1.7	2.1
	2.4	2.2
	1.8	1.9
	1.4	1.3
	2.7	2.8
	2.1	2.3
	1.9	2.0
	2.3	2.4
	1.8	*1.8*
Standard deviations	0.375	0.400

$$\text{F ratio} = \frac{(0.400)^2}{(0.375)^2} \frac{0.160}{0.141} = 1.13$$

Degrees of freedom of larger variance estimate = 12 − 1 = 11
Degrees of freedom of smaller variance estimate = 12 − 1 = 11
From Table 9.2, then the F-test = 2.9
F ratio < F-test

There is not a statistically significant difference in the variances and the t-test can be utilized. (Note: If we had found the ratio $F_{ratio} = S_1^2 / S_2^2 > F_{0.05,\, df1,\, df2}$ where S_1^2 and S_2^2 are based on degrees of freedom df_1 and df_2, respectively, then the result is significant at the 5% level and we have reasonable evidence that H_0 is untrue.

When n_1 and n_2 are large, an alternative procedure to the above F-test is to test the significance of a difference in standard deviations using the standard error of difference methodology and the fact that the variance for the distribution of sample standard deviation is $\frac{\sigma^2}{2n}$ (Note: It is the standard deviation and not the mean and thus the reasoning for the "2" in the denominator) and the fact that the variance of a difference is equal to the sum of the variances.

9.4.2.1 Pooled Variance

Assume that two variances from the two monitoring locations are not of the same magnitude but that the F-test did not indicate rejection of the hypothesis, indicating the two samples are not significantly different. Each sample furnishes an estimate, $\hat{S}^2$, of the population variance, σ^2. Assuming the variances have not proven to be significantly different, the pooled variance is estimated from

$$\hat{S}^2 = \frac{(n_1 - 1)S_1^2 + (n_2 - 1)S_2^2}{n_1 + n_2 - 2} \tag{9.14}$$

where n_1 is the number of samples in the first dataset and n_2 is the number of samples in the second dataset.

The two variances are pooled to get the best estimate of a population variance that the t-test assumes is the same for both populations.

The standard error of difference of the two means is then

$$S_m = \sqrt{\hat{S}^2\left(\frac{1}{n_1} + \frac{1}{n_2}\right)} \tag{9.15}$$

and the t-test is then calculated as

$$t^* = \frac{\overline{x}_1 - \overline{x}_2}{\sqrt{\frac{n_1 + n_2}{n_1 n_2}\left(\frac{(n_1 - 1)S_1^2 + (n_2 - 1)S_2^2}{n_1 + n_2 - 2}\right)}} \tag{9.16}$$

with the number of degrees of freedom calculated as $df = n_1 + n_2 - 2$.

Example 9.11 Using the benzene concentration information in Example 9.9, determine whether the concentrations measured at MW2 are statistically significantly larger than those measured at MW1 at 5% level of significance.

	Mean (μg/l)	Standard deviation (μg/l)
MW1	1.99	0.375
MW2	2.08	0.400

The F-test in Example 9.9 has already demonstrated there is not a statistically significant difference in the variance. Thus, the pooled variance is

$$\hat{S}^2 = \frac{11(0.375)^2 + 11(0.400)^2}{22} = 0.150$$

The standard error of the difference is $S_m = \sqrt{0.150\left(\frac{1}{12} + \frac{1}{12}\right)} = 0.158.$

The t-statistic $t^* = \frac{2.08 - 1.99}{0.158} = 0.57$.

The degrees of freedom df = 12 + 12 − 2 = 22 and α = 5% gives t_c = 1.72 and $t^* < t_c$, indicating there is no basis to indicate there is a difference in the benzene concentrations at MW1 and MW2.

Example 9.12 Monitoring results for pH at two locations in the wastewater treatment plant are indicated below.

Monitoring point	pH values
MW1	7.2, 7.5, 7.1, 7.4, 7.7
MW2	6.7, 7.3, 6.7, 7.1

Is there a significant difference between the pH monitoring results at the two monitoring locations?

Since pH values can be significantly larger or smaller, a two-sided test for statistical significance is needed.

Monitoring point	Mean	Standard deviation	Sample size
MW1	7.38	0.24	5
MW2	6.95	0.30	4

$$\text{F ratio} = \frac{(0.30)^2}{(0.24)^2} = 1.56 < 6.6 \quad (\text{from Table 9.2})$$

The pooled variance is $\hat{S}^2 = \frac{4(0.24)^2 + 3(0.30)^2}{5 + 4 - 2} = 0.071.$

The standard error of difference of the two means is

$$S_m = \sqrt{0.071\left(\frac{1}{5} + \frac{1}{4}\right)} = 2.40$$

The t-test statistic $t^* = \frac{|7.38 - 6.95|}{0.179} = 2.40.$

The degrees of freedom = 5 + 4 − 2 = 7.

For α = 5% and a two-sided test, t_c = 2.37. Thus $t^* > t_c$ or 2.40 > 2.37, which indicates sufficient difference to reject the null hypothesis. Thus, we conclude there is a statistically significant difference in the pH levels at the two monitoring locations.

In summary, the t-test for pH and similar monitoring parameters is constructed in the same manner except we utilize the absolute value in the numerator and use the two-tailed table for determination of t_c.

Example 9.13 From a sampling program, we summarize the results obtained from two monitoring programs.

Observation	Location A	Location B
1	2.6	3.1
2	4.2	3.6
3	3.0	3.1
4	3.2	4.4
5	2.9	3.6
6	4.1	4.7
Mean	3.33	3.75
Standard deviation	0.66	0.67
COV	COV = 0.66/3.33 = 0.19 (ok for normality assumption)	COV = 0.67/3.75 = 0.18 (ok for normality assumption)

a) Is there a statistically significant difference between the results as monitored at locations A and B? Assume $\alpha = 5\%$.
b) If we use the mean and variances calculated from the above data to predict results of additional samples at the same locations, how many samples would we need in order to show a statistically significant difference? What is a primary assumption upon which this calculation is based?

Part a

Because they do not have the exact same variance, we begin by calculating the ratio of the variance as

$$\text{F-ratio} = \frac{\text{Greater estimate of the variance of the population}}{\text{Lesser estimate of the variance of the population}} = \frac{0.67^2}{0.66^2} = 1.03$$

We now go to Table 9.2 in the text and find F-test = 5.0 (where the number of degrees of freedom for a sample of n items is equal to (n – 1)). Because the F-ratio < F-test, there is not a statistically significant difference between the variances, and the t-test can be utilized.

We can now pool the variances to get the best estimate of a population variance that the t-test assumes is the same for both populations:

$$\hat{S}^2 = \frac{(n_1-1)S_1^2+(n_2-1)S_2^2}{n_1+n_2-2} = \frac{(6-1)0.66^2+(6-1)0.67^2}{6+6-2} = 0.44$$

The standard error of the difference of the two means is then

$$S_m = \sqrt{\hat{S}^2\left(\frac{1}{n_1}+\frac{1}{n_2}\right)} = \sqrt{(0.44)\left(\frac{1}{6}+\frac{1}{6}\right)} = 0.38$$

The t-statistic is then calculated as $t^* = \frac{|3.33-3.75|}{0.38} = 1.11$.

We could also get the t-statistic using

$$t^* = \frac{\bar{x}_1-\bar{x}_2}{\sqrt{\frac{n_1+n_2}{n_1 n_2}\left(\frac{(n_1-1)S_1^2+(n_2-1)S_2^2}{n_1+n_2-2}\right)}} = \frac{|3.33-3.75|}{\sqrt{\frac{6+6}{6\times 6}\left(\frac{(6-1)0.66^2+(6-1)0.67^2}{6+6-2}\right)}} = 1.11.$$

From Table A.4 with the degrees of freedom df = 6 + 6 − 2 = 10 and α = 5% for a two-tailed test gives $t_c = 2.228$ (once again because we are only checking if there is a difference between the two sites, it is a two-tailed test). Since $t^* < t_c$, there is insufficient evidence of a difference between the results at location A and location B.

Part b
We can work in Microsoft Excel to find the number of additional samples at each location (need to find the first number of samples that would give $t^* > t_c$ (just try $n_1 = 7$, $n_2 = 7$, calculate the t-statistic and compare it against a new critical t-value from Table A.4 for df = 7 + 7 − 2 = 12).

From Excel we find that when $n_1 = 21$ and $n_2 = 21$, we have

$$\hat{S}^2 = \frac{(n_1 - 1)S_1^2 + (n_2 - 1)S_2^2}{n_1 + n_2 - 2} = \frac{(21-1)0.66^2 + (21-1)0.67^2}{21 + 21 - 2} = 0.442$$

The standard error of the difference of the two means is then

$$S_m = \sqrt{\hat{S}^2\left(\frac{1}{n_1} + \frac{1}{n_2}\right)} = \sqrt{(0.442)\left(\frac{1}{21} + \frac{1}{21}\right)} = 0.205$$

The t-statistic is then calculated as

$$t^* = \frac{|3.33 - 3.75|}{0.205} = 2.04$$

From Table A.4 with the degrees of freedom df = 21 + 21 − 2 = 40 and α = 5% for a two-tailed test gives $t_c = 2.021$. Since $t^* > t_c$, we would conclude that there is a statistically significant difference between the results obtained from location A and location B.

9.4.3 Effect of Nonnormality on the Hypothesis Test

The t-test is premised on the assumption that the data are normally distributed. It is fairly robust to deviations from this assumption, but the appropriateness of the assumption should be assessed. A simple indication of normality (although only approximate) involves the utilization of the coefficient of variation test (from Section 5.3).

Example 9.14 Benzene concentrations as measured at two monitoring locations are as listed below.

Monitoring location	Benzene concentrations (μg/l)
MW1	1.6, 1.9, 2.3, 1.7, 2.4, 1.8, 1.4, 2.7, 2.1, 1.9, 2.3, 1.8
MW2	1.8, 1.9, 2.5, 22.2, 2.2, 1.9, 1.3, 2.8, 2.3, 2.0, 2.4, 1.7

Is there a statistically significant difference in the concentrations at the two monitoring locations? Sample statistics are as follows.

Monitoring location	Mean (μg/l)	Standard deviation (μg/l)	Coefficient of variation
MW1	1.99	0.38	0.19
MW2	3.75	5.82	1.55

Noteworthy is the sizable magnitude of the fourth monitored value at MW2 (22.2), which results in a substantial increase in the mean and the standard deviation. The coefficient of variation of the results at MW2 is 5.82/3.75 = 1.55, indicating a failure to pass the test of normal distribution assumption of normality.

Alternatives for testing the assumption of normality are indicated in Section 5.3.

In the event of the data deviating from the assumption of normality, the usual practice is to transform the data so that the data can be described by the normal distribution. The transformations may include the log transformation, square root transformation, and the arcsine transformation. Alternatively, if an analyst cannot identify a transformation, an alternative approach is to use a distribution-free procedure.

9.4.4 Assumption of Independence

The t-test model assumes that the observations are the result of a random sampling and that each observation is an independent random sample from the parent population. An example of this type of problem occurs when the collection of a single groundwater sample is divided into aliquots for laboratory analyses. The laboratory results are not independent of one another.

Example 9.15 The air quality as quantified by 16 samples from one area of a city showed a mean of 107 and a standard deviation of 10, while the air quality of 14 samples from another area of the city showed a mean of 112 and a standard deviation of 8. Is there a significant difference between the air quality at the two locations at significance levels of (i) 0.01 and (ii) 0.05?

If μ_1 and μ_2 denote the population mean and air quality from the two areas, respectively, we have to decide between the hypotheses:

H_o: $\mu_1 = \mu_2$ and there is essentially no difference between the air quality.
H_1: $\mu_1 \neq \mu_2$ and there is a significant difference between the air quality.

Under hypothesis H_o

$$t^* = \frac{\bar{x}_1 - \bar{x}_2}{\hat{S}\sqrt{\frac{1}{n_1} + \frac{1}{n_2}}} \quad \text{where } \hat{S} = \sqrt{\frac{n_1 S_1^2 + n_2 S_2^2}{n_1 + n_2 - 2}} \tag{9.17}$$

Thus,

$$\hat{S} = \sqrt{\frac{16(10)^2 + 14(8)^2}{16 + 14 - 2}} = 9.44$$

and

$$t^* = \frac{112 - 107}{9.44\sqrt{\frac{1}{16} + \frac{1}{14}}} = 1.45$$

a) Using a two-tailed test at the 0.01 significance level, we would reject H_o if t^* were outside the range $-t_{0.995}$ to $t_{0.995}$, which for $(n_1 + n_2 - 2) = 16 + 14 - 2 = 28$ degrees of freedom is the range −2.76 to 2.76. Thus, we cannot reject H_o at the 0.01 significance level.
b) Using a two-tailed test at the 0.05 significance level, we would reject H_o if t^* were outside the range $-t_{0.975}$ to $t_{0.975}$, which for 28 degrees of freedom is the range −2.05 to 2.05. Thus, we cannot reject H_o at the 0.05 significance level.

We conclude that there is no significant difference in the air quality of the two locations.

9.4.5 Examples of t-Test Applications

Consider three sets of data: A_1, A_2, and A_3. The individual values are listed in Table 9.4.

a) *A standard t-test evaluation*
   Comparison of A_1 and A_2

   i) The coefficient of variation is less than 1.0, which indicates an assumption of the normal distribution is probably reasonable.
   ii) F-test is

$$F = \frac{(1.58)^2}{(1.45)^2} = 1.19$$

From Table 9.2, for $n_1 = 11$ and $n_2 = 11$ and thus, degrees of freedom of 10 and 10:

$F_{crit} = 3.0.$

$1.19 < 3.0$ so the two variances may be considered the same.

   iii) Pooled variance is

$$\hat{S}^2 = \frac{10(1.45)^2 + 10(1.58)^2}{20} = 2.30$$

and the standard error of difference is

$$S_m = \sqrt{2.30\left(\frac{1}{11} + \frac{1}{11}\right)} = 0.98$$

$t^* = \dfrac{6.26 - 4.98}{0.98} = 1.30$ and degrees of freedom $= 11 + 11 - 2 = 20$ and 5% level of significance, and

Table 9.4 Sample monitoring data.

	Monitoring locations		
	A_1	A_2	A_3
Monitoring data	4.8	5.9	6.6
	6.8	7.3	6.3
	4.2	7.2	21.9
	4.7	7.5	6.0
	7.0	8.1	5.8
	6.5	2.2	7.2
	3.9	5.6	5.7
	5.1	6.2	2.8
	2.1	5.8	5.2
	5.6	7.1	5.4
	4.1	6.0	6.1

$t_c = 1.812.$

A comparison of t* and t_c gives

$1.30 < 1.812$

which indicates there is not sufficient evidence to indicate a statistically significant difference.

b) *t-Test where a data transformation is needed*

Consider now a comparison of A_1 and A_3. The sampling results at A_3 are similar in many respects to those at location A_2, with the exception of the third reported value, which is much larger. Assuming that the large value is real and not a laboratory or sampling error, the analytical procedure is carried out as follows:

i) As with (a)(i) above, the coefficient of variation test is passed.
ii) The F-test gives

$$F = \frac{(5.01)^2}{(1.45)^2} = 11.9$$

From Table 9.2, 11.9 > 3.0, and thus the variances cannot be considered to be samples from the same population. One approach to the problem involves transforming the data by taking logarithms. Note that the average and standard deviations of the transformed data are as indicated below (i.e. the mean of the log-transformed data is not the same as the log of the mean).

	Statistics of the log-transformed data	
Sampling location	**Mean**	**Standard deviation**
A_1	1.56	0.34
A_2	1.84	0.48

The F ratio now becomes

$$F = \frac{(0.48)^2}{(0.34)^2} = 1.99$$

Since F < F_{crit}, i.e. 1.99 < 3.0, the two variances may be considered the same.

iii) The pooled variance is $\hat{S}^2 = 0.173$

and the standard error of difference is $S_m = 0.177$ and the t-test becomes

$$t^* = \frac{1.84 - 1.56}{0.177} = 1.57$$

Since $t_c = 1.58$ [from (a)(iii)]

$t^* < t_c$

indicating there is insufficient evidence to reject the null hypothesis or insufficient evidence to indicate there is a significant difference between the means.

9.5 Acceptance and Rejection Regions

In conducting statistical inference tests, incorrect conclusions sometimes occur. An acceptance rate is assumed, but what that really means is that an error is made a specified percentage of the time.

In reality, because the population means are being estimated from small samples, the comparison of these means involves the risk of making one of two possible errors, either a type I or a type II error. We attempt to achieve a proper balance between two conditions: (i) the risk the procedures will falsely indicate that, for example, a landfill is causing background values or concentration limits to be exceeded, when in fact the landfill is not causing a problem (false positive), and (ii) the risk that the procedures will fail to indicate that background values or concentration limits are being impacted when that is, in fact, the situation that a landfill is contaminating the groundwater (false negative).

Consider Table 9.5 and the following discussion as a demonstration:

Null hypothesis H_o: No contamination at the compliance well.
Alternative hypothesis H_A: Contamination has occurred.

The four possibilities are indicated in Figure 9.5 where the left side of the diagram represents the decision and along the top is truth or reality; thus,

i) We can accept the hypothesis and the hypothesis H_o is true and no error is made.
ii) We can reject H_o (accept H_A, the alternative hypothesis) but, in fact, H_o is true. This is a type I error, denoted by α, which represents the rejection of the null hypothesis when it is true. This error is also called a false positive (e.g. a "source" is identified as impacting the groundwater when, in fact, it is not impacting the groundwater or saying something is present when it is not).
iii) We can reject H_o and H_A is true and thus, no error is made.
iv) We can accept H_o, and H_A is true. This is a type II error, denoted by α, and represents the acceptance of the null hypothesis when it is false, or not true. We accept H_A when in fact H_A should be rejected. This error is also called a false negative (e.g. a "source" is identified as nonpolluting when, in fact, it is polluting or saying something is not present when it really is present). The result of this type of error, for example, is that a remedial measure is not implemented on a contaminating source that is, in fact, degrading the groundwater quality.

Since the type II error has environmental consequences, we frequently tend to be rather conservative.

Table 9.5 Types of errors in statistical testing.

		Truth	
		H_o	H_1
Decision	Accept H_o	No error	Type II error
		Probability = 1 − α	Probability = α
	Accept H_1	Type I error	No error
		Probability = α	Probability = 1 − α = power

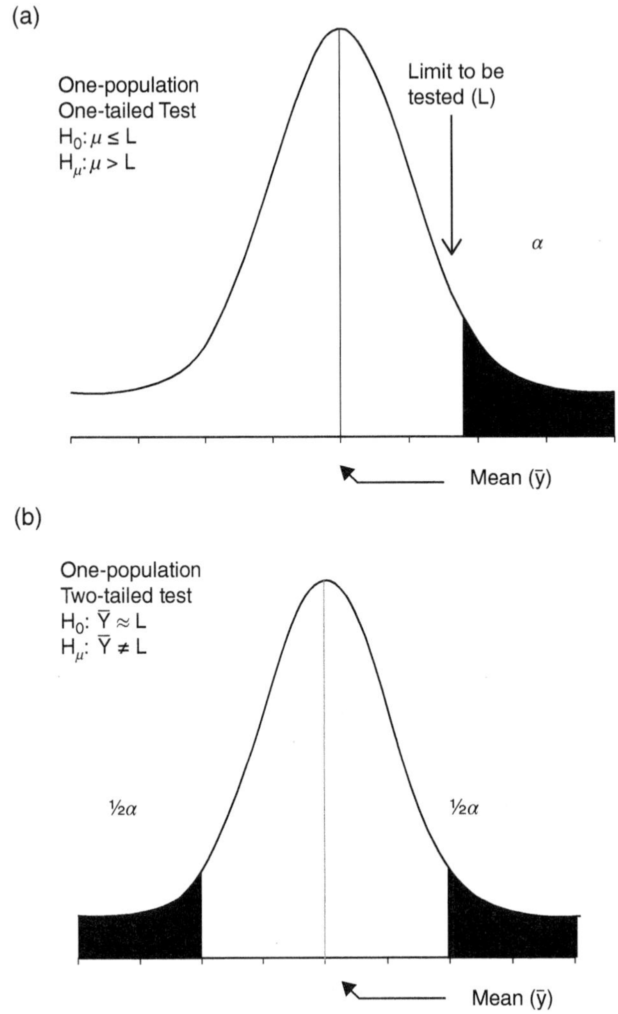

Figure 9.5 (a) A one-sided test of significance. (b) A two-sided test of significance.

In quantifying the likelihood of these errors, the level of significance, α, represents the probability of making a type I error. By specifying $\alpha = 0.05$, this indicates that in the long term, the type I error will be made 5% of the time. However, the two types of errors, α and β, are not independent. By making α smaller (e.g. $\alpha = 0.01$), we make the false positive error less often but increase the probability of making the false negative or type II error. The smaller we make the significance level, α, the more confidence we shall have that the null hypothesis is really false on the occasions when we reject it. The price we pay for this increased assurance is that the smaller we make α, the more difficult it becomes to disprove the hypothesis that is false.

For a given scenario, α and β are linked, that is, they cannot be minimized simultaneously. To examine this trade-off in greater detail, consider the following example: After carrying out a significance test, we have some evidence on which to decide whether or not to reject the null hypothesis. Assume $\alpha = 5\%$, $\sigma = 150\,\text{mg/l}$ of chloride, sample size n = 25, and

$$H_o : \mu = 1250\,\text{mg/l}$$

and

$$H_A : \mu \neq 1250\,\text{mg/l}$$

Then $t^* = \dfrac{\overline{x} - 1250}{150/\sqrt{25}}$

If H_o is true, this corresponds to

$$\text{Probability}\left(t^* > 1.64\right) = 0.05$$

Thus, H_o is rejected at the 5% level if the observed value of t* is greater than 1.64. Therefore, 1.64 is the critical value of t*. This is equivalent to rejecting H_o if $\overline{x} > 1250 + 1.64(150/\sqrt{25}) = 1299$. We have chosen the critical value in such a way that the probability of an error of type I or false positive rate is 5%.

In general, the probability of getting a significant result when H_o is true is denoted by α. In words, there is only a 5% probability of getting a mean chloride concentration greater than 1299 mg/l.

9.6 Power of the Discrimination Tests

If we wanted to be more certain, we could make α = 0.01% or 1% (this would correspond with making the type I error smaller, the situation in which we would find that the source is impacting the groundwater when it is actually doing so). Then, if H_o is true, we have

$$\text{Probability}\left(\frac{\overline{x} - 1250}{150/\sqrt{25}} < 2.33\right) = 0.99 \tag{9.18}$$

so that the critical value of t* is 2.33, i.e. $\overline{x} < 1250 + \dfrac{2.33(150)}{\sqrt{25}} = 1319.9.$

As expected, we need to have a larger mean (1320) to reject the null hypothesis when α = 1% than when α = 5% (1299).

If we have a specific alternative hypothesis and a given value of α, we can also find the probability of an error of type II, or β.

If H_o is tested with α = 0.05, then the critical value of $\overline{x}$ is 1299. If the observed value of $\overline{x}$ exceeds the critical value, then the null hypothesis will be correctly rejected. However, a type II error will occur if the observed value of $\overline{x}$ is less than the critical value, so that there is a probability that the null hypothesis, H_o, is accepted when it is false. If the above alternative hypothesis is true, we have

$$\text{Probability }\left(\overline{x} < 1299\right) = \text{prob}\left[\frac{\overline{x} - 1319.9}{150/\sqrt{25}} < \frac{1299 - 1319.9}{150/\sqrt{25}}\right] \tag{9.19}$$

$$\begin{aligned} &= \text{prob}\left[t^* < -0.70\right] \\ &= 1 - 0.82 = 0.18 \end{aligned} \tag{9.20}$$

and this is the probability of an error of type II.

Alternatively, if $\alpha = 0.01\%$ or 1%, the critical value is 1250 + 2.33 (150) = 1319.9. If the above alternative hypothesis is actually true, the probability of getting a nonsignificant result is given by

$$\begin{aligned}\text{Probability } (\bar{x} < 1319.9) &= \text{prob}\left[\frac{\bar{x} - 1319.9}{150/\sqrt{25}} < \frac{1319.9 - 1319.9}{150/\sqrt{25}}\right] \\ &= \text{prob}\left[t^* < 0\right] \\ &= 0.5\end{aligned} \tag{9.21}$$

Thus, with the lower level of significance (0.05 decreased to 0.01), there is much higher chance of making a type II error (0.18 from [9.20] increased to 0.5 from [9.21]).

The foregoing example shows that the two types of errors are dependent on one another. If the critical value is increased in order to reduce the probability of a type I error, then the probability of a type II error will increase.

For a specific alternative hypothesis, the power of a significance test is obtained by calculating the probability that H_o is rejected when it is false (that is, the correct decision is made). Then

$$\begin{aligned}\text{Power} &= 1 - \text{probability (error of type } II) \\ &= 1 - \beta\end{aligned} \tag{9.22}$$

or, in summary form,

α	β	Power
0.05	0.18	0.82
0.01	0.500	0.500

Figure 9.6 summarizes the alternative configurations for the α and β errors depicted in 9.6. The power of a statistical test is defined as the probability $(1 - \beta)$ that the test will successfully detect a statistically significant difference. Alternatively, it is the probability of rejecting a null hypothesis when, in fact, it is false and should be rejected.

A major impact with strict regulations is that they tend to minimize the false negative rates (i.e. the probability that you conclude that there is contamination when there is) by increasing the false positive rates (i.e. the probability that you conclude that there is contamination when there is none). Simply selecting a test based on its false negative rate is a concern, because a test with a high false positive rate will necessarily have a low false negative rate. In the extreme case in which the false positive rate is 100%, the false negative rate is 0%.

A reasonable trade-off between type I and type II errors is desirable. We can develop power curves to examine the trade-offs. For a given sample power curves, allow selection of an optimal test in the sense that it achieves its intended false positive (e.g. 5%) while simultaneously achieving a reasonable false negative rate (e.g. 5%).

In summary, the power of a test may be increased in one or more of the following ways:

a) By reducing the variance of the measurements to be tested.
b) By increasing the size of the difference to be detected.
c) By increasing the number of samples.
d) By using a higher significance level. For example, a test that uses a significance level of 0.10 will have a higher power (i.e. a lower chance of a type II error) but also a higher probability of a type I error than will a test using the 0.05 level of significance. The power may be increased by raising the significance level but at the risk of increasing the probability of incorrectly rejecting the null hypothesis.

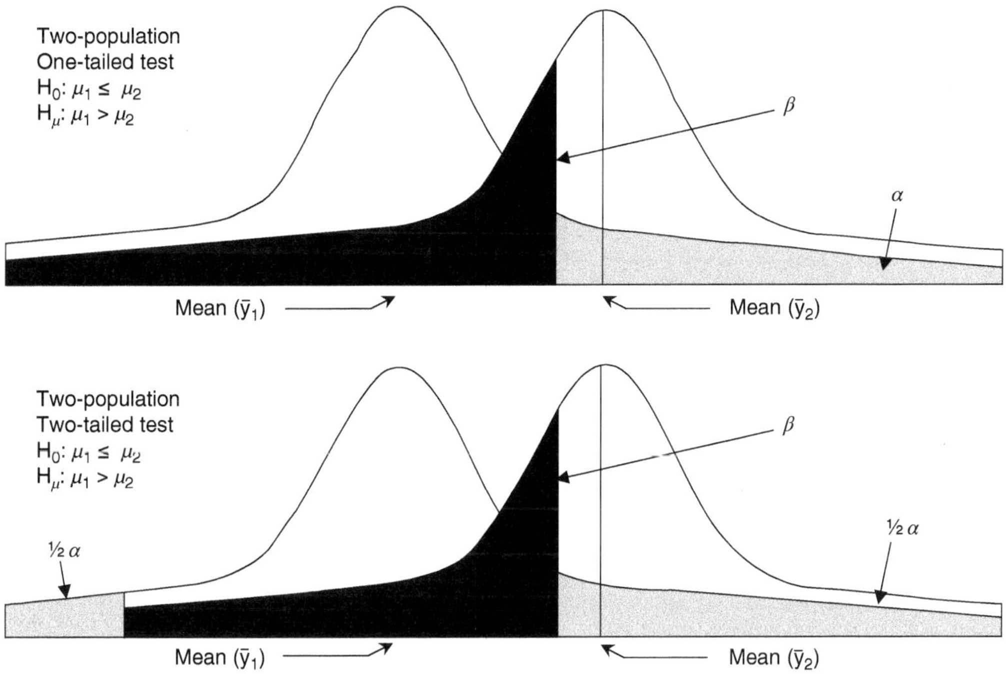

Figure 9.6 Probability of error in statistical testing.

Approximate power may be found early for three differences in the two population means (U.S. EPA, 1994):

a) At a difference of zero, the power is α, the significance level.
b) When the difference in mean concentration equals the critical value of the t-statistic (from the table) times the standard error of the difference (the denominator of the form of the t-statistic that is being used), the power is approximately 50%.
c) When the difference in mean concentration is twice that in (b), the power is approximately $100(1 - \alpha)\%$.

The power of the statistical test for detecting differences tends to increase as the alternative is farther from the null hypothesis.

9.6.1 Power of the t-Test

For given sample sizes, n_1 and n_2, significance level (α) and hypothesized difference between population means, d, the power of the t-test may be estimated (after Zar, 1982) by computing

$$t_{\beta(1)} = \frac{d}{\sqrt{2\hat{S}^2 / n}} - t_{\alpha/2} \tag{9.23}$$

where $\hat{S}^2$ is the pooled variance, $t_{\alpha(2)}$ is the critical value of t (for $n_1 + n_2 - 2$ degrees of freedom), and n is the size of each sample. If the two samples are of different sizes, n_1 and n_2, then

$$n = \frac{2n_1 n_2}{n_1 + n_2} \tag{9.24}$$

Finally, $t_{\beta(1)}$ denotes a "one-tailed" t, the value of t that delineates the proportion β in one tail of the distribution.

Example 9.16 If the difference between the population means is 0.41 mg/l, what is the probability that the t-test would conclude the null hypothesis when $n_1 = 12$, $n_2 = 17$, $\hat{S}^2 = 0.31$, and $t_{\alpha(2)} = 2.052$.

$$\text{Then } n = \frac{2(12)(17)}{12+17} = 14.07$$

$$\text{and } t_{\beta(1)} = \frac{0.41}{\sqrt{\frac{2(0.31)}{14.07}}} - 2.052 = -0.10$$

To estimate β, consider what proportion of a normal curve is ≥ -0.10; the probability of $t_{\beta(1)} \geq -0.10$ (estimated from Table A.1) is 0.54. Therefore, the power of the test is estimated as $1 - \beta = 1 - 0.54 = 0.46$. This indicates to the analyst that there is only a 46% chance of detecting a difference of 0.45 mg/l between the two population means. Increased power in testing may be achieved by increasing the sample size, n, and/or increasing a, the probability of a type I error.

9.7 Extensions of the t-Test

The limitations/assumptions implicit within the t-test preclude utilization of the test on a number of problems of environmental quality. As alternatives, a number of modified t-tests have been developed:

i) Satterthwaite's modified t-test (1964)
ii) Cochran's approximation to the Behrens–Fisher t-test (Behrens, 1929; Cochran, 1964)
iii) Paired sample t-test

These alternative tests have focused on relaxation of the assumptions implicit in the standard t-test, but caution in their application must be exercised since the tests are approximate, which means the tests have corresponding limitations.

9.7.1 Satterthwaite's Modified t-Test

Satterthwaite's test represents a modified t-test that allows unequal variance between the samples being compared (after Iman and Conover, 1983). The remaining assumptions necessary for utilization of the t-test must still be satisfied, namely, independence between observations and the assumptions that the populations are normally distributed.

When the variances are unequal, $S_1^2 \neq S_2^2$, the variances cannot be pooled as per Equation (9.14). However, the formula for the variance of $\bar{x}_1 - \bar{x}_2$ in independent samples still holds:

$$S^2_{\bar{x}_1-\bar{x}_2} = \frac{S_1^2}{n_1} + \frac{S_2^2}{n_2} \tag{9.25}$$

Satterthwaite's test then computes the t-statistic from

$$t^* = \frac{|\bar{x}_1 - \bar{x}_2|}{\sqrt{\frac{S_1^2}{n_1} + \frac{S_2^2}{n_2}}} \tag{9.26}$$

However, t* does not follow the t-distribution when $\mu_1 = \mu_2$. Two different forms of the distributions of t* have been identified (after Behrens (1929) and Fisher and Yates (1957)). In Satterthwaite's model (1946), an approximate number of degrees of freedom is assigned to t* so that the ordinary t-tables may be used. The approximate number of degrees of freedom are determined from

$$df = \frac{\left(\frac{S_1^2}{n_1} + \frac{S_2^2}{n_2}\right)^2}{\frac{\left(\frac{S_1^2}{n_1}\right)^2}{n_1 - 1} + \frac{\left(\frac{S_2^2}{n_2}\right)^2}{n_2 - 1}} \tag{9.27}$$

which is, in turn, used to obtain the critical "t" value, t_c, from the standard t-tables for a selected level of significance a. The value of df to use in the ordinary t-tables is "rounded down" (e.g. 16.8 becomes 16). As apparent from the procedure, Satterthwaite's approximation works by adjusting the degrees of freedom.

9.7.2 Cochran's Approximation to the Behrens–Fisher t-Test

An alternative to Satterthwaite's approximate t-test, to examine whether there is a statistically significant difference in the environmental quality at two locations when the variances are not equal, is Cochran's approximation to the Behrens–Fisher test (CABF). The CABF test functions by relaxing the requirement that the variance of the data at both sampling locations be of the same magnitude by adjusting the degrees of freedom. Cochran's test computes the t-statistic from

$$t^* = \frac{\left|\bar{x}_1 - \bar{x}_2\right|}{\sqrt{\frac{S_1^2}{n_1} + \frac{S_2^2}{n_2}}} \tag{9.28}$$

The critical t-statistic t_c is computed from

$$t_c = \frac{\frac{S_1^2}{n_1} t_1 + \frac{S_2^2}{n_2} t_2}{\frac{S_1^2}{n_1} + \frac{S_2^2}{n_2}} \tag{9.29}$$

where

$$S_1 = \sqrt{\frac{\Sigma\left(x_{i1} - \bar{x}_1\right)^2}{n_1 - 1}}$$

$$S_2 = \sqrt{\frac{\Sigma\left(x_{i2} - \bar{x}_2\right)^2}{n_2 - 1}}$$

where t_1 and t_2 are determined from t-tables for df of $n_1 - 1$ and $n_2 - 1$, respectively, for a specified level of significance, α.

Example 9.17 Consider the analysis in Section 9.4.5(b) where a data transformation was utilized because the variances could not be considered to be samples from the same population. As an alternative to the data transformation, the CABF test can be employed.

Specifically, using Equation (9.28),

$$t^* = \frac{7.10 - 4.98}{\sqrt{\frac{(4.53)^2}{2} + \frac{(1.45)^2}{11}}} = \frac{2.12}{1.43} = 1.48$$

The critical t-statistic is

$$t_c = \frac{\frac{(4.53)^2}{11}(1.81) + \frac{(1.45)^2}{11}(1.81)}{\frac{(4.53)^2}{11} + \frac{(1.45)^2}{11}} = 1.81$$

Since $t^* < t_c$ (1.48 < 1.81) there is insufficient evidence to indicate a significant difference between the sampling results at A_1 and A_3.

The difference between Cochran's test and Satterthwaite's test is in the calculation of the degrees of freedom, df. Cochran's test weighs the individual t values for each sample, while Satterthwaite's test calculates a new degree of freedom and then uses the t-tables.

As noted above, CABF allows relaxation of the equality of variance requirement. However, there are situations in which the assumptions implicit within the basic t-test are *not* violated and the utilization of the approximate test in such a situation will imply different findings from those intended. This concern was described in McBean et al. (1988). For example, when $n_1 = 3$ (the downgradient location has three sample results) and $n_2 = 4$ (the upgradient has four samples), with the variance ratio $S_1^2 / S_2^2 = 4$, the t-test allows a deviation of $|\bar{x}_1 - \bar{x}_2|$ of 2.28 before a significant difference is identified (a = 5% if a one-tailed test). Translation of the allowable deviation by Cochran's test into an equivalent deviation corresponds to $|\bar{x}_1 - \bar{x}_2| = 3.56$, which, in turn, gives a significance level of 1.3%. In other words, if Cochran's test is used instead of the t-test, the level of significance is not 0.05, but 0.013.

Cochran's test and Satterthwaite's test are both approximate formulas for the t-test. Conditions exist in which both approximate tests give substantially different results from the t-test. The following recommendations are made on the approximate t-tests (after McBean et al., 1988):

i) When the number of samples at the two locations being compared are similar in magnitude, the difference between the tests is minimal.
ii) When the variances are determined not to be different in the statistical significance sense (using the F-test), the t-test should be utilized. Cochran's test is more stringent than the t-test when

$n_1 > n_2$ for $\frac{S_1^2}{S_2^2} > 1$ and less stringent when $n_1 < n_2$ for $\frac{S_1^2}{S_2^2} > 1$.

Cochran's test is more stringent than the t-test when

$n_1 < n_2$ for $\frac{S_1^2}{S_2^2} < 1$ and less stringent when $n_1 > n_2$ for $\frac{S_1^2}{S_2^2} < 1$.

When the variances are determined to be different (using the F-test), Cochran's test should be used since the t-test assumptions are violated.

9.7.3 Paired t-Test

The paired t-test is a statistical test that can be used to determine if data from two separate treatments or environmental media fall within the same population or are "equivalent." For example, Anderson et al. (1993) used two pilot plants to examine such features as chlorine versus ozone and alum dosage versus silicate dosage as alternatives for the treatment of water. If the two sides of the treatment plant are equivalent in terms of impact on the treatment of water, the expected difference between two data pairs would be zero.

Paired sample t-tests are used when the sample populations are not independent. This may occur when successive sampling of the same water samples takes place upstream and downstream of a contaminant source. This situation occurs in the situation depicted in Figure 9.2a. A single element of water may be considered to move from an upstream point through a zone of possible contamination to a point downstream, such that it is virtually the same sample and therefore not independent. The paired t-test represents a variation on the standard t-test.

For some types of correlated data, it is possible to define the population being sampled in order to remove or minimize the effect on the test outcome (e.g. cyclical variations in both background and monitored data are not included in the calculation of the standard error). This is done by pairing the individual observations from the correlated populations. Once the differences in the pairs are calculated, the differences are treated as a single random independent sample. With the pairwise test, there is one-half the degrees of freedom, but the standard error is smaller.

The aim of pairing is to make the comparison more accurate by having the numbers of any pair as similar as possible except in the "treatment" difference. For example, the variable upon which the pairing takes place should predict accurately the concern by which the effects of the "treatments" are to be judged.

Pairwise testing is also appropriate when there is a sinusoidal variation in the input. For sinusoidally varying data, the variance will be large, and the t-value low. However, the variance is large because of a recurring feature, which is predictable, not random. Seasonal fluctuations in water chemistry are particularly apparent in shallow aquifers where groundwater quality may vary with recharge cycles (e.g. Pettyjohn, 1976; Spalding, 1984; Pekny et al., 1989). For these types of situations, the data must be combined to yield a new population.

When comparing two treatment methods, which occurs when two alternative procedures are used to analyze the same samples, it may happen that experiments are carried out in pairs. The difference between each pair of measurements is of interest.

This can be tested when we test whether both samples are equally affected by the external effects:

$$t^* = \frac{\bar{D} - \delta}{\frac{S_D}{\sqrt{n}}} \tag{9.30}$$

where δ may be taken as $\delta = 0$ or $\delta \neq 0$.

The paired t-test involves the derivation of a new population. The differences, D_i, for each of the individual pairs are assumed to be distributed about a mean, μ_D. The deviations $D_i - \mu_D$ may be due to various causes. In the analysis, the deviations $D_i - \mu_D$ are assumed to be normally and independently distributed with population mean zero. When these assumptions hold, the sample mean difference $\bar{D}$ is normally distributed about μ_D, with standard error $\sigma_D / \sqrt{n}$ where σ_D is the standard deviation of the population of differences. Since the value S_D is seldom known, the sample estimate is utilized as

$$S_D = \sqrt{\frac{\Sigma\left(D_i - \bar{D}\right)^2}{n-1}}$$

Hence, $S_{\bar{D}} = S_D / \sqrt{n}$ is an estimate of $\sigma_{\bar{D}}$ based on (n – 1) degrees of freedom:

$$t^* = \frac{\left|\bar{D} - \mu_D\right|}{S_{\bar{D}}} \tag{9.31}$$

If the H_o is true, the distribution of t* will be a t-distribution with (n – 1) degrees of freedom. The t-distribution may be used to test the null hypothesis that $\mu_D = 0$.

Similarly, the confidence interval for the mean is $\bar{D} \pm t_\alpha S_{\bar{D}}$. If the observation (difference) falls inside the tolerance interval, one concludes that no significant shift in level has occurred.

Note that with pairing we have fewer degrees of freedom (the t-value used in the test of significance or in computing confidence limits would be smaller with independent samples).

Example 9.18 Determine if there is a statistically significant difference in the arsenic concentrations between the quality at the two monitoring locations (twice/year of sampling). To examine this question, subtract the concentrations at the two locations to end up with a single set of concentration changes. The test may then be conducted as an examination of whether the changes are significantly different from zero.

		Monitoring results (μg/l)		
Date of sampling	Sample number	Station 1	Station 2	Difference in concentration (μg/l)
July	1	37	43	6
January	2	28	33	5
July	3	33	37	4
January	4	26	25	−1
July	5	41	45	4
January	6	22	23	1
July	7	33	32	−1
January	8	24	27	3
July	9	31	32	1
January	10	23	21	−2
Mean		$\bar{x} = 29.8$	31.8	2.0
Standard deviation		S = 6.30	8.11	2.79

a) Consider first the test as a standard t-test determination. We begin by calculating the ratio of the variances as

$$F = \frac{(8.11)^2}{(6.30)^2} = 1.52$$

According to Table 9.2, there is a not a statistically significant difference between the variances.

The pooled variance estimate is $\hat{S} = \dfrac{9(6.30)^2 + 9(8.11)^2}{18} = 50.00$

The standard error of the mean = $S_m = \sqrt{50.0\left(\dfrac{1}{10} + \dfrac{1}{10}\right)} = 3.25$

and the t-value $t^* = \dfrac{31.8 - 29.8}{3.25} = 0.61$
and $t_c = 1.73$

where t_c is determined using 18 degrees of freedom.

Since $t^* < t_c$, the analyses indicate there is no significant difference between the two monitoring locations. To leave it at this point would be clearly an incorrect approach. The reason that the test fails to detect a significant difference is the seasonality (the values are considerably higher during the July period, as opposed to the January period). However, when the seasonality is retained in the dataset, the variance is high and the t^* value is low.

b) Consider now the test as one of difference between the two monitoring locations.

The result is $t^* = 2.0 / \left(2.79 / \sqrt{10}\right) = 2.27$

$t_c = 1.83$

where t_c is determined using 10 – 1 = 9 df and α = 5%.

Since $t^* > t_c$, these findings indicate that there is a significant difference.

Example 9.19 The surface water quality monitoring results obtained from measuring both upstream and downstream of a potential nonpoint source are included as columns II and III in the table below. Some remedial technologies were implemented in October/November 1980 and the water quality monitoring data are as characterized by column V (McBean, 1986), as measured in 1981/1982. Of interest are two questions:

i) Is the source contributing significantly to the river?
ii) Did the remedial technologies significantly improve the water quality?

Upstream and downstream water quality monitoring records:

	Preremedial records		Postremedial records	
Date of sampling	Upstream measurements (mg/l)	Downstream measurements (mg/l)	Date of sampling	Downstream measurements (mg/l)
I	II	III	IV	V
10/79	0.29	4.3	10/81	0.53
11/79	12	16	11/81	1.5
12/79	0.32	6.1	12/81	1.3
1/80	—	—	1/82	—
2/80	—	—	2/82	—
3/80	0.49	2.66	3/82	2.1
4/80	0.14	3.0	4/82	1.1
5/80	1.58	4.42	5/82	—
6/80	1.77	5.74	6/82	1.8
7/80	1.07	1.40	7/82	1.2
8/80	0.07	1.49	8/82	0.64
9/80	0.14	2.3	9/82	1.1
Mean	1.69	4.74		1.25
Std. deviation	3.67	4.28		0.50

Note: "—" means no measurement available.

Using Satterthwaite's approximate test, an examination of the upstream and downstream concentration finds

$\bar{x}_1 = 1.69$	$\bar{x}_2 = 4.74$
$n_1 = 10$, $df_1 = 9$	$n_2 = 10$, $df_2 = 9$
$\frac{S_1^2}{n_1} = 1.35$	$\frac{S_2^2}{n_2} = 1.83$

$$t^* = 1.71$$

$$df = 17.5, \quad \text{which is taken } as\, 17$$

Finally, for a one-sided test (from the standard t-tables)

$$t_{c,0.05} = 1.74$$

Since $t_{c,0.05} > t^*$, these results indicate a statistically significant change has not been identified at the 95% level. However, a visual comparison of the upstream/downstream data in columns I and II clearly demonstrates that the downstream water quality is at a lesser water quality level.

For the type of correlation existing between upstream and downstream points, the pairing of individual observations and then examining only the differences between the observations is appropriate. Once the differences in the pairs are calculated, they are treated as a single random independent sample.

Using the paired test

$$\bar{D} = 2.95$$

$$S_D = 1.55$$

and

$$t^* = 6.0$$

For 9 degrees of freedom $t_c = 1.83$ for 5% and a statistically significant impact ($t^* > t_c$), indicating the nonpoint source has impacted the river.

Impact of the remedial technology

Using Satterthwaite's approximation in comparing pre- and postremedial measurements gives

$$t^* = 2.56$$

$$t_c = 1.7$$

which indicates the remedial measures have had a statistically significant impact.

Example 9.20 Consider the problem of comparing two methods for finding the percentage of iron in a compound. Six different samples were analyzed by both methods and the results are as given below.

Sample number	Method A	Method B
1	13.3	13.4
2	17.6	17.9
3	4.1	4.1
4	5.7	5.1
5	7.6	7.0
6	12.9	12.8

Is there a significant difference between the two methods of analysis? Note that it would be incorrect to calculate the average percentage for method A and method B and proceed in the standard t-test because the variations between compounds will overwhelm any difference there may be between the two methods. Instead, the correct approach is to proceed with comparing differences as follows.

Sample number	Difference in concentration
1	0.1
2	0.3
3	0
4	−0.6
5	−0.6
6	−0.1

If the two methods gives similar results, the above differences should be a sample of six observations from a population with a mean of zero.

The statistics of the difference in concentration are:

$$\bar{D} = -0.15$$

$$S_D = 0.37$$

$$t^* = \frac{-0.15}{\frac{0.37}{\sqrt{6}}} = -0.99$$

$t_c = 2.015$ for df $= 6 - 1 = 5$ and a $= 5\%$.

Since $t^* < t_c$, there is not sufficient evidence to indicate a difference exists between the two methods of analysis.

9.7.4 Summary of Alternative Tests

A series of alternative statistical significance tests exist, each of which involves a series of assumptions. In utilizing a particular test for a specific application, it is important to be cognizant of the limitations/approximations implicit in the individual procedures. All of the tests described in this chapter are parametric tests, with parametric meaning the value of concentration is utilized. Later chapters will deal with nonparametric procedures, where the nonparametric procedures use relative rankings and not the concentrations. Less stringent assumptions will be necessary for some of these procedures.

Both Cochran's and Satterthwaite's tests are approximate formulas for the t-test that allow relaxation of the quality of variances requirement. However, these are situations in which the assumptions implicit in the basic t-test are violated, and the utilization of the approximate tests in such situations may imply different findings from those intended (McBean and Rovers, 1990).

9.8 References

Anderson, W., Douglas, I., Van Den Oever, J. et al. (1993). Experimental techniques for pilot plant evaluation. Presented at the AWWA Water Quality Technology Conference, Miami, FL (7–11 November 1993).

Behrens, W.V. (1929). Ein beitrag zur fehlerberechnung bei wenigen beobachtungen. *Landwirtschaftliche Jahrbucher* 68: 807–837.

Cochran, W. (March 1964). Approximate significance levels of the Behrens-Fisher test. *Biometrics* 20: 191–195.

Draper, N.R. and Smith, H. (1981). *Applied Regression Analysis*. New York: John Wiley & Sons, Inc.

Fisher, R.A. (1925). *Statistical Methods for Research Workers*. London: Oliver and Boyd.

Fisher, R.A. and Yates, F. (1953). *Statistical Tables for Biological, Agricultural and Medical Research*, 6e, 53. Edinburgh: Oliver and Boyd.

Fisher, R.A. and Yates, F. (1957). *Statistical Tables*, 5e. Edinburgh: Oliver and Boyd.

Iman, R.A. and Conover, W.J. (1983). *A Modern Approach to Statistics*. New York: Wiley.

Kerlinger, F.N. and Pedhazur, E.J. (1973). *Multiple Regression in Behavioral Research*. New York: Holt, Rinehart and Winston Inc.

Koch, G.S. and Link, R.F. (1970). *Statistical Analysis of Geologic Data*. New York: John Wiley & Sons Inc.

Krumbein, W.C. and Graybill, F.A. (1965). *An Introduction to Statistical Models in Geology*. New York: McGraw-Hill Inc.

Kufs, C.T. (1992). Statistical modeling of hydrogeologic data – part I: regression and ANOVA models. *Ground Water Monitoring Review* Spring: 120–130.

McBean, E. (1986). Alternatives for identifying statistically significant differences. In: *Statistical Aspects of Water Quality Monitoring*, Developments in Water Science, vol. 27 (ed. A. El-Shaarawi and R. Kwiatkowski). Delft: Elsevier Science Publishers.

McBean, E., Kompter, M., and Rovers, F. (1988). A critical examination of approximations implicit in Cochran's procedure. *Ground Water Monitoring Review* Winter: 83–87.

McBean, E.A. and Rovers, F.A. (1985). Analysis of variances as determined from replicates versus successive samplings. *Ground Water Monitoring Review* Summer: 61–64.

McBean, E. and Rovers, F. (1990). Flexible selection of statistical discrimination tests for field-monitored data. In: *Ground Water and Vadose Zone Monitoring*, ASTM STP 1053 (ed. D. Nielsen and A. Johnson), 256–265. Philadelphia, PA: American Society for Testing and Materials.

Milliken, G.A. and Johnson, D.E. (1984). *Analysis of Messy Data*. New York: Van Nostrand Reinhold Co.

Pekny, V., Skorepa, J., and Vrba, J. (1989). Impact of nitrogen fertilizers on ground-water quality – some examples from Czechoslovakia. *Journal of Contaminant Hydrology* 4: 51–67.

Pettyjohn, W.A. (1976). Monitoring cyclic fluctuations in ground-water quality. *Ground Water* 14 (6): 472–479.

Satterthwaite, F. (1964). An approximate distribution of estimates of variance components. *Biometric Bulletin* 2: 110.

Spalding, M.E. (1984). Investigation of temporal variations and vertical stratification of ground water nitrate-nitrogen in the Hall County special use area, Nebraska Water Resource Center, Project G854-06, 33pp.

U.S. EPA (April 1994). Statistical training course for groundwater monitoring data analysis, EPA 530-R-93-003.

Zar, J.H. (1982). Power and statistical significance impact evaluation. *Ground Water Monitoring Review* Summer: 33–35.

9.9 Problems

9.1 (a) Do the sampling results listed below indicate that the treatment plant is not meeting its discharge permit (assume one-sided at 0.975)?

Trichloroethylene concentration data (μg/l)

9.5	
21.8	
32.9	
8.1	
18.04	Average
11.64	Std. deviation
135.56	Variance

(b) Explain in words what the results in (a) above mean, including 0.975.

9.2 There is a number of assumptions on which the hypothesis is based for purposes of determining if there is statistically significant difference between two datasets. What are the assumptions?

9.3 You may find that when preparing a site-specific risk assessment (SSRA), the numbers you calculate exceed the generic criteria and tables. Does that mean that you may conclude your analysis and report to the client, or is there merit in undertaking a next step (and if there is next step), and what step would that be?

9.4 There are three assumptions on which the hypothesis test is based for purposes of determining if there is a statistically significant difference between two datasets. List the assumptions, and if they are not appropriate for your application, how do you change your analysis so that you can test whether there is a statistical significance between two datasets?

9.5 Listed below are successive data points that you have assembled, which describe the air quality adjacent to a landfill. You need to prepare for a public meeting an estimate of the air quality of benzene to which they are exposed from emissions from the refuse and be 95% confident that you are correct in terms of your estimate. While you have some measurements, you want to be able to indicate the estimate of the mean is within ±15 μg/m^3.

Sample number	Measured benzene concentration	Cumulative mean	Cumulative standard deviation
1	15		
2	45	30.0	21.21
3	23	27.7	15.53
4	89	43.0	33.19
5	78	50.0	32.73
6	23	45.5	31.28
7	80	50.4	31.39
8	112	58.1	36.31
9	52	57.4	34.03
10	48	56.5	32.22

9.6 The remediation of a contaminated site has involved the construction of a grout curtain as depicted in the figure below (this involves excavation of a trench and then placement of cement into the trench to create a "curtain" that separates the groundwater on both sides of the trench). You are wondering if there is leakage under the grout curtain, so you have measured the groundwater levels at locations A and B at 12 p.m. (noon) each day. You are finding that the water levels from these measurements are highly correlated (when one is high, the other tends to be high, and when one is low, the other tends to be low).

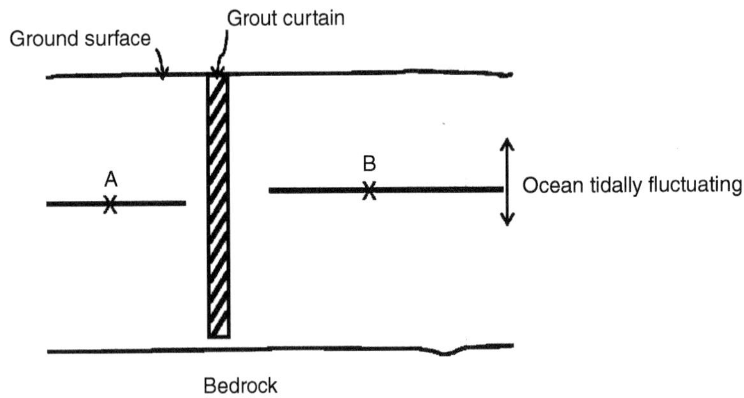

The success of your remediation technology depends on a quantity C, which is the cumulative effect of the variances at A and B (i.e. C = A + B). Demonstrate how you would calculate the variance of C, and indicate what you think might be the magnitude of the variance of C.

10

Multiple Comparisons Using Parametric Analyses

Testing is often the only way to answer our questions, but it does not produce unassailable, universal truths that should be carved on stone tablets. Instead, testing produces statistics which must be interpreted.

—Robert Hooke

10.1 Introduction

The focus in Chapter 9 was on single comparisons using various forms of the t-test to test for significance of the difference between monitoring records. However, for situations such as depicted in Figure 10.1 in which a number of comparisons are to be made (for the situation illustrated, comparisons of a number of upgradient and downgradient wells), we are interested in not one statistical comparison as per Chapter 9 but many comparisons. One approach then involves testing whether concentrations at one or more monitoring locations are statistically significantly different from other monitoring locations (e.g. do the concentrations at the compliance well significantly exceed the mean background concentration level at the upgradient well?).

An appropriate statistical analysis procedure to simultaneously analyze whether differences in mean concentrations between monitoring locations, or groups of monitoring locations, are statistically significant is through use of the analysis of variance (ANOVA), for example, to examine whether there is significant contamination at one or more downgradient monitoring locations 1 through 4 as depicted in Figure 10.1. ANOVA is a method by which the total variation in a set of data may be reduced to components associated with possible sources of variability, whose relative importance is of interest.

ANOVA is the name given to a wide variety of flexible and powerful statistical procedures. All of these procedures compare the means of different groups of observations to determine whether there are any significant differences among the groups, and if differences are identified, additional procedures may be used to determine where the differences lie.

ANOVA represents a technique closely related to regression. In ANOVA, the independent variables represent group membership (in contrast to discriminate analysis in which the dependent variable represents group membership). While regression techniques are generally used to evaluate data trends, the purpose of the ANOVA model is to detect differences in the average values of the dependent values of the dependent variables in the various groups.

Risk Assessment: Procedures and Protocols, First Edition. Edward A. McBean.

Companion website: www.wiley.com/go/McBean/RiskAssessProcedure_1e

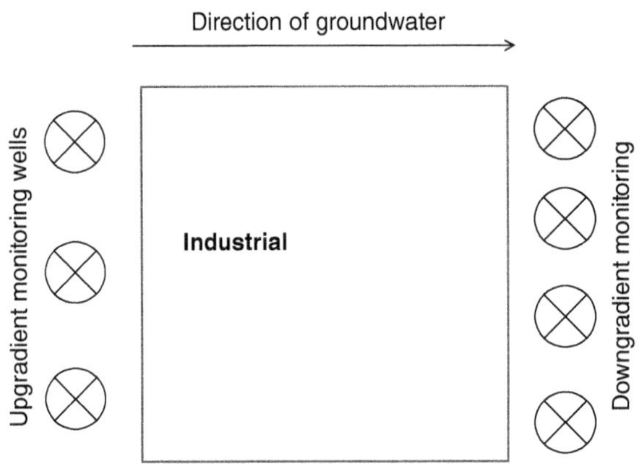

Figure 10.1 Plan view of upgradient and downgradient monitoring well locations.

Example 10.1 The types of questions to which ANOVA can be applied include:

a) Considering the data for one time period at several wells to determine if there is significant difference between different groupings of wells.
b) Considering the monitoring results from several wells over several time periods, assuming no seasonality is present, to determine if changes are occurring over time.

Note that in the examples as described above, the data are assumed to not exhibit seasonality or trends, which allows an analysis by the so-called one-way ANOVA. If the data show evidence of seasonality (observed, for example, by plotting the data over time), a trend analysis or perhaps a two-way ANOVA may be appropriate. The ramifications of a one-way ANOVA versus a two-way ANOVA will be considered in Section 10.2.3.

Two general types of ANOVA exist, namely, parametric and nonparametric. Both methods of ANOVA are appropriate for application to a wide array of environmental quality problems. However, parametric ANOVA is the focus of this chapter and nonparametric ANOVA is dealt with in Chapter 12. Of the two, the parametric ANOVA is the preferred procedure, but if the percentage of nondetects is greater than approximately 15%, then nonparametric one-way ANOVA must be utilized (where ranks are used as opposed to parametric values, e.g. concentrations).

The basis of the ANOVA concept was first developed by R. Fisher, and the F-test (which is the key to ANOVA) was named in his honor. The F-test makes a comparison of variances to determine whether there is a significant difference.

Since this chapter represents an extension of Chapter 9, from the single comparison test to the multiple comparison test, a reasonably good understanding of Chapter 9 is, therefore, presumed in the text below. As well, for ease of explanation, the problem context for describing the specifics of the calculations will be directed to comparison of upgradient and downgradient monitoring well locations as depicted in Figure 10.1.

10.2 Analysis of Variance (ANOVA)

10.2.1 Development of the Null Hypothesis

The mathematical procedures involved in ANOVA are best demonstrated by example. The null hypothesis is made that the means (e.g. the mean water quality for two sets of wells for a particular constituent) are equal and are, therefore, all simultaneously estimates of the same population. Rejecting the null hypothesis means that a statistically significant difference is said to be identified.

10.2.2 Multiple Comparisons and Statistical Power

Applying the same statistical significance test to each comparison is acceptable if the number of comparisons is small, but when the number of comparisons is moderate to large, the false positive rate can be quite high. This means that if a sufficient number of tests are run, there will be a significant chance that at least one test will indicate a significant difference exists (i.e. contamination has occurred, even if no actual contamination has occurred).

To avoid this problem, one could lower the type I error, but, for a given statistical test, in general, a lower false positive rate causes a lower power of the test to detect a real difference (e.g. contamination at the downgradient well). If the statistical power drops too much, real contamination will not be identified when it occurs.

To establish a recommended standard for the statistical power of a testing strategy, it must be understood that the power is not a single number, but rather a function of the level of contamination actually present. For more tests, the higher the level of contamination, the higher the statistical power and likewise, the lower the contamination level, the lower the power. As such, for example, when increasingly contaminated groundwater passes a particular well, it becomes easier for the statistical test to distinguish background levels from the contamination, i.e. the power is a function of the contamination level.

The power of any testing method can be increased merely by relaxing the false positive rate or letting α become larger than 5%.

10.2.3 One-Way ANOVA and Two-Way Tests of ANOVA

The one-way ANOVA uses a row to represent each sample observation and a separate column to represent each grouping to be compared (e.g. groups of groundwater monitoring wells). The two-way ANOVA involves the addition of a third factor, the blocking variable, which is used to associate some of the variation in the sample results. Thus, as an approach, we classify the monitoring locations into P mutually exclusive groups (where $P \geq 3$). If there are only two groups of monitoring wells, the two-sample test information from this chapter is used. The testing procedures described in this chapter need at least three to four samples per group. Each of the P groups should have at least three observations, and the total sample size, n, should be large enough so that $n - P \geq 5$.

Assumptions implicit in the ANOVA procedure include the data with normally distributed residuals with equal variances across P groups of wells, where the residual is defined as

$$\text{Residual} = \text{observed} - \text{predicted} \tag{10.1}$$

The hypothesis tests with parametric ANOVA usually assume that the errors (residuals) are normally distributed with constant variance. These assumptions can be checked by saving the residuals (the difference between the observations and the values predicted by the ANOVA model). Therefore, the procedures involve the calculation of the residuals and then testing the normality of the residuals. The testing of normality involves the pooling of all residuals together and checking the distribution using, for example, the Shapiro–Wilk test (see Section 5.3.6). The ANOVA procedure then tests equality of variances across the groups. Assuming equality of variance, we then compute for each source of variability in turn, (i) the sum of squares, and (ii) the number of degrees of freedom.

The total variation of the observations is partitioned into two components, one measuring the variability between the group means, $\overline{x}_1$, $\overline{x}_2$,...,$\overline{x}_n$, and the other measuring the variation within each group.

10.3 Testing for Homogeneity of Variance

Assumptions within the parametric ANOVA procedures include the assumption that the different groups (e.g. different wells) have approximately the same variance. If this is not the case, the power of the F-test (its ability to detect differences among the group means) is reduced. Mild differences in variance are acceptable, but the effect of differences in the variances becomes noticeable when the largest and smallest group variances differ by a ratio in excess of 4 and become severe when the ratio is 10 or more (Milliken and Johnson, 1984a).

10.3.1 Box Plots

Probably the simplest procedure to check the variances across the P groups is to use box–whisker plots (described in Section 4.7.2). A box plot is constructed for each well group, and the boxes are compared to see if the assumption of equal variances is reasonable. Three horizontal lines are drawn for each well grouping – one line each at the lower and upper quartiles and another at the median concentration. If the box length for each group is less than three times the length of the shortest box, the sample variances are probably sufficiently close to assume equal group variances. If the group variances appear to be different, the data should be further checked using Levene's test. If Levene's test is significant (see Section 10.3.2), the data may need to be transformed or a nonparametric procedure considered.

10.3.2 Levene's Test

Levene's test is a more formal procedure than box plots for testing the homogeneity of variance. Levene's test is not sensitive to nonnormality in the data and has been shown to have power nearly as great as Bartlett's test (Section 10.3.3) for normally distributed data and power superior to Bartlett's for nonnormal data (Milliken and Johnson, 1984a).

To conduct Levene's test, first compute the new variable:

$$Z_{ij} = \left| x_{ij} - \overline{x}_i \right| \tag{10.2}$$

where x_{ij} represents the jth value for the ith well and $\overline{x}_i$ is the ith well mean. The values Z_{ij} represent the absolute values of the residuals. A standard one-way ANOVA is then computed on the variables Z_{ij}. If the F-test is significant, the hypothesis of equal group variances is rejected.

Example 10.2 Use Levene's test for testing homogeneity of variance for monitoring data collected at four individual wells.

Sampling time	Sulfide concentration (mg/l)			
	Monitoring well 1	Monitoring well 2	Monitoring well 3	Monitoring well 4
1	29.4	16.1	7.8	27.9
2	12.6	4.7	12.4	21.9
3	18.2	36.1	3.6	12.6
4	*31.3*	*32.1*	*18.4*	*36.7*
Group mean $\bar{x}_i$	22.9	22.3	10.6	24.8

Step 1: Calculate the group mean for each well ($\bar{x}_i$).

Step 2: Compute the absolute residuals $Z_{ij} = \left|x_{ij} - \bar{x}_i\right|$ in each well and the well means of the residuals ($\bar{Z}_i$).

Sampling time	Absolute residuals			
	Well 1	Well 2	Well 3	Well 4
1	6.5	6.2	2.8	3.1
2	10.3	17.6	1.8	2.9
3	4.7	13.8	7.0	12.2
4	*8.4*	*9.8*	*7.8*	*11.9*
Well mean ($\bar{Z}_i$) =	7.48	11.85	4.85	7.53
Overall mean ($\bar{Z}$) =	7.93	Variance of the residual S_z^2 =	20.0	

Step 3: Compute the sums of squares for the absolute residuals:

$$SS_{Total} = (n-1)S_z^2 = (16-1)(20.0) = 300.0$$

$$\begin{aligned} SS_{Wells} &= \Sigma n_i \bar{Z}_i^2 - n\bar{Z}^2 = 4(7.48)^2 + 4(11.85)^2 + 4(4.85)^2 + 4(7.53)^2 - 16(7.93)^2 \\ &= 223.8 + 561.7 + 94.1 + 226.8 - 1005.5 \\ &= 100.9 \end{aligned}$$

$$SS_{Error} = SS_{Total} - SS_{Wells} = 300.0 - 100.9 = 199.1$$

Step 4: Construct an analysis of variance table to calculate the F statistic. The degrees of freedom (df) are equal to (number of groups – 1) = 4 – 1 = 3 df and (number of samples – number of groups) = 16 – 4 = 12 df.

Source	Sum of squares	df	Mean square	F-ratio
Between wells	100.9	3	33.6	2.02
Error	199.1	12	16.6	
Total	300.0	15		

Step 5: The tabulated value of $F_{0.05}$ = 3.49 with 3 and 12 df is larger than the F ratio of 2.02, and so the assumption of equal variances cannot be rejected.

(Note: If the F ratio had been larger than the F statistic, the assumption of equal variances would be rejected. In this situation, the most likely next step would be to log-transform the data and retest the data).

10.3.3 Bartlett's Test

Another means of testing whether a number of population variances of normal distributions are equal involves use of Bartlett's test. Recall that homogeneity of variances is an assumption made in the ANOVA when comparing concentrations of constituents between background and compliance monitoring locations.

Bartlett's test examines whether the sample data provide evidence that the well groups have different variances. However, Bartlett's test is sensitive to nonnormality in the data and may give misleading results unless one knows in advance that the data are approximately normal (Milliken and Johnson, 1984a). With long-tailed distributions, the test too often rejects equality (homogeneity) of the variances.

Test Procedure – Assume that data from k monitoring well groupings are available and that there are n_i data points for monitoring group i.

Step 1: Compute the k sample variances:

$$S_i^2 = \frac{\sum_{j=1}^{n_i}\left(x_{ij} - \bar{x}_i\right)^2}{n_i - 1} \quad \text{for all i groups} \tag{10.3}$$

Each variance has associated with it, $f_i = n_i - 1$ degrees of freedom.

Take the natural logarithms of each variance as $\ln\left(S_1^2\right),\ldots,\ln\left(S_k^2\right)$.

Step 2: Compute the test statistic:

$$\chi^2 = f\ln\left(S_p^2\right) = \sum_{i=1}^{k} f_i \ln\left(S_i^2\right) \tag{10.4}$$

where

$$f = \sum_{i=1}^{k} f_i = \left(\sum_{i=1}^{k} n_i\right) - k \tag{10.5}$$

In words, f is the total sample size minus the number of well groups. As well,

$$S_p^2 = \frac{1}{f}\sum_{i=1}^{k} f_i S_i^2 \tag{10.6}$$

is the pooled (weighted variance across wells groups).

Step 3: Using the chi-square table, find the critical value for χ^2 with (k – 1) degrees of freedom at a predetermined significance level (e.g. 5%). If the calculated value χ^2 is larger than the tabulated value, then conclude that the variances are not equal at that significance level.

Example 10.3 Utilize Bartlett's test to determine if there are statistically significant differences in the variances for the six monitoring well records for iron.

	Iron concentrations (μg/l)					
Sample number	**MW1**	**MW2**	**MW3**	**MW4**	**MW5**	**MW6**
1	812	712	675	556	623	472
2	694	695	490	503	591	499
3	774	—	526	—	—	526
4	858	—	499	—	—	403
n_i	4	2	4	2	2	4
$f_i = n_i - 1$	3	1	3	1	1	3
S_i	69.4	12.0	86.4	37.5	22.6	52.8
S_i^2	4819.7	144.5	7459	1404	512	2790
$f_iS_i^2$	14459	144.5	22377	1404	512	8370
$\ln(S_i^2)$	8.48	4.97	8.92	7.25	6.24	7.93
$f_i \ln(S_i^2)$	25.44	4.97	26.75	7.25	6.24	23.79

$$f = \sum f_i = 3+1+3+1+1+3 = 12$$

$$S_p^2 = \frac{1}{f}\sum_{i=1}^{k} f_i S_i^2 = \frac{1}{12}\left(14\ 459.1 + 144.5 + 22\ 377 + 1\ 404 + 512 + 8\ 370\right)$$
$$= 3938.9$$

$$\ln\left(S_p^2\right) = 8.28$$

$$\chi^2 = f \ln\left(S_p^2\right) - \sum f_i \ln\left(S_i^2\right)$$
$$= 12(8.28) - (25.44 + 4.97 + 26.75 + 7.25 + 6.24 + 23.79)$$
$$= 99.34 - 93.37 = 5.97$$

The critical χ_c^2 value with 6 − 1 = 5 degrees of freedom at 5% level is 11.1. Since 5.97 < 11.1, we cannot reject the null hypothesis, i.e. the test is not significant at the 5% level, suggesting that the variances are not significantly different.

It is noteworthy that the differences in the variances between the various monitoring groups are quite large and yet the null hypothesis is not rejected. With very small datasets, the differences in the variances have to be very large for the differences to be considered statistically significant. As a demonstration of this, consider the redefinition of the example where n_i for each of the six monitoring wells is taken to be five times as large, but the remaining values are as indicated previously. Then,

n_i	20	10	20	10	10	20
f_i	19	9	19	9	9	19
S_i	69.4	12.0	86.4	37.5	22.6	52.8
S_i^2	4819.7	144.5	7459	1404	512	2790
$f_iS_i^2$	91574	1300.5	141721	12636	4608	53010
$\ln(S_i^2)$	8.48	4.97	8.92	7.25	6.24	7.93
$f_i \ln(S_i^2)$	161.1	44.73	169.5	65.25	56.16	150.7

$$f = \sum_{i=1}^{6} f_i = 84$$

$$S_p^2 = \frac{1}{f}\sum_{i=1}^{k} f_i S_i^2 = 3629$$

$$\ln\left(S_p^2\right) = 8.20$$

$$\chi^2 = 41.39$$

The critical χ_c^2 is 11.1 and is less than the calculated 41.39; the null hypothesis is rejected.

If the results were shown to have different (heterogeneous) variances, then a logarithmic transformation of the data should be utilized and Bartlett's test redone. On the other hand, unequal variances among monitoring data could be a direct indication of well contamination, since the individual data could come from different distributions (i.e. different means and variances). The next step, given the significant difference in the variance, is to test which variances are different. In the case of two sets of data, the test of equality of variances becomes the F-ratio test described in Chapter 9. Bartlett's test simplifies in the case of equal sample sizes for each of the groups, i.e. for $n_i = n$ for $i = 1, \ldots, k$. The test used then is Cochran's test where Cochran's test focuses on the largest variances and compares it to the sum of all the variances (U.S. EPA, 1989). Technical aids for the procedures under the assumption of equal sample sizes are given by Nelson (1987).

10.4 ANOVA Procedure

The mathematical steps of the ANOVA procedure involve the following: Assume there are P well groupings with n_i data points (concentrations) at the ith well. These data can be from either a single sampling period or from more than one. The statistical analyst can check for seasonality before proceeding by plotting the data over time.

Step 1: Arrange the $N = \sum_{i=1}^{P} n_i$ monitoring results in a data table as indicated below and calculate the residuals.

Well number	Monitored results				Well mean	Residuals		
1	x_{11}	x_{12}	x_{13}	$x_{1,n1}$	$\overline{x}_1$	R_{11}	R_{12}	R_{13}
2	x_{21}	x_{22}	x_{23}	$x_{2,n2}$	$\overline{x}_2$	R_{21}	R_{22}	R_{23}
i	x_{i1}	x_{i2}	x_{i3}	$x_{i,ni}$	$\overline{x}_i$	R_{11}	R_{12}	R_{13}
p	x_{p1}	x_{p2}	x_{p3}	$x_{p,np}$	$\overline{x}_p$	R_{11}	R_{12}	R_{13}
				Grand mean = $\overline{x}$				

The well means are calculated as

$$\overline{x}_i = \frac{\sum_{j=1}^{n_i} x_{ij}}{n_i} \tag{10.7}$$

$$\text{The grand mean calculated as } \overline{x} = \frac{\sum_{i=1}^{p}\sum_{j=1}^{n_i} x_{ij}}{N} \tag{10.8}$$

where N is the total number of observations, $N = \sum_{i=1}^{p} n_i$.

Step 2: Compute the sum of squares of difference *between* well means and the grand mean:

$$SS_{wells} = \sum_{i=1}^{p} n_i \left(\bar{x}_i - \bar{x}\right)^2 \qquad (10.9)$$

This sum of squares has (p – 1) degrees of freedom associated with it and is a measure of the variability *between* wells.

Step 3: Compute the corrected total sum of squares:

$$SS_{total} = \sum_{i=1}^{p}\sum_{j=1}^{n_i}\left(x_{ij} - \bar{x}\right)^2 \qquad (10.10)$$

This sum of squares has (N – 1) degrees of freedom associated with it.

Step 4: Compute the sum of squares of differences of observations *within* wells from the well means. This is the sum of squares due to error and is obtained by subtraction:

$$\begin{aligned} SS_{error} &= SS_{total} - SS_{wells} \\ &= \sum_{i=1}^{p}\sum_{j=1}^{n_i}\left(x_{ij} - \bar{x}\right)^2 \\ &= \Sigma\Sigma R_{ij}^2 \end{aligned} \qquad (10.11)$$

It has associated with it (N – P) degrees of freedom and is a measure of the variability *within* wells.

Step 5: Set up and ANOVA table (where MS is the mean square).

Source of variation	Sum of squares	Degrees of freedom	Mean squares	F
Between wells	$SS_{wells} = \Sigma\left(\bar{x}_i - \bar{x}\right)^2$	P – 1	$MS_{wells} = SS_{wells}$ (P – 1)	$F = \frac{MS_{wells}}{MS_{error}}$
Error (within wells) (residual variation)	$SS_{error} = \Sigma\Sigma\left(x_{ij} - \bar{x}_i\right)^2$	N – P	$MS_{error} = SS_{error}/(N - P)$	
Total	SS total $\sum_{i=1}^{p}\sum_{j=1}^{n_i}\left(x_{ij} - \bar{x}\right)^2$	N – 1		

The two mean squares, MS_{wells} and MS_{error}, are determined by dividing the appropriate sum of squares by the appropriate number of degrees of freedom.

If the "between sample variation" is significantly greater than the "within sample variation," then the potential exists that the samples were not, in fact, drawn from the same population, but from populations whose average values differed. If this proves to be the situation, in addition to the "within population variation," there exists also a "between population variation."

Step 6: To test the hypothesis of equal means for all P wells, compute F = MS_{wells}/MS_{error}. Compare this statistic to the tabulated F statistic with (P – 1) and (N – P) degrees of freedom. If the calculated F value exceeds the tabulated value, reject the hypothesis of equal well means. Otherwise conclude there is no significant difference between the concentrations at the P wells (groupings).

If a significant difference exists, at least one pair of well means is different. To find out which compliance well exceeds background, multiple comparisons must be calculated.

In the case of a significant F, one must determine which of the compliance well(s) is contaminated. This is done by comparing each compliance well with the background well(s). Concentration differences between a pair of background wells and compliance wells or between a compliance well and a set of background wells are called *contrasts* in the ANOVA and multiple comparison framework (U.S. EPA, 1989).

Step 7: Determine if the significant F is due to differences between background and compliance wells (involves computation of Bonferroni t-statistics as described below). Compute the Bonferroni t-statistic (on each compliance well).

Assume that of the P wells, u are background and m are compliance wells (i.e. u + m = P). Then m differences (m compliance wells are each compared with the average of the background wells) need to be computed and tested for statistical significance. If there are more than five downgradient wells, the individual comparisons are done at the comparisonwise significance level of 1%, which make the experimentwise significance level greater than 5%.

If more than five comparisons are involved, perform each comparison at 1% significance level. If less than five, divide by the number of comparisons.

Obtain the total sample size of all u background wells:

$$n_{up} = \sum_{i=1}^{u} n_i \tag{10.12}$$

Compute the average concentration from the u background wells:

$$\overline{x}_{up} = \frac{1}{n_{up}} \sum_{i=1}^{n} \overline{x}_i \tag{10.13}$$

Compute the m differences between the average concentrations from each compliance well and the average background wells:

$$\overline{x}_i - \overline{x}_{up} \quad \text{for all } i = 1, \ldots, m$$

Compute the standard error of each difference as

$$SE_1 = \left[MS_{error} \left(\frac{1}{n_p} + \frac{1}{n_p} \right) \right]^{0.5} \tag{10.14}$$

where MS_{error} is determined from the ANOVA matrix and n_1 is the number of observations at well i (i.e. they are all tested at once). Obtain the t-statistic $t = t_{N-P,\ 1-\alpha/2}$ from Bonferroni's t-table with $\alpha = 0.05$ and df = N − P degrees of freedom. Then compute the m quantities $D_1 = SE_i \times t$ for each compliance well i. If m > 5, use the entry for $t_{N-P\ 1-0.01}$ (i.e. the entry at m = 5). If the difference $\overline{x}_i - \overline{x}_{up}$ exceeds D_i, then the ith compliance well is concluded to have significantly higher concentrations than the average background wells. The comparison needs to be performed for each of the m compliance wells individually. The test is designed so that the overall experimentwise error is 5% when there are not more than five compliance wells.

Example 10.4 The sulfide concentration data from Example 10.2 plus data for two additional monitoring wells are listed below. If monitoring wells 1 and 2 are upgradient, determine if any of the monitoring wells 3 through 6 are significantly different.

Monitoring well	Monitoring data				Mean concentration $\bar{x}_i$	Standard deviation S_i
1	29.4	12.6	18.2	31.3	22.9	8.96
2	16.1	4.7	36.1	32.1	22.3	14.5
3	17.8	22.4	13.6	28.4	20.6	6.35
4	27.9	21.9	12.6	36.7	24.8	10.1
5	29.8	32.1	41.0	28.9	33.0	5.53
6	39.6	46.2	42.0	39.9	41.9	3.04
					$\bar{x} = 27.6$	

Complete the between-well sum of squares:

$$\begin{aligned} SS_{wells} &= 4(22.9-27.6)^2 + 4(22.3-27.6)^2 + 4(20.6-27.6)^2 + 4(24.8-27.6)^2 + 4(33.0-27.6)^2 \\ &\quad + 4(41.9-27.6)^2 \\ &= 88.4 + 112.4 + 196 + 31.4 + 116 + 818 \\ &= 1362 \end{aligned}$$

with six wells – 1 = 5 degrees of freedom total.

Compute the corrected total sum of squares:

$$\begin{aligned} SS_{total} &= (29.4-27.6)^2 + (12.6-27.6)^2 + (18.2-27.6)^2 + \cdots \\ &= 2793 \end{aligned}$$

with 24 observations – 1 = 23 degrees of freedom.

Obtain the within-well or error sum of squares by subtraction:

$$SS_{error} = SS_{total} - SS_{wells} = 2793 - 1362 = 1431$$

with 24 observations – 6 wells = 18 degrees of freedom.

The calculated F statistic is 3.4. The tabulated F value with 5 and 18 df at the $\alpha = 0.05$ level of significance is 2.77 (see Table 9.2). Since the calculated value exceeds the tabulated value, the hypothesis of equal means must be rejected. Subsequent analyses must be carried out to determine where the difference exists.

Compute the Bonferroni t-statistic. There are four downgradient or compliance wells, so m = 4 comparisons are involved:

$n_{up} = 8 =$ the total number of samples in the upgradient or background wells.

$\bar{x}_{up} = 22.6 =$ the average concentration of upgradient wells.

Source of variation	Sums of squares	Degrees of freedom	Mean squares	F
Between wells	1362	5	$\frac{1362}{5} = 272.4$	$\frac{272.4}{79.5} = 3.4$
Error (within wells)	1431	18	$\frac{1431}{18} = 79.5$	
Total	2793	23		

Compute next the differences between the four compliance wells and the average of the two background wells:

$$\bar{x}_3 - \bar{x}_{up} = 20.6 - 27.6 = -7.0$$
$$\bar{x}_4 - \bar{x}_{up} = 24.8 - 27.6 = -2.8$$
$$\bar{x}_5 - \bar{x}_{up} = 33.0 - 27.6 = 5.4$$
$$\bar{x}_6 - \bar{x}_{up} = 41.9 - 27.6 = 14.3$$

Compute the standard error of each difference. Note that for this case, since the number of observations is the same for all compliance wells, the standard errors for the four differences are all the same:

$$SE_i = \left[79.5\left(\frac{1}{8} + \frac{1}{4}\right)\right]^2 = 5.46 \quad \text{for } i = 3, \ldots, 6$$

From Table A.15, the critical t with (24 – 6) = 18 degrees of freedom, m = 4, and $\alpha = 0.05$ gives t = 2.4.

Compute the quantities D_i as

$$D_i = SE_i \times t = 5.46 \times 2.4 = 13.1 \quad \text{for } i = 3, \ldots, 6$$

Only for $\bar{x}_6 - \bar{x}_{up} = 14.3$ exceeds the critical value of 13.1. Consequently, there is significant evidence of contamination at monitoring well 6.

10.5 Two-Way ANOVA

One-way ANOVA was used in the previous sections to examine differences between a number of monitoring locations. In the one-way ANOVA, the variation of the observations is partitioned into two components, one measuring the variability between the group means and the other measuring the variation within each group. Alternatively, in the two-way ANOVA, term [1] of Equation (10.15) below, measures the variance between treatments, term [2] measures the variation between blocks, and term [3] measures the residual variation. The algebraic identity is

$$\sum_{i,j}\left(x_{ij} - \bar{x}\right)^2 = \underbrace{c\sum_{i=1}^{r}\left(\bar{x}_i - \bar{x}\right)^2}_{[1]} + \underbrace{r\sum_{j=1}^{c}\left(\bar{x}_j - \bar{x}\right)^2}_{[2]} + \underbrace{\sum_{i,j}\left(x_{ij} - \bar{x}_i - \bar{x}_j - \bar{x}\right)^2}_{[3]} \tag{10.15}$$

10.6 Iterations and Data Transformations

The ANOVA is a useful test, but its use does entail a series of assumptions, which sometimes do not apply in a particular application. As a result, data transformations are often performed to obtain approximate normality and to stabilize the variance of ANOVA residuals across different groups. Avoidance of heteroscedasticity or "unequal variances" is of interest because approximately equal variances are required for a parametric ANOVA to give valid results. This homogeneity (equality) of variances of residuals across monitoring locations is the most important assumption in parametric ANOVA. If this assumption is not met, the power of the F-test, that is, its ability to detect differences among the group means, is reduced. Checking the equality of variance is important for the analysis of variance procedure is often more sensitive to unequal variances than to moderate departures from normality. The intent of this test is to determine whether group variances are equal or significantly different.

If the residuals from the transformed data do not meet the parametric ANOVA requirements, then nonparametric approaches to ANOVA are a potential approach (see Chapter 12 for a description of the nonparametric or Kruskal–Wallis test. The Kruskal–Wallis procedure utilizes the ranks of the observations as opposed to the monitored values or magnitudes). Alternatively, use of a data transformation to stabilize or equalize the variances may be utilized. Levene's test of homogeneity of group variances is not as sensitive to departures from normality as Bartlett's test. Bartlett's test checks equality, or homogeneity, of more than two variances with unequal sample sizes. The F-test is a special situation when there are only two groups to be compared.

In the case of equal sample sizes but more than two variances to be compared, the user may want to use Hartley's (or maximum F-ratio) test (Nelson, 1987).

10.7 Concerns with Multiple Comparisons

Although the error rate of any single test is low, if a sufficiently large number of comparisons are completed, the chance of making at least one mistake (false positive error) overall is high. Miller et al. (2001) showed mathematically that the chance of one or more false positives, when performing several statistical tests, can never be greater than the sum of the chances of false positives on each of the individual tests.

In general, $\Pr\{\text{at least one false positive}\} = 1 - (1-\alpha)^{wc}$

where w is the number of monitoring locations and c is the number of constituents being tested for. For example, 100 tests run each at 5% type I error rate, and we would expect on average

$$100 \times 0.05 = \text{Five false positives}$$

The probability of at least one false positive = $1 - (0.95)^{100}$ = 99.4%. In other words, for an assessment for a large monitoring program, it is likely we would see at least one significant test result during statistical testing.

The Bonferroni technique (see Section 8.4) is conservative, in that the overall significance level (risk of at least one false positive) is generally smaller than the nominal rate although it will not be excessively so if the number of tests is small and the tests are not highly correlated with each other.

The use of t-tests can be carried out with critical values adjusted to account exactly for a known correlation structure between tests. When the correlations are induced by using the same denominator and background mean for tests with means of several downgradient wells, it is appropriate to use *multiple* comparisons with control (MCC) tests, of which the best known is Dunnett's test (1955).

10.8 Summary

ANOVA models are typically used to detect differences in observed phenomena rather than predict new values of a dependent variable. In large networks, even the parametric ANOVA has a difficult time finding the "needle in a haystack." The reason for this is that the ANOVA

F-test combines all downgradient wells simultaneously so that any "clean wells" are mixed together with one or more contaminated wells.

References to ANOVA techniques include Winer (1971), Hays (1973), Kerlinger and Pedhazur (1973), Milliken and Johnson (1984b), and Kufs (1992). Multiple discriminate analysis determines statistically, by an ANOVA, those linear combinations of variables that best discriminate among the different variables (e.g. Chappell, 1976).

10.9 References

Chappell, V.G. (1976). Determining watershed sub-areas with principal component analysis. *Water Resources Bulletin* 12 (6): 1133–1139.

Dunnett, C.W. (1955). A multiple comparisons procedure for comparing several treatments with a control. *Journal of American Statistics Association* 50: 1096–1121.

Hays, W.L. (1973). *Statistics for the Social Sciences*, 2e. New York: Holt, Rinehart and Winston.

Kerlinger, F.N. and Pedhazur, E.J. (1973). *Multiple Regression in Behavioral Research*. New York: Holt, Rinehart and Winston, Inc.

Kufs, C.T. (1992). Statistical modeling of hydrogeologic data – part I: regression and ANOVA models. *Groundwater Monitoring and Remediation* Spring: 120–130.

Milliken, G.A. and Johnson, D.E. (1984a). *Analysis of Messy Data*, Designed Experiments, vol. 1. Belmont, CA: Lifetime Learning Publications.

Milliken, G.A. and Johnson, D.E. (1984b). *Analysis of Messy Data*. New York: Van Nostrand Reinhold Co.

Nelson, L.S. (1987). Upper 10 percent, 5 percent and 1 percent points of the maximum F-ratio. *Journal of Quality Technology* 19: 165–167.

U.S. EPA (April 1989). Statistical analysis of ground-water monitoring data at RCRA facilities. Interim Final Guidance. Washington, DC: Office of Solid Waste, Waste Management Division.

Winer, B.J. (1971). *Statistical Principles in Experimental Design*. New York: McGraw-Hill Book Co.

10.10 Problems

10.1 You have identified there is a leaking gasoline storage tank, a tank that has been leaking for a sufficient period of time to have created a groundwater plume (remember that gasoline "floats" on the surface of groundwater as gasoline is lighter in density than water), as depicted in the figure below.

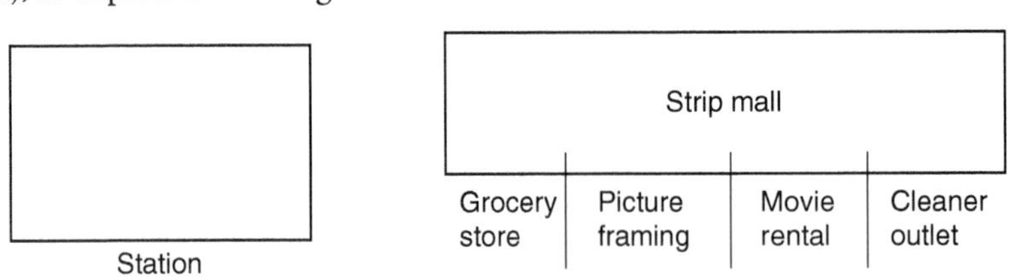

The adjacent commercial complex, as indicated in the figure, is a strip mall (e.g. contains a small grocery store, a picture framing store, a movie rental shop, and a dry cleaning outlet store). Develop a flowchart to indicate the data you would collect to allow you to calculate whether there is a significant human health risk related to the shops in the strip mall.

Monitoring location	Benzene concentrations (ug/l)						Statistics		
							Mean	Std. dev.	Skewness
	14 January	21 January	28 January	4 February	11 February	18 February			
A	28	14	12	22	17	26	19.8	6.52	0.09
B	7	6	4	6	7	4	5.7	1.37	−0.52
C	8	8	8	6	1	6	6.2	2.71	−1.79
A–B	21	8	8	16	10	22	14.2	6.40	0.30
A–C	20	6	4	16	16	20	13.7	6.98	−0.72
B–C	−1	−2	−4	0	6	−2	−0.5	3.45	1.64

10.2 There are three locations at which you have mercury concentration in groundwater. The number of samples at A equals 12, the number at B equals 14, and the number at C equals 16.

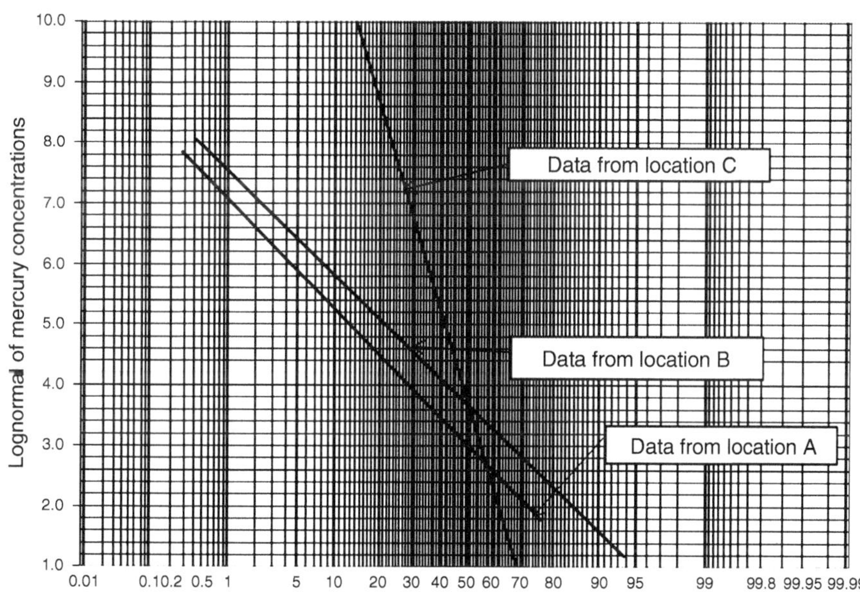

a) What are the geometric mean concentrations at A, B, and C?
b) Determine whether concentrations at locations A and B are statistically significantly different. Show the steps of your evaluation.
c) It would be more challenging to test the hypothesis whether A and C are statistically significantly different. Why?

10.3 Two locations on a stream separated by 0.5 km (a short distance) were monitored, yielding the results in the table below. Without transforming the data (i.e. use the data as is), determine whether any source is contributing between the two sampling locations such that there is a statistically significant difference between the two monitoring locations.

Chloride concentration (mg/l)	
Upstream location	**Downstream location**
0.35	1.2
2.90	2.8
0.37	1.6
0.48	0.50
0.74	0.55
0.61	0.65
0.36	0.90

Is the paired t-test a reasonable approach or can you use the standard t-test?

10.4 From a sampling program, we summarize the results obtained from two monitoring programs as defined below:
Location A: 2.6, 4.2, 3.0, 3.2, 2.9, 4.1
Location B: 3.1, 3.6, 3.1, 4.4, 3.6, 4.7

a) Is there a statistically significant difference between the results as monitored at Locations A and B? Assume $\alpha = 5\%$.
b) If we use the mean and variances calculated from the above data to predict results of additional samples at the same locations, how many samples would we need in order to show a statistically significant difference? What is a primary assumption upon which this calculation is based?

10.5 The following data have been obtained upstream and downstream of an agricultural field with significant seasonality in the runoff pattern of nitrogen from the field.

Sample	Location A	Location B
1	0.06	0.07
2	0.08	0.09
3	0.13	0.15
4	0.24	0.15
5	0.08	0.12
6	0.05	0.07
7	0.06	0.06
8	0.12	0.11
9	0.15	0.16

Use the paired t-test to determine if there is a statistically significant difference between the sampling results at the two monitoring locations.

10.6 You are assessing monitoring data obtained from sampling a treatment process (location B is downstream from location A). The results of the monitoring are as detailed in the table below.

A	B
3.6	4.1
3.7	3.8
3.9	3.9
3.7	4.3
3.73	4.03 Mean
0.13	0.22 Std. dev.

a) Is there a statistically significant difference between the results as monitored at locations A and B? Assume $\alpha = 5\%$.
b) What assumptions are you relying upon in your evaluation in (a)?
c) Just before you are about to submit your report, you are told the sample locations A and B are not independent sampling points Do not do the calculations, but indicate the calculations you would undertake to determine whether the treatment process is having an impact on the water quality, given this revised information.

10.7 What are the three major assumptions the data must satisfy to allow utilization of the hypothesis test to compare two sets of monitoring data?

10.8 Wastewater from a chemical plant needs to be treated to a larger extent than has historically been utilized. You have identified two possible treatment train sequences as depicted in the figure below, which might be effective. You arranged for one of your lab techs to complete the operation of the treatment alternatives and to collect samples at intervals, over a 6-week period, for benzene. The resulting values are summarized in the table below.
a) You first task is to evaluate whether the treatment process is effective at removal of the benzene. Undertake this task assuming 95% significance.

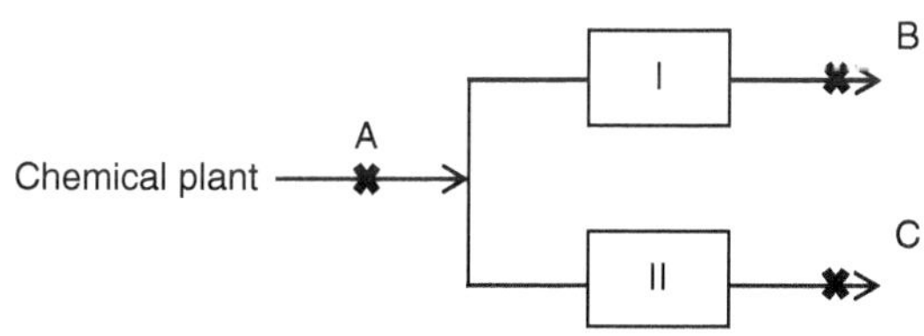

Monitoring location	Benzene concentrations (ug/l)					
	14 January	21 January	28 January	4 February	11 February	18 February
A	28	14	12	22	17	26
B	7	6	4	6	7	4
C	8	8	8	6	1	6
A–B	21	8	8	16	10	22
A–C	20	6	4	16	16	20
B–C	−1	−2	−4	0	6	−2

Statistics		
Mean	**Std. dev.**	**Skewness**
19.8	6.52	0.09
5.7	1.37	−0.52
6.2	2.71	−1.79
14.2	6.40	0.30
13.7	6.98	−0.72
−0.5	3.45	0.64

b) Are the two treatment sequences providing the same degree of benzene removal, with 95% confidence?

11

Testing Differences Between Monitoring Records When Censored Data Records Exist

The public has become used to conflicting opinion... many have come to feel that for every Ph.D., there is an equal and opposite Ph.D.

—Tim Hammonds, Food Marketing Institute

11.1 Introduction

The public is expressing a strong interest in decreasing their exposures to environmental contaminants. This considerable interest has arisen in part due to widespread reporting by the media about environmental quality concerns. As well, increasing stringency is being required in regulations to limit exposure risks. The result has been expansions in monitoring programs to include many more constituents. The stringency of environmental quality standards being promulgated is such that the maximum concentration limit (MCL) allowable is at, or very near to, the detection capability of laboratory instrumentation for some constituents. The consequence is many monitoring programs contain numerous censored data entries, or concentrations are reported as "less than" a method detection level (MDL).

The degree of censoring of environmental quality data is variable from one sampling program to another. Different laboratory procedures have varying limitations in their measurement capabilities. Other sources of difficulty arise due to matrix interferences, where the presence of one constituent influences the ability to quantify the concentrations of another constituent. A greater discussion of the rationale for different magnitudes of detection level is presented in Section 11.2; of present concern is an appreciation that the actual concentration of censored data can lie anywhere between zero and the method detection limit. In the absence of a specified magnitude, estimation of the means and standard deviations becomes difficult, and calculations using the t-test are less defensible. Of interest is how to deal with "less than" detection level values.

One approach to allow calculations of the statistics (e.g. the mean and the standard deviation) is to ignore all censored values. This gives a biased overestimate of the mean and a biased underestimate of the standard deviation.

Example 11.1

a) Assume the actual air quality concentrations of benzo(a)pyrene are 0.4, 1.2, 2.6, 0.3, 1.7, and 0.6 $\mu g/m^3$, and yet the available instrumentation cannot measure at levels less than 1. The resulting vector of concentrations would be reported as <1, 1.2, 2.6, <1, 1.7, and <1. Determine the impact on estimates of the mean and the standard deviation using the simple procedure of truncating the "less than" values.

Risk Assessment: Procedures and Protocols, First Edition. Edward A. McBean.

Companion website: www.wiley.com/go/McBean/RiskAssessProcedure_1e

The estimated mean and standard deviation using the actual concentrations are as follows:

For n = 6, $\overline{x} = 1.13$ and S = 0.89

Using the abbreviated dataset (truncating the censored data),

For n = 3, $\overline{x}_{trun} = 1.83$ and S = 0.71

The result is that the mean has increased and the standard deviation has decreased.

To ignore the values below the detection level, by truncating the dataset, is to ignore information. Another simple approach is to set the less than detection level magnitudes equal to the detection level, but this approach has similar faults to that of the preceding approach.

Example 11.1 (continued)

Redo the calculations for $\overline{x}$ and S with less thans set equal to the method detection level.

For n = 6 and censored dataset equal to the detection limit, $\overline{x} = 1.42$ and S = 0.64.

Other approaches exist to deal with censored data; these include the maximum likelihood method and regression analyses, but these approaches are more involved mathematically (e.g. see Sharma et al., 1995a, b). Clearly, some approach must be employed in the statistical analyses of environmental quality data, and the information must be extracted from the data when the problem of censored data exists; the appropriateness of the procedure to utilize in a particular circumstance is likely problem specific.

The objective in this chapter is to indicate some of the pros and cons associated with a number of alternative approaches to deal with censored data, particularly those applicable to brief datasets. Each of the approaches will be seen to have their own advantages and disadvantages. Different alternatives associated with different circumstances will be examined to identify a consensus as to which procedure to apply and when.

The focus in this chapter is on approaches that retain the parametric magnitudes of the data. Thus, these tests will utilize replacement of less than values by some parametric value. Chapter 12 deals with the approaches that are nonparametric or ranking approaches.

11.2 Alternative Types of Censoring

Various laboratory procedures have limitations in their measurement capabilities resulting in data censoring. In current environmental monitoring records, the frequency of occurrence of censored data may be very high. Reasons for this include the following:

1) Concerns about environmental quality are becoming more prevalent. Current sampling programs involve dramatically increased numbers of constituents over those required, for example, 10 years ago. Since many programs utilize a standard set of constituents for which analyses are required, many of the constituents are at very low levels.
2) As noted in Chapter 2, many environmental quality datasets are skewed to the left because concentration data are constrained on the low side (concentrations cannot be negative) but not on the upper side.
3) Improved analytical instrumentation now provides measurement capability of water sample constituents in the μg/l range and lower, while our awareness of the toxicological effects of various contaminants has widened.

The result is that many of the environmental quality datasets have left-censored data as depicted schematically in Figure 11.1.

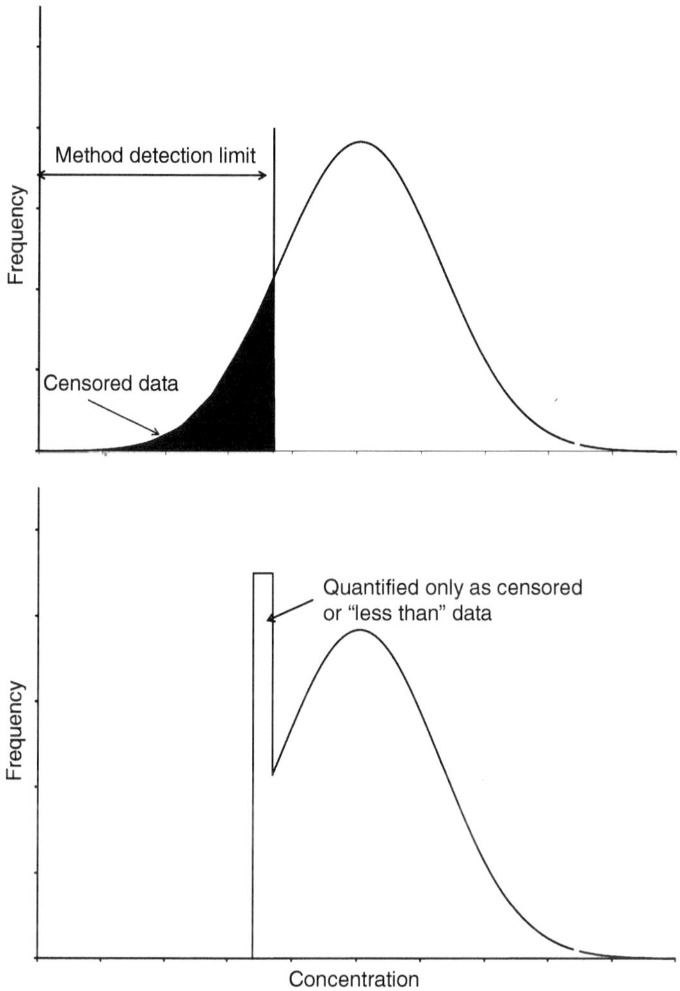

Figure 11.1 Influence of left censoring of concentration data.

It is important to recognize that uncertainty in reported values arises in the instrumentation for a variety of reasons:

1) Some degree of variability in test data is inevitably present in any chemical test, no matter how carefully conducted.
2) The results of detection limit calculations are less certain when the data have been generated by a number of different laboratories and/or over time where the technology for measurements has evolved. The effect of different standards, reagents, analysts, instruments, and lab protocols often result in biases among labs and variable detection limits.
3) Multiple measurements of the same sample by the same laboratory will give different results, which may lead to conflicting determinations.

All of these features relate to QA/QC where QA is defined as quality assurance, a definitive plan for laboratory operation that specifies the measures used to produce data of known precision and bias, and QC refers to quality control where a set of measures is utilized to ensure that the process is under control.

There have been numerous investigations of methods for estimating parameters for censored data drawn from assumed specified parent distributions such as the normal or lognormal (e.g. David, 1981). There have been far fewer studies of the application of these methods to environmental data for which parent distributions are unknown and sample sizes are small. When the frequency of data censoring is low, the approach utilized may not make a significant difference, but as the frequency of data censoring increases, the reliability of statistical analyses that can be applied to censored data decreases.

The complexity of statistical analyses increases substantially when multiple levels of detection limits exist in environmental quality records. As an example of this situation, see the values listed in Table 11.1 obtained from one of the monitoring wells at the Operating Industries Landfill site in California. The variation in detection levels as noted in these values is not unusual in monitoring data now being reported.

The analysis process normally detects and measures data below the reporting limit, but the difficulty is the quantification; the presence of the constituents is known but not the magnitude. The most frequently used of these limits of measurement are the detection and quantitation limits, but others exist. There are several detection and quantitation limit acronyms commonly used by laboratories, as apparent from Table 11.2.

Further complications in the statistical analysis of data arise with the data qualifiers. Examples of data qualifiers include those listed in Table 11.3.

Table 11.1 Example of a groundwater quality monitoring record with multiple detection values.

Date of sampling	Cadmium concentration (μg/l)
7 March 1985	<5
27 May 1985	<3.6
13 August 1985	<1.9
9 September 1986	6
5 February 1987	<0.2
13 August 1987	<2.4
14 October1987	<4
14 October 1987	5.4
12 May 1988	12.6
28 September 1988	<3.1
9 August 1989	<4
7 November 1989	<5
5 February 1990	<2.4
30 April 1990	<5
20 June 1991	<3
15 May 1992	<3.7
19 July 1993	<1.5

Source: CH2M-Hill (1994).

Table 11.2 Indication of detection and quantitation limits.

Limit of detection (LOD)	The lowest concentration that can be determined to be statistically different from a blank
Method detection limit (MDL)	The minimum concentration of an analyte that can be determined with 99% confidence that the true value is greater than zero. The measure is suitable only for detecting the presence of a pollutant is actually present. The MDL is estimated on the basis of ideal laboratory conditions with ideal analyte samples and does not account for matrix or other interferences encountered when analyzing specific, actual field samples
Instrument detection limit (IDL)	Smallest signal above background noise that an instrument can detect reliably. This measure is based on a signal-to-noise ratio and takes into account no chemical or other preparatory steps involved in an analysis
Limit of quantitation (LOQ)	The concentration above which quantitative results can be obtained with a specified degree of confidence
Method quantification limit (MQL)	The minimum concentration of a substance that can be measured and reported
Practical quantitation limit (PQL)	The lowest level that can be reliably achieved within specified limits of precision and accuracy during routine laboratory operating conditions. The "PQL" is the concentration at which laboratories should be able routinely to determine quantitatively how much of a pollutant is present. For this reason, the PQL should be taken as the most reasonable upper bound for nondetect concentrations. PQL is time and laboratory independent for regulatory purposes, e.g. numerous current NPDES permits and cleanup levels for US sites include limits for PCBs at or even below the PQL. Compliance with such limits may be the result of random chance as much as environmental diligence (Scroggin, 1994)

Source: After Anne (1992).

Table 11.3 Explanation of some of the data qualifiers commonly used in relation to monitoring data.

Data qualifier	
"U"	This qualifier indicates that the compound was analyzed for, but not detected at, the reported concentration and is commonly referred to as "not detected." This means that the reported concentration is the quantitation limit (or reporting limit). The signal characteristic for the constituent being analyzed could not be observed or distinguished from "background noise" during laboratory analyses. For example, 3 μg/l of cadmium means that the sample contains less than 3 μg/l of cadmium
"B"	This qualifier means the chemical was identified in the sample and in the method blank analyses with the sample, which means that all or part of the reported concentration may be due to laboratory contamination. For example, consider bis(2-ethylhexyl) (or "bis" for the sake of simplicity) phthalate reported as 2B μg/l. bis is a common contaminant associated with sample collection apparatus and an analysis of a blank sample (one that is collected for checking of the lab QA/QC) is 1.5 μg/l. One should suspect the bis in the sample is due to laboratory contamination, not environmental contamination. It is left to the judgment of the data user, not the laboratory
"J"	This flag indicates that the reported concentration is an estimated value. It may be referred to as "estimated" for one of a number of reasons including the concentration is a tentatively identified compound, the quantitation is below the reporting limit, or the sample exceeded the holding time
	For example, if the sample quantitation limit is 10 μg/l but a concentration of 3 μg/l is calculated, the value may be reported 3 J

The detection limit is the lower limit at which you can differentiate that measurements are quantitatively meaningful (Taylor, 1987). Additional comments include that the method of detection limit (MDL) is similar and in some cases the same as the limit of detection (LOD). The only difference is the LOD is defined with a sample blank, whereas the MDL is defined with either a sample blank or with one or several matrix samples.

The practical quantitation limit (PQL) is usually equal to the MDL multiplier by a factor of 3–5 (National Archives and Records Administration, 1987) but varies for different chemicals and different samples.

The most important interpretive features regarding data qualifiers are, in summary, as follows:

1) Below the MDL – The results do not demonstrate the presence of the chemical.
2) Between the MDL and PQL – The constituent is present, but we are unsure of the magnitude.
3) Above the PQL – Take the concentration as given.

11.3 Alternative Procedures for Statistical Analysis of Environmental Quality Datasets

A number of procedures for examining data records with varying degrees of censorship have been published. The majority of these procedures assume the data are singly censored, e.g. <5 as opposed to multiply censored datasets (e.g. the monitoring record consists of concentrations reported as <5 as well as <15). Gilbert and Kinnison (1981), Gilliom and Helsel (1986), Gleit (1985), Helsel and Gilliom (1986), and Owen and deRouen (1980) all examine procedures for estimating parameters based on singly censored datasets.

11.3.1 Simple Substitution Methods

Consider the difficulties that arise when using the t-test for data with censored data. From Chapter 9, recall that a simple form of the t-test evaluation for the purposes of explanation is written as

$$t^* = \frac{\overline{x}_1 - \overline{x}_2}{S/\sqrt{n}} \tag{11.1}$$

where $\overline{x}_1$ and $\overline{x}_2$ are the means, S is the standard deviation, and n is the number of samples. When the data are constant or nearly so, this leads to a zero or near-zero standard deviation. When the dataset contains a high percentage of censored values and these censored values are transformed to some constant value, the standard deviation in (11.1) approaches zero. As S approaches zero, the calculation of t^* in (11.1) either requires a division by zero or leads to ill-conditioning, a situation in which very small changes in the value of a single observation produce enormous changes in the value of t^*.

The development of this problem as a result of a simple transformation to some constant value may occur as indicated in Example 11.2.

Example 11.2 Assume that the true concentrations are as follows, namely, 3.6, 4.2, 1.7, 5.0, and 4.1, but that the instrumentation has an MDL of 5.0. The resulting reported concentrations would be <5, <5, <5, 5.0, and <5. Compute the standard deviations for both situations.

For actual data, S = 1.24.

For censored data, S = 0.0 for the simple substitution rule of replacing the "less thans" with values "equal to."

Extension of the situations in which the less thans are set equal to is the *simple substitution methods*. These methods set all censored observations equal to zero, or some fraction of the detection limit, e.g. the MDL, 1/2 MDL, or zero.

Example 11.3 If the monitoring record consists of the following eight values, compute the mean and standard deviation

$$< 5, 7, 8, 12, < 5, < 5, 6, 9$$

using the simple substitution rule.

For the values reported as <MDL, use both 0 and MDL to calculate a mean and standard deviation and report the value of mean as a range.
For values reported as <MDL taken as zero

$$\overline{x} = 4.75 \quad \text{and} \quad S = 4.43$$

For <MDL taken as equal to the MDL

$$\overline{x} = 7.13 \quad \text{and} \quad S = 2.47$$

The mean is between the interval 4.75 and 7.13 and the standard deviation is in the range of 2.47–4.43.

If a small proportion of the observations are not detected, one approach is to replace the censored data with a small number, e.g. MDL/2. The US EPA recommends the MDL/2 approach if less than 15% of all samples are nondetect, but these simple substitution methods tend to perform poorly in statistical tests when the nondetect percentage is substantial (Gilliom and Helsel, 1986). Any of these simple substitution procedures provide a degree of quantification but may seriously affect subsequent utilization of the parameters in the t-test. Hashimoto and Trussell (1983) compared several estimators of the mean for censored water quality data. Their examples illustrate the bias caused by three commonly used methods: discarding censored observations, setting all censored observation equal to zero, or assigning the detection limit to all censored observations.

11.3.2 Test of Proportions

The test of proportions is a procedure that does not account for potentially different magnitudes among the concentrations of detected values. Rather, each sample is treated as a 1 or 0 depending on whether the measured concentration is above or below the detection limit. The test of proportions relies on a normal probability approximation to the binomial distribution of 0s and 1s and assumes that the sample size is reasonably large. Generally, if the proportion of detected values is denoted by P, and the sample size is n, then the normal approximation is adequate, provided that nP and n(1 − P) both are greater than or equal to 5.

The test of proportions is appropriate for use when the proportion of quantified values is small to moderate (e.g. 10–50%) or if more than 50% of the data are below detection but at least 10% of the observations are quantified.

A test of proportions might be used, for example, if none of the background well observations are above the MDL, but if all of the compliance well observations are above the detection limit, one would suspect contamination.

Assuming the findings at two monitoring locations are being examined for significant difference, the calculation steps involved in the procedure are as follows:

Step 1: Determine w, the number of background well samples in which the compound was detected. Let n be the total number of samples obtained from the background wells (u is the number of background or upgradient samples). Compute the proportion of detects as

$$P_u = w / n \tag{11.2}$$

Step 2: Determine v the number of downgradient well samples in which the compound was detected, out of m total samples:

$$P_d = v / m \tag{11.3}$$

Step 3: Compute the standard error of the difference in proportions:

$$S_D = \left\{ \left[\frac{w+v}{n+m} \right] \left[1 - \frac{w+v}{n+m} \right] \left[\frac{1}{n} + \frac{1}{m} \right] \right\}^{0.5} \tag{11.4}$$

and form the statistic

$$Z = \frac{P_u - P_d}{S_D} \tag{11.5}$$

Step 4: Compare the absolute value of Z from (10.5) to a value (e.g. 97.5%) from the standard normal distribution. For the 97.5%, $Z_c = 1.96$. If the absolute value of Z exceeds $Z_c = 1.96$, this provides statistically significant evidence at the 5% significance level (i.e. a two-sided test) that the proportion of compliance well samples where the compound was detected exceeds the proportion of background well samples where the compound was detected.

If the percentage of nondetects is very high and/or the sample size is fairly small, the test of proportions may lead to inaccurate results.

Example 11.3 (continued) Five samples out of 26 upgradient samples showed detection levels, whereas 8 out of 33 downgradient samples showed detection. Is there a statistically significant difference between the upgradient and downgradient monitoring results?

$$P_u = 5 / 26 = 0.192$$

$$P_d = 8 / 33 = 0.242$$

$$S_D = \left\{ \left[\frac{5+8}{26+33} \right] \left[1 - \frac{5+8}{26+33} \right] \left[\frac{1}{26} + \frac{1}{33} \right] \right\}^{0.5}$$
$$= 0.017$$

$$Z = \frac{0.192 - 0.242}{0.017} = -2.94$$

$$|Z| = 2.94 > 1.96 = Z_c$$

Therefore, there is a statistically significant difference between the results at the upgradient and downgradient monitoring locations.

11.3.3 Plotting Position Procedure

Use of probability paper provides a simple means of estimating the mean and standard deviation in the presence of censored data (McBean and Rovers, 1984). The limitation of this method arises in that sufficient detected data must exist to convince the analyst that the correct type of probability paper is being utilized and that a reasonable line can be fit to the plotted data. This procedure involves fitting the probability distribution to the data above the detection limit and then extrapolating the line to, in turn, estimate the mean and standard deviation, as depicted schematically in Figure 11.2. Potential candidate distributions include the normal (see Chapter 5) and lognormal (see Chapter 6) since the probability plotting papers are readily available and the distributions are generally appropriate to environmental quality data. For alternative distributions, see the discussion in Chapter 7.

The distributional assumptions for parametric procedures can be rather difficult to check when a substantial fraction of nondetects exist. To utilize this procedure, the following steps are utilized:

Step 1: Rank-order the detect data and, utilizing a plotting position formula such as the Weibull formula, plot the data on probability paper:

Rank	Plotting position
(m)	$m/(n+1)$

where m is the rank of the sample and n is the total number of samples.

Step 2: Extrapolate the fitted line and read the mean and standard deviation from the graph (the mean concentration corresponds to the 50th percentile and the standard deviation corresponds to the difference between the 84th and the 50th percentile, since the standard deviation for the normal distribution corresponds to approximately 34% of the area under the normal distribution curve).

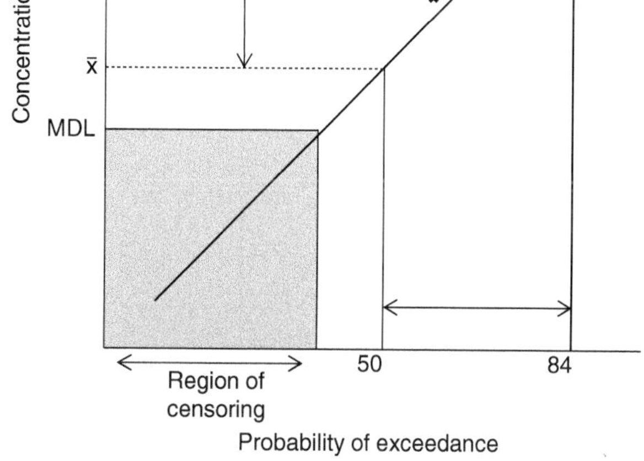

Figure 11.2 Extrapolation of fitted line from probability paper to estimate the mean and standard deviation.

To be successful, a sufficient portion of the data must be in excess of the detection limit to establish that the data are reasonably described by the assumed probability distribution. As well, there must be sufficient quantities of information from which the fitted line can be drawn. For example, with a brief dataset, one large value can substantially bias the resulting line (McBean and Rovers, 1984). Mage (1982) describes several techniques for plotting ranked observations on probability paper and drawing lines.

A significant advantage of the probability plotting procedure is that it gives a visual assessment of the data. The analyst can downplay the effect of an individual data point if there is a valid rationale for presuming the particular point is a data outlier. The disadvantages of probability plots are that there is no commonly accepted rule for plotting the data and drawing lines by eye and there is no simple objective way to judge how well the data points conform a straight line (Mage, 1982).

Example 11.4 Consider the monitoring values as summarized below. Use the plotting position procedure to estimate the mean and standard deviation. Monitoring data are as follows.

I	II	III	IV
Monitoring data	**Rank**	**Rank-ordered data**	**Plotting position**[a]
<5	1	10.2	0.077
10.2	2	9.2	0.153
9.2	3	7.8	0.231
<5	4	7.2	0.308
<5	5	5.9	0.385
7.2	6	5.4	0.462
5.3	7	5.3	0.538
<5	8	<5	0.615
7.8	9	<5	0.692
<5	10	<5	0.769
5.4	11	<5	0.846
5.9	12	<5	0.923

[a] $m/(n+1)$.

The plot of the data in columns III and IV is illustrated in Figure 11.3. From the dashed line, the mean is estimated to be 5.4 μg/l. The 84th percentile is 8.8 with the result that the standard deviation is estimated as $S = 8.8 - 5.4 = 3.4$ μg/l.

11.3.4 Cohen's Test

Another procedure for estimating the mean and variance of a censored normal distribution was presented by Cohen (1961). Cohen's test may be applied up to the situation of 90% nondetects. This approach assumes that the observed data (detects and nondetects) come from the same normal or lognormal population but that the nondetect values have been censored at their detection limits. The premise of the technique is based on the censored probability model where we estimate new or adjusted mean and standard deviations.

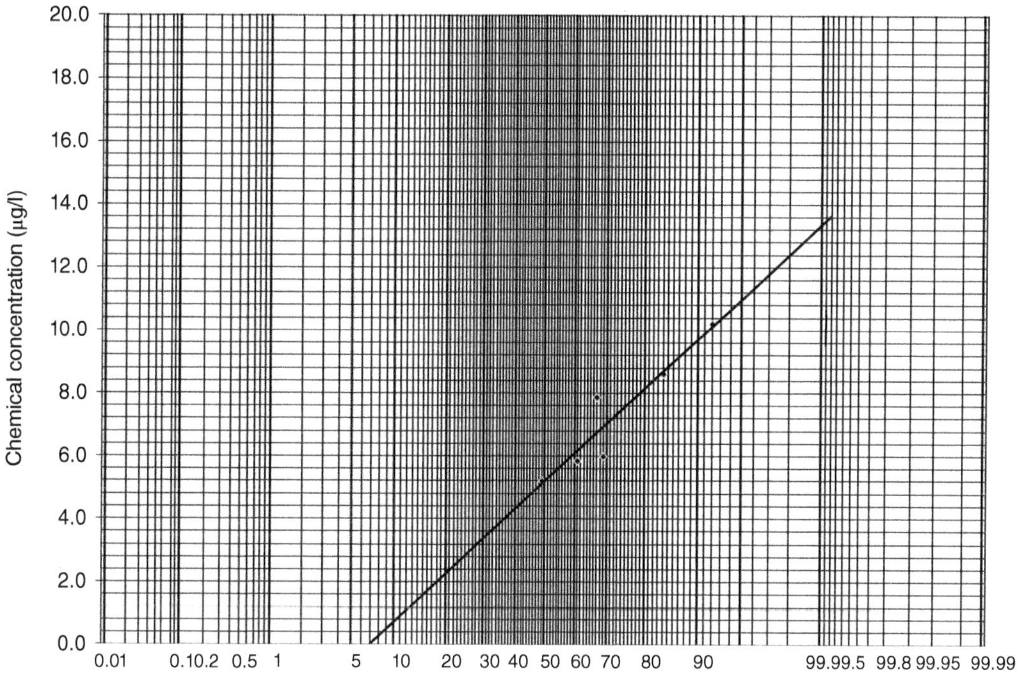

Figure 11.3 Plotting position procedure to estimate the mean and standard deviation.

Cohen's method adjusts the sample mean and sample standard deviation to account for data below the detection limit, assuming the data are normally distributed and the detection limit remains the same.

Step 1: Compute the sample mean $\overline{x}_d$ from the data above the detection limit as follows:

$$\overline{x}_d = \frac{1}{m}\sum_{i=1}^{m} x_i \tag{11.6}$$

Step 2: Compute the sample variance S_d^2 from the data above the detection limit as follows:

$$S_d^2 = \frac{\sum_{i=1}^{m}\left(x_i - \overline{x}\right)^2}{\left(m-1\right)} \tag{11.7}$$

Step 3: Compute the two parameters h and γ as

$$h = \frac{n-m}{n} \tag{11.8}$$

$$\gamma = \frac{S_d^2}{\left(\overline{x} - DL\right)^2} \tag{11.9}$$

where DL is the detection limit.

Use these to determine the value of the parameter $\overline{x}$ from the Table A.11.

Step 4: Estimate the corrected sample mean, which accounts for the data below the detection limit as

$$\bar{x} = \bar{x}_d - \bar{\lambda}(\bar{x}_d - DL) \quad (11.10)$$

and

$$S = \left(S_d^2 + \bar{\lambda}(\bar{x}_d - DL)^2\right)^{0.5} \quad (11.11)$$

Example 11.5 The monitoring record for benzene consists of the following:

$$10.1, <5, <5, 8.9, 6.2, 5.6, <5\,\mu g/l$$

Use Cohen's method to estimate the sample mean and standard deviation.

Step 1: Compute the sample mean $\bar{x}_d$ for the data above the detection limit:

$$\bar{x}_d = \frac{10.1 + 8.9 + 6.2 + 5.6}{4} = 7.7$$

Step 2: Compute the sample standard deviation for data above the detection limit:

$$S_d^2 = 4.62$$

Step 3: Compute $h = \frac{7-4}{7} = 0.429$

Compute $\gamma = \frac{4.62}{(7.7-5)^2} = 0.633$

$$\bar{\lambda} = 0.817 \quad \text{(interpolating from Table A.11)}$$

Step 4: Estimate the corrected sample mean:

$$\bar{x} = 7.7 - 0.817(7.7 - 5) = 5.49$$

Correct the standard deviation as

$$S = \left[4.62 + 0.817(7.7 - 5)^2\right]^{0.5} = 3.25$$

Cohen's method provides maximum likelihood estimates of the mean and variance of a censored normal distribution. Cohen's adjustment may not give valid results if the proportion of nondetects exceed 50%. McNichols and Davis (1988) found that the false positive rate associated with the use of t-tests based on Cohen's method rose substantially when the fraction of nondetects was greater than 50%. This occurred because the adjusted estimates of the mean and standard deviation are more highly correlated as the percentage of nondetects increases, leading to less reliable statistical tests.

11.3.5 Aitchison's Method

When at least 10% of the groundwater samples have a measurable (detected) value of a particular constituent, the mean and variance of the distribution can be approximated using a method due to Aitchison (1955). Like Cohen's method, the assumption of a particular probability model for the data leads to adjusted estimates of $\bar{x}$ and S. However, Aitchison's adjustment is constructed on the assumption that the nondetect samples are free of contamination, so that all nondetects may be regarded as zero concentrations. To compute Aitchison's adjustment

(Aitchison, 1955), it is assumed that the detected samples follow an underlying normal distribution or if the detects are found to be lognormally distributed, the entire computation can be carried out using the logarithms of the data instead. Let d be the number of nondetects and n the total number of samples (detects and nondetects). Then, if $\overline{x}_d$ and S_d denote, respectively (where these are calculated as per Equations [11.6] and [11.7]), the same sample mean and standard deviation of the detected values, the adjusted overall mean $\hat{x}$ (where the ^ indicates the sample mean is an estimate) is

$$\hat{x} = \left(1 - \frac{d}{n}\right)\overline{x}_d$$

and the adjusted overall standard deviation, $\hat{S}$, is:

$$\hat{S} = \left[\left(\frac{n-(d+1)}{n-1}\right)(S_d)^2 + \frac{d(n-d)(\overline{x}_d)^2}{n(n-1)}\right]^{0.5}$$

Example 11.6 Utilizing the same data as in Example 11.5, estimate the sample mean and standard deviation employing Aitchison's method:

$$d = 3 = \text{number of nondetects}$$

$$n = 7$$

$$\overline{x}_d = 7.7$$

$$S_d = \sqrt{4.62} = 2.15$$

$$\hat{x} = (1 - 3/7)7.7 = 4.40$$

$$\hat{S} = \left[\frac{7-4}{6}(2.15)^2 + \frac{3}{7}\left(\frac{4}{6}\right)(7.7)^2\right]^{0.5} = 4.39$$

It is interesting to compare these adjusted statistics with those determined using Cohen's method.

	Cohen's method	Aitchison's method
Mean	5.49	4.40
Standard deviation	3.25	4.39

The lower mean of Aitchison's method in comparison with Cohen's method is due to the assumption within Aitchison's method that nondetect samples are free of contamination.

Example 11.7 Utilize Aitchison's method to adjust the estimate of the mean and standard deviation of the data listed below:

$$<5, 5, <5, 7.2, 12, 66, 6.1, 5.9\ \mu g/l$$

and then compute the 95% upper tolerance level.

Apparent from the detection data listed is that the data distribution is skewed (the value of "66"). Therefore, the data should be log-transformed prior to analysis.

Reported value	Log-transformed (natural logarithm)
<5	—
6	1.79
<5	—
<5	—
7.2	1.97
12	2.48
66	4.19
6.1	1.81
5.9	1.77

$$\overline{x} = 2.34$$

$$S_d = 0.946$$

$$\hat{x} = \left(1 - \frac{3}{9}\right)(2.34) = 1.56$$

Reversing the transformation gives $e^{1.56} = 4.76$

$$\hat{S} = \left[\frac{9-4}{8}(0.946)^2 + \frac{3}{9}\left(\frac{6}{8}\right)(2.34)^2\right]^{0.5} = 1.39$$

The upper 95% tolerance interval (two-sided test) is n = 9, degrees of freedom = 9 − 1 = 8, and the t-test.

First, using the log-transformation

$$TI = 1.56 \pm 2.31(1.39) = 11.95$$

In arithmetic terms

$$TI = e^{11.95} = 154.8\ \mu g/l$$

11.3.6 Maximum Likelihood Procedure

The maximum likelihood estimator (MLE) procedure as a way of incorporating the effect of "less thans" in censored datasets is based on an assumption of the underlying distribution of the entire dataset. MLE estimators of distributional parameters can then be derived from the uncensored observations. This procedure is useful for data with censored values but is considerably more complicated than the preceding alternatives and is best accomplished when there are large datasets. For small sample sizes, MLE are not necessarily minimum variance, unbiased estimates.

Owen and deRouen (1980) addressed the problem of estimating a mean from censored air contaminant data. They used Monte Carlo techniques to evaluate the performance of MLE methods derived for the lognormal distributions. The technical literature also includes Holland and Fitz-Simmons (1982) who describe a computer program for fitting statistical distributions to air pollution data using the maximum likelihood estimation. Sharma et al. (1995a, b) utilize the MLE for parameter estimation in air pathways migration models.

There are several methods for handling regression with censored data. These include the iterative least squares (ILS) method (Schmee and Hahn, 1979), the linear unbiased estimate method (Nelson and Hahn, 1973), and the MLE method (Dempster et al., 1979; Haas and Jacangelo, 1993). The ILS method is simpler to implement than the MLE method and has good statistical convergence properties (Schmee and Hahn, 1979). The ILS method was developed for parameter estimation of a linear model with right censored data that arise from life tests.

11.4 Multiple Detection Limits

There has been very little reported research in the environmental quality field using techniques to handle multiply censored datasets (Millard and Deveral, 1988). However, the continuing evolution of analytical chemistry techniques has brought about the ability to detect increasingly smaller concentrations of chemicals in the environment such that a fairly common feature is now one of occurrence of multiple detection limits.

There are at least three possible causes of multiple detection limits:

1) The limit of detection of a particular analyte depends upon the method that is used to measure it. There may be more than one method available, and each method may be optimal (have the smallest percent measurement error) in a certain range of analyte concentration. For example, the protocol may call for method 1 to be used if the specific conductance is above a certain threshold and method 2 if a specific conductance is below a threshold value.
2) A second cause of multiple detection limits involves the process of dilution, resulting in varying detection limits.
3) Decreasing detection limits over time as the measurement techniques improve.

Utilization of standard parametric statistical methods such as t-tests and multiple regression becomes difficult with multiply censored datasets.

Statisticians in the field of survival analysis and life testing have developed numerous techniques for analyzing multiply censored datasets (e.g. Kalbfleisch and Prentice, 1980). Multiple correlation procedures may be useful in this problem to infill or estimate some of the data.

11.5 References

Aitchison, J. (1955). On the distribution of a positive random variable having a discrete probability mass at the origin. *Journal of the American Statistical Association* 50: 901–908.

Anne, D.C. (May 1992). Know your detection and quantitation limits. *Pollution Engineering* 1: 68–71.

CH2M-Hill. (1994). Remedial Investigation Report for Operating Industries, Inc. Landfill, Monterey Park, California, report to U.S. EPA, San Francisco, CA, 25 October 1994.

Cohen, A. (1961). Tables for maximum likelihood estimates from singly truncated and singly-censored samples. *Technometrics* 3: 535–541.

David, H.A. (1981). *Order Statistics*. New York: John Wiley & Sons, Inc.

Dempster, A.P., Laird, N.M., and Rubin, D.B. (1979). Maximum likelihood from incomplete data via the EM algorithm. *Journal of the Royal Statistical Society, Series B* 39 (1): 1–38.

Gilbertk, R. and Kinnison, R.R. (1981). Statistical methods for estimating the mean and variance from radionuclide data sets containing negative, unreported or less-than values. *Health Physics* 40: 377.

Gilliom, R.J. and Helsel, D.R. (1986). Estimation of distributional parameters for censored trace level water quality data 1. Estimation techniquest. *Water Resources Research* 22: 135–146.

Gleit, A. (1985). Estimation for small normal data sets with detection limits. *Environmental Science and Technology* 19: 1201–1206.

Hashimoto, L.K. and Trussell R.R. (1983). Evaluating water quality data near the detection limit. Paper presented at the Advanced Technology Conference, American Water Works Association, Las Vegas, NV (5–9 June 1983).

Haas, C. and Jacangelo, J.G. (1993). Development of regression models with below-detection data. *Journal of Environmental Engineering* 119: 214–230.

Helsel, D.R. and Gilliom, R.J. (1986). Estimation of distributional parameters for censored trace level water quality data 2. Verification and applications. *Water Resources Research* 22 (February): 147–155.

Holland, D. and Fitz-Simmons, T. (1982). Fitting statistical distributions to air quality data by the maximum likelihood method. *Atmospheric Environment* 16 (5): 1971–1076.

Kalbfleisch, J.D. and Prentice, R.L. (1980). *The Statistical Analysis of Failure Time Data*. New York: John Wiley.

Mage, D.T. (1982). An objective graphical method for testing normal distributional assumptions using probability plots. *The American Statistician* 36 (2): 116–120.

McBean, E.A. and Rovers, F.A. (1984). Alternatives for handling detection limit data in impact assessments. *Ground Water Monitoring Review* Spring: 42–44.

McNichols, R.J. and Davis, C.B. (1988). Statistical issues and problems in ground water detection monitoring at hazardous waste facilities. *Ground Water Monitoring Review* Fall: 135–150.

Millard, S.P. and Deveral, S.J. (1988). Nonparametric statistical methods for comparing two sites based on data with multiple nondetect limits. *Water Resources Research* 24 (12): 2087–2098.

National Archives and Records Administration. (1987). Federal Register, Volume 52, No. 130, 8 July 1987.

Nelson, W. and Hahn, J. (1973). A Problem in the statistical comparison of measuring devices. 15 (4).

Owen, W.J. and DeRouen, T.A. (1980). Estimation of the mean for lognormal data containing zeroes and left-censored values, with applications to the measurement of worker exposure to air contaminants. *Biometrics* 36 (4): 707–719. doi: 10.2307/2556125.

Schmee, J. and Hahn, G.J. (1979). A simple method for regression analysis with censored data. *Technometrics* 21: 417–432.

Scroggin, D.G. (1994). Detection limits and variability in testing methods for environmental pollutants: misuse may produce significant liabilities. *Hazardous Waste and Hazardous Materials* 11 (1): 1–4.

Sharma, M., Thomson, N.R., and McBean, E.A. (1995a). Linear regression analyses with censored data: estimation of PAH washout ratios and dry deposition velocities to a snow surface. *Canadian Journal of Civil Engineering* 22 (4): 819–833.

Sharma, M., McBean, E.A., and Thomson, N.R. (1995b). Maximum likelihood method for parameter estimation with below-detection data. *ASCE – Journal of the Environmental Engineering Division* 11: 776–784.

Taylor, J.K. (1987). *Quality Assurance of Chemical Measurements*. Chelsea, MI: Lewis Publishers Inc.

11.6 Problems

11.1 What is the purpose of Cohen's test and what is the primary assumption implicit in the test?

11.2 For the following water quality data, estimate the mean and standard deviation using Cohen's test procedure:

$1.7, < 5, 63, 8.3, < 5, 5.5, < 5, 6.7$

For information on this procedure, see Section 11.3.4 of the textbook. Works best when 15–50% of the data are nondetects.

11.3 Your professional team has collected four samples from a river, one per quarter over the last year. The arsenic concentrations are as listed in the table below. The river is used as a water source that is used for drinking by cottage dwellers located along the river.

a) Utilize the probability plot method to estimate the 95% exceedance level (i.e. the concentration below which you expect 95% of the arsenic samples to be less than).
b) Are your data reasonably described by the normal distribution? How do you know?
c) What is the variance of the data, as determined from the probability plot method? Indicate on the diagram how you determined you value of the variance.
d) Compute the probability of the residents developing cancer as a result of exposure to the arsenic, assuming you take the 95% exceedance level, and they spend a total of 35 days/year, every year, at their cottage, and drink 3 l of water per day when at their cottage. Assign other necessary values as needed describing your reasoning for your assigned values you select.

Sample	As concentration (μg/l)
1	71
2	55
3	52
4	27

11.4 For the following water quality data, estimate the mean and standard deviation using Cohen's test procedure:

8.2,5.3,< 5,< 5,10.1,9.3,7.6,< 5

For information on this procedure, see Section 11.3.4. Works best when 15–50% of the data are nondetects.

11.5 For the following water quality data, estimate the mean and standard deviation using Aitchison's test:

8.2,5.3,< 5,< 5,10.1,9.3,7.6,< 5

11.6 For the following water quality data, estimate the mean and standard deviation using the probability plot procedure:

1.7,< 5,63,8.3,< 5,5.5,< 5,6.7

For information on this procedure see Section 11.3.3. Works best when 15–50% of the data are nondetects.

11.7 For the following water quality data, estimate the mean and standard deviation using Aitchison's test procedure:

1.7,< 5,63,8.3,< 5,5.5,< 5,6.7

11.8 Use the Mann–Whitney test to compare the two groups.

Control	Treatment
12	4
11	9
9	1
6	13
12	10
9	18
10	17
9	14
6	12
8	16

11.9 Background and downgradient monitoring results are summarized below. Determine if there is a significant difference between the background and monitoring results.

	Downgradient		
Background	MWA	MWB	MWC
12	26	13	26
10	8	18	12
8	11	17	13
14	12	15	17
16	13	18	21
14	18	21	22

11.10 Use the Kruskal–Wallis test to determine whether there is evidence of contamination at the site, given the data as listed below.

	Downgradient		
Background	MWA	MWB	MWC
<10	15	16	17
<10	<10	<10	13
11	<10	<10	<10
13	18	12	<10
15	26	<10	21
11	20	21	20

12

Nonparametric Procedures

12.1 Introduction

Previous chapters have focused on the available procedures for addressing hypothesis testing questions and the identification of trends. However, as noted in Chapter 10, there are significant difficulties with these procedures when a substantial percentage of censored data exist within the dataset and/or when there is a failure to adhere to assumptions implicit within the parametric tests. The result is that many existing environmental quality databases are not amenable to analyses by the parametric methods described previously. Many techniques are not suitable because of missing data, censored data, and changing laboratory techniques. As well, many of these procedures cannot be applied to data that are clearly nonnormal in distribution.

A group of tests have been devised in which no assumptions are made about the distribution of the observations. For this reason, the tests are referred to as distribution-free. An array of these alternative tests exists, collectively referred to as "nonparametric" tests. The fundamental characteristic of these procedures is that the ranks of the data are utilized instead of data values. The analyses of data on ranks are a direct parallel of the more traditional parametric methods. For data that do not fulfill the necessary assumptions for the parametric analyses, the nonparametric methods are as powerful or, in fact, more powerful than the equivalent parametric tests.

A large body of technical literature exists based on analyses of the ranks of the data. The mathematics involved in nonparametric tests are generally very straightforward. As well, quite different from the parametric procedures examined in Chapters 9 through 11, for the nonparametric procedures, there is practically no knowledge needed whatsoever about the distribution of the population. Thus, in the event that one or more of the assumptions implicit in the t-test are violated (e.g. there is a significant difference between the variances), an alternative class of procedures involving nonparametric analyses may be utilized.

The nonparametric methods usually have the additional advantage that they require less burdensome calculations since they are generally not related specifically to the parameters of a given distribution. The major disadvantage of nonparametric methods is that they may be wasteful of information and usually they have a smaller efficiency than the corresponding parametric methods, provided that the assumptions of the standard (parametric) methods can be met. Hamby (1994) discusses many of the features of the nonparametric tests. In large normally distributed samples, the nonparametric tests have an efficiency of approximately 95% relative to Student's t-test (Hodges and Lehman, 1946), and in small normally distributed samples, the signed-rank test (see Section 12.2.2) has been shown to have an efficiency slightly higher than this (Klotz, 1963). With nonnormally distributed data, the efficiency of the nonparametric tests relative to the t-test never falls below 86% in large samples and may be greater than 100% for distributions that have long tails (Hodges and Lehman, 1946).

Risk Assessment: Procedures and Protocols, First Edition. Edward A. McBean.

Companion website: www.wiley.com/go/McBean/RiskAssessProcedure_1e

Example 12.1 The concentrations measured at a groundwater monitoring location are as follows: 13.2, 12.1, 2.1, 2.0, and 1.9. Assign the ranks for these five measurements.

The assignments of the ranks are as follows.

Concentration	Rank
13.2	1
12.1	2
2.1	3
2.0	4
1.9	5

All subsequent statistical analyses are now done using the "rank" information implying equal spacing between successive values. Note that the relative proximity information (e.g. the relative grouping of the data into two high and three relatively low values) is lost. This loss of information content is, however, the rationale for subsequently not having to be concerned with the probability distribution of the concentration data.

If two or more observations are tied, each of the observations is assigned the mean of the ranks they would jointly occupy.

Rank or nonparametric tests are generally tests that do not require any assumptions other than independent samples from continuous populations. During the last 40 years, the role played by nonparametric tests has undergone considerable change. In the 1950s and 1960s, it was quite common to characterize nonparametric methods as "rough and ready." This attitude has changed where now practically every introductory statistics text contains a chapter concerning nonparametric or distribution-free or rank methods (Noether, 1981).

The nonparametric tests include the Wilcoxon–Mann–Whitney test, the Kruskal–Wallis test, the Wilcoxon signed-rank test, the Friedman test, Spearman's rho, and others. Since the tests are relatively quickly completed, the rank tests are highly useful for the investigator who is doubtful that the data can be regarded as normal.

12.2 Single Comparison Procedures

The single comparison procedures are for testing the null hypothesis that the two samples come from identical continuous populations against the alternative that the populations have different means. A highly efficient class of nonparametric tests of this and similar hypothesis is based on rank sums; that is, the observations are assigned ranks according to their order of magnitude and the tests are performed on the basis of various sums of these ranks.

12.2.1 Mann–Whitney Test

The Mann–Whitney test or U test is a highly efficient nonparametric test based on rank sums to compare the means of two independent samples. The Wilcoxon rank sum test for two groups' test is essentially the same as the Mann–Whitney test. The Mann–Whitney U statistic is a linear transformation of the Wilcoxon test and therefore an equivalent statistic. Probability theory does not depend on the distribution type when operating on ranks (Conover, 1980). The data in each of the two samples are first ordered from lowest to highest values, and then the combined dataset is ranked. The ranks are then summed for each group.

The null hypothesis is that the two samples are taken from a common population so that there should be no consistent difference between the two sets of rankings. Denoting the respective sample sizes by "n" and "m" and the sum of the ranks occupied by the first sample by R_1, it can be shown that the mean and the variance of the sampling distribution of the statistic are as follows:

$$U = nm + \frac{n(n+1)}{2} - R_1 \tag{12.1}$$

which are given by

$$\bar{U} = \frac{nm}{2} \tag{12.2}$$

and

$$S_U^2 = \frac{nm(n+m+1)}{2} \tag{12.3}$$

If both n and m are greater than 8, the distribution of the U statistic can be approximated closely by a normal distribution and hence the test can be based on the statistic:

$$Z = \frac{U - \bar{U}}{S_U} \tag{12.4}$$

With knowledge of Z, conclusions can then be made as to whether the null hypothesis of identical population can or cannot be rejected.

If one sum of the single-sample ranks is sufficiently larger than the other, then the two sample means are determined to be significantly different.

Example 12.2 Concentrations at two locations were monitored as tabulated below.

MW1	1.3	0.9	0.8	0.2	0.4	0.6	0.1	5.1	0.2	
MW2	1.7	3.5	7.8	0.9	0.7	2.6	0.2	1.5	15.3	0.7

Utilize the Mann–Whitney test to determine if there is a statistically significant difference between the monitoring results at the two locations.

First, arrange the 19 observations according to size, retaining the sample identity of each observations. Then, we assign the ranks 1, 2, 3, ... and 19 to the observations, as shown below.

Monitoring well	MW1	MW1	MW1	MW2	MW1	MW1	MW2	MW2	MW1
Observation	0.1	0.2	0.2	0.2	0.4	0.6	0.7	0.7	0.8
Rank	1	3	3	3	5	6	7.5	7.5	9

Monitoring well	MW1	MW2	MW1	MW2	MW2	MW2	MW2	MW1	MW2
Observation	0.9	0.9	1.3	1.5	1.7	2.6	3.5	5.1	7.8
Rank	10.5	10.5	12	13	14	15	16	17	18

Monitoring well	MW2
Observation	15.3
Rank	19

Note that if two or more observations are tied in rank, we assign each of the observations the mean of the rank they jointly occupy.

If we denote the respective sample sizes by n_1 and n_2 and the sum of the ranks occupied by the first sample by R_1, it can be shown that the mean and the variance of the sampling distribution of the statistic

$$U = n_1 n_2 + \frac{n_1(n_1+1)}{2} - R_1 = 9 \cdot 10 + \frac{9 \cdot 10}{2} - 66.5 = 68.5$$

are given by

$$\bar{U} = \frac{n_1 \cdot n_2}{2} = \frac{9 \cdot 10}{2} = 45.0$$

and

$$SU^2 = \frac{n_1 n_2 (n_1 + n_2 + 1)}{12} = \frac{9 \cdot 10 \cdot 12}{12} = 150$$

Thus

$$Z = \frac{68.5 - 45.0}{\sqrt{150}} = 1.93$$

Since this value falls between −1.96 and 1.96, the critical values for a two-sided alternative (95%) and $\alpha = 0.05$ (see Table A.1), we conclude that the null hypothesis of identical populations cannot be rejected.

12.2.1.1 Use of the Mann–Whitney Test to Test Equality of Variance

If the ranks of the data are assigned in a different way from that outlined above, the same U statistic can also be used to test the null hypothesis of identical populations against the alternative that the populations have unequal dispersions or variances. For this type of test, the ranks are assigned "from both ends toward the middle" by assigning rank 1 to the smallest observation, ranks 2 and 3 to the largest and second largest observation, and so on. All other aspects of this test for unequal dispersions are identical with those of the Mann–Whitney U test.

12.2.2 Spearman's Rank Correlation Coefficient

Spearman's rank correlation coefficient procedure is a very simple ranking technique. Consider the situation where two laboratories are completing parallel analyses on groundwater quality. The results from analyses of 10 different tests (repetitions for the same constituent) are shown in Table 12.1 where the entries indicate the relative quality (as determined by each laboratory) for each of the samples. Therefore, for example, laboratory 1 identified sample 2 as having the worst water quality, sample 1 as the second worst, and so on. Of concern is, do the laboratories show evidence of agreement among themselves in regard to ranking? This can be addressed by calculating Spearman's rank correlation coefficient as

$$R = 1 - \frac{6\Sigma d^2}{n^3 - n} \tag{12.5}$$

Table 12.1 Ranking in accordance with Spearman's rank correlation coefficient.

Sample	1	2	3	4	5	6	7	8	9	10
Laboratory 1	2	1	3	4	6	5	8	7	10	9
Laboratory 2	3	3	2	4	6	7	5	9	10	8
Rank difference = d	1	2	−1	0	0	2	−3	2	0	−1
Square of the difference = d^2	1	4	1	0	0	4	9	4	0	1

where n is the number of samples ranked and d is the rank difference.

For the values in Table 12.1

$$\Sigma d^2 = 24$$

so

$$R = 1 - \frac{6 \cdot 24}{1000 - 10} = 0.85 \quad (12.6)$$

The rank correlation coefficient procedure has been designed so that when the two rankings are identical, R has the value +1 and when the rankings are as greatly in disagreement as possible (i.e. one ranking is exactly the reverse of the other), R has the value −1.

To examine whether this measure of agreement could have arisen by chance, the significance of R may be tested (provided n > 10):

$$t^* = R\sqrt{\frac{n-2}{1-R^2}} \quad (12.7)$$

with (n − 2) degrees of freedom.

In the example,

$$t^* = 0.85\sqrt{\frac{10-2}{1-(0.85)^2}} = 4.6$$

with 8 degrees of freedom so that the level R = 0.85 did not likely arise by chance. This demonstrates agreement in the findings of the laboratories.

12.2.3 Sign Test for Paired Observations

This test represents an alternative to the paired t-test. For example, if a sequence of x's and y's correspond to an observation and a model's prediction, respectively, then individual values

$$x_1, x_2, \ldots, x_n \quad \text{and} \quad y_1, y_2, \ldots, y_n$$

The sign test does not depend on the size of numerical values of the differences but only on their sign or rank order. The assumption of normality is not required in order for this test to be valid. If the difference between the values is expected to equal zero, we would expect approximately one-half of the differences to be negative. If a small proportion of the differences are either negative or positive, we suspect that there might be a real difference but the magnitude of the proportion is the question.

The true proportion of positive (or negative) signs equal to p = 1/2. Hence, in n independent trials in which a positive or negative sign for each pair is determined,

$$b(x) = \frac{n!}{x!(n-x)!}\left(\frac{1}{2}\right)^n \quad (12.8)$$

With paired observations, the sign test provides a simple test of the null hypothesis that x and y have a common distribution (as opposed to the alternative that x is less than y or x is greater than y).

To examine this, consider a new variable z defined as

$$Z_i = x_i - y_i \tag{12.9}$$

If the null hypothesis is true, then

$$\Pr\left[Z_i > 0 \middle| Z_i \neq 0\right] = \Pr\left[Z_i < 0 \middle| Z_i \neq 0\right] = 0.5$$

The probabilities in [ ] are made conditional on $Z_i \neq 0$ to emphasize that pairs for which $Z_i = x_i - y_i = 0$ can be disregarded.

A reasonable test statistic is, S, the number of Z_i's that are greater than zero. If the alternative hypothesis is true, a smaller value of S would occur than if the null hypothesis were true. If there are m Z_is not equal to zero ($m < n$), then with the null hypothesis, S has a binomial distribution and

$$\Pr\left[S < s\right] = \sum_{k=0}^{S} \frac{m!}{k!(m-k)!}\left(\frac{1}{2}\right)^m \tag{12.10}$$

12.3 Multiple Comparison Procedures

The Kruskal–Wallis H test is utilized for testing whether k samples come from identical populations against the alternative that the populations have unequal means when there are three or more samples. This test represents a nonparametric alternative to the ANOVA test. The null hypothesis of the H test is that the samples have been taken from populations with identical distributions and any differences between the samples are due to chance variation inherent in the process of random sampling. Hence, this is utilized for testing whether k samples come from identical populations against the alternative that the populations have unequal means.

12.3.1 Kruskal–Wallis Test (or Nonparametric ANOVA)

The Kruskal–Wallis H test for deciding whether k independent samples come from identical populations is conducted in a manner similar to the U test. As with the U test, the observations are ranked jointly, and if R_1 is the sum of the ranks occupied by the n_i observations of the ith sample, the test is based on the statistic

$$H = \frac{12}{n(n+1)}\left[\Sigma \frac{R_1^2}{n_1} + \frac{R_2^2}{n_2} + \cdots + \frac{R_k^2}{n_k}\right] - 3(N+1) \tag{12.11}$$

where $n_1 + n_2 + \cdots + n_k$ are the number of observations in each of the k groups.

When $n_i > 5$ for all i and the null hypothesis holds, the distribution of the H statistic is well approximated by the chi-square distribution with $k - 1$ degrees of freedom. There should be at least three groups with a minimum sample size of 3 in each group.

The Kruskal–Wallis H test is utilized when the data or the residuals have been found to be significantly different from normal and when a log-transformation fails to adequately normalize the data. This test does not imply assumptions about the underlying distribution.

In a one-way nonparametric ANOVA, the assumption under the null hypothesis is that the data from each monitoring location come from the same continuous distribution and hence have the same mean concentration of a constituent. The Kruskal–Wallis test is equivalent except that the test is performed using ranks, as opposed to concentrations.

If the monitoring locations are found to differ, post hoc comparisons are needed to determine if contamination is present.

Define the number of groups by k and the number of observations in each group by n_i with N being the total number of well observations. Let x_{ij} denote the jth observation in the ith group, where j runs from 1 to the number of observations in the group, n_i and i = 1, …, k groups.
The steps of the procedure are then as follows:

Step 1: The analysis is initiated by ranking all the observations from the least to the greatest from the combined groups to be analyzed. Let R_{ij} denote the rank of the jth observation in the ith group. As a convention, denote the background monitoring location(s) as group 1. Ties are given the average rank of the tied values.
Step 2: Compute the sum of the rank values for each group and the average rank within each group (k groups). The sum of the ranks for the ith group is defined as R_i, and the average rank for each $\overline{R}_i = R_i / n_i$.
Step 3: Compute the average rank of the overall dataset.
Step 4: Calculate the Kruskal–Wallis test statistic, H, as

$$H = \frac{12}{N(N+1)}\left[\frac{R_1^2}{n_1} + \frac{R_2^2}{n_2} + \cdots + \frac{R_k^2}{n_k}\right] - 3(N+1) \tag{12.12}$$

where $n_1, n_2, \ldots, n_k$ are the number of observations in each group.
Step 5: Compare H with appropriate chi-square critical value χ^2, with degrees of freedom − k − 1 where k is the number of groups. If H is greater than χ^2 this indicates there is a significant difference between at least two of the well groups. The null hypothesis is rejected if the computed value exceeds the tabulated critical value.
Step 6: If the H statistic is significant, then individual comparisons are computed between the background (group 1) and compliance wells. If the computed value exceeds the value from the chi-square table, compute the critical difference for well comparisons to the background, assumed to be group 1.
Step 7: Compute the differences of average ranks from each group to the average rank of background.
Step 8: Calculate the critical difference

$$C_i = Z_{\alpha|(k-1)}\left[\frac{N(N+1)}{12}\right]^{0.5}\left[\frac{1}{n_1} + \frac{1}{n_i}\right]^{0.5}$$

where $Z_{\alpha|(k-1)}$ is the upper $\alpha|(k-1)$ percentile from the standard normal distribution, with zero mean and unit variance, and $\alpha = 5\%$.
The null hypothesis of the H test is that the samples have been from populations with identical distributions. Any differences between the samples are due to variations inherent in the process of random sampling.
Step 9: Compare each rank difference to C. The differences of the average ranks for each group to the background are compared with the critical values found in step 8 to determine which wells indicate evidence of contamination. That is, compare $R_i - R_1$ to C_1 for i taking the values 2 through k (remembering that group 1 is the background).

Example 12.3 Suppose that the experiments with the results are shown below. Note that tied observations are again assigned the mean of the ranks they jointly occupy.

MW1	0.2	0.3	0.4	0.5	1.7	1.9	2.0
MW2	0.8	1.1	1.3	1.9	2.5	7.8	
MW3	0.7	0.9	8.2	12.0	12.1	15.3	
MW4	0.1	0.1	0.3	0.5	2.9	12.8	

The observations in the first sample set occupy ranks 3, 4.5, 6, 7.5, 14, 15.5, and 17 so that $R_1 = 67.5$. Similarly, the observations in the second sample occupy ranks 10, 12, 13, 15, 5, 18, and 20 so that $R_2 = 88.5$. Continuing, $R_3 = 111.0$ and $R_4 = 58.0$. Substituting into the formula for H, we obtain

$$H = \frac{12}{25 \cdot 26}\left(\frac{67.5^2}{7} + \frac{88.5^2}{6} + \frac{111^2}{6} + \frac{58^2}{6}\right) - 3(26) = 6.4$$

And if we compare this figure with 7.815 (from Table A.12, the value of $\chi^2_{0.05}$ with 3 degrees of freedom), we find that the null hypothesis cannot be rejected. In other words, we cannot reject the null hypothesis that the sample comes from identical populations against the *alternative that the population means are not all equal.*

Example 12.4 The data below indicate the benzene concentrations at three locations in a manufacturing facility. The null hypothesis is that the three samples are from the sample population.

Location 1		Location 2		Location 3	
Concentration	Rank	Concentration	Rank	Concentration	Rank
23	7	11	3	110	18
15	4	17	5	84	16
42	12	31	11	85	17
8	1	27	9.5	45	13
10	2	63	14	64	15
18	6	27	9.5	25	8
$R_1 = 32$		$R_2 = 52$		$R_3 = 87$	
$n_1 = 6$		$n_2 = 6$		$n_3 = 6$	

The data are ranked overall from lowest to highest. Identical values are given the mean of the ranking they would otherwise have received.

The sum of the ranks is noted for each of the sampling locations. It is then calculated as

$$\begin{aligned} H &= \frac{12}{N(N+1)}\sum_{i=1}^{3}\frac{R^2}{n_i} - 3(N+1) \\ &= \frac{12}{18 \cdot 19}\left(\frac{32^2}{6} + \frac{52^2}{6} + \frac{87^2}{6}\right) - 3(19) \\ &= 8.90 \end{aligned}$$

When sampling locations all contain more than five individual values, the sampling distribution of H is virtually identical to the sampling distributions of chi-square with 2 degrees of freedom (Table A.12), which is 5.99. Since the calculated H is larger than the critical value, the null hypothesis is rejected. In other words, it is safe to assume that the benzene concentrations are different in at least at some of the three sampling locations.

Next calculate the critical value to compare with locations 2 and 3 with 1:

$$C_i = Z_{\alpha|(k-1)}\left[\frac{N(N+1)}{12}\right]^{1/2}\left[\frac{1}{n_1} + \frac{1}{n_i}\right]^{1/2}$$

There are two comparisons to be made. For $\alpha = 0.05$, we find the upper 0.05/2 = 0.025 percentile of the standard normal distribution to be 1.96 (from Table A.1). The critical values for sampling locations C_2 is

$$C_2 = 1.96\left[\frac{18(19)}{12}\right]^{1/2}\left[\frac{1}{6}+\frac{1}{6}\right]^{1/2}$$
$$= 3.45$$

The critical value for sampling locations C_3 is also 3.45. Compute the differences.

Differences		Critical value
Between 1 and 2	52 − 32 = 20	3.45
Between 1 and 3	87 − 32 = 55	3.45

Comparing the differences with the critical differences indicates that both sampling locations 2 and 3 are significantly different from location 1.

12.3.2 Special Consideration of the Kruskal–Wallis Test

Equation (12.12) should be modified if there are any tied ranks in the dataset. To allow for tied ranks, the calculated value of H is divided by a correction factor:

$$1-\frac{\Sigma\left(t^3-t\right)}{n^3-n} \tag{12.13}$$

where t is the number of individual values involved in each set of tied ranks. The summation $\Sigma(t^3 - t)$ must be calculated for each set of tied ranks.

Example 12.5a Use the Kruskal–Wallis test to determine whether there is evidence of contamination at the site, given the monitoring records listed below.

	Concentrations					
	Background wells			Compliance wells		
Sampling time	Well 1	Well 2	Well 3	Well 4	Well 5	Well 6
1	<5	<5	8.1	<5	<5	<5
2	<5	<5	<5	12.1	13.4	<5
3	<5	6.9	<5	<5	8.6	15.6
4	7.7	<5	<5	8.0	<5	18.1

Step 1: Compute the overall percentage of nondetects. In this case, nondetects account for 15/24 = 63% of the data. More than 15% of the records are "less thans," and therefore parametric ANOVA should not be utilized. Use the Kruskal–Wallis test instead, pooling all three background wells into one group and treating each compliance well as a separate group.

Step 2: Compute ranks for all the data including the tied observations (e.g. nondetects) as in the following table. Note that each nondetect is given the same mid rank, equal to the average of the first 14 unique ranks.

Concentrations						
	Background wells			Compliance wells		
Sampling time	Well 1	Well 2	Well 3	Well 4	Well 5	Well 6
1	8.0	8.0	19	8.0	8.0	8.0
2	8.0	8.0	8.0	21	22	8.0
3	8.0	16	8.0	8.0	20	23
4	17	8.0	8.0	18	8.0	24
Rank sum		$R_1 = 124$		$R_4 = 55$	$R_5 = 58$	$R_6 = 63$
Rank mean		$\overline{R_1} = 10.3$		$\overline{R_4} = 13.8$	$\overline{R_5} = 14.5$	$\overline{R_6} = 15.8$

Step 3: Calculate the sums of the ranks in each group (R_i) and the mean ranks in each group ($\overline{R}_i$).
Step 4: Compute the Kruskal–Wallis statistic H using the formula

$$H = \frac{12}{N(N+1)} \sum_{i=1}^{k} \frac{R_i^2}{n_i} - 3(N+1)$$

where N is the total number of samples, n_i is the number of samples in the ith group, and k is the number of groups:

$$N = 24, k = 4$$

$$H = \frac{12}{24(25)} \left(\frac{124^2}{12} + \frac{55^2}{4} + \frac{58^2}{4} + \frac{63^2}{4} \right) - 3(25)$$
$$= 0.20(1281 + 756 + 841 + 993) - 75 = 2.42$$

Step 5: Compute the adjustment for ties. There is only one group of distinct tied observations, containing 15 samples. The adjusted Kruskal–Wallis statistic is given by

$$H' = \frac{2.42}{1 - \left(\frac{15^3 - 15}{24^3 - 24} \right)} = 3.20$$

Step 6: Compare the calculated value of H' to the tabulated chi-square value with (k – 1) = (# of groups – 1) = 4 – 1 df, $\chi^2_{3,0.05} = 7.81$. Since the observed value of 3.20 is less than the chi-square critical value, there is insufficient difference between the well groups. Post hoc pairwise comparisons are unnecessary.

Example 12.5b Consider now the situation as if part (a) of the example had not been carried out and the monitoring results were obtained as indicated in the modified table below, the change being the addition of results from well 7.

Sampling time	Background wells	Well 7
1	Values as listed in Example 12.5a	24.9
2		18.2
3		29
4		24.9

Step 1: The overall percentage of nondetects is 16/28 = 57%. More than 15% are nondetect and therefore use the Kruskal–Wallis test.
Step 2: Computation of ranks.

Sampling time	1	2	3	4	5	6	7
1	8.0	8.0	19	8.0	8.0	8.0	26.5
2	8.0	8.0	8.0	21	22	8.0	25
3	8.0	16	8.0	8.0	20	23	28
4	17	8.0	8.0	18	8.0	24	26.5
Rank sum =		12.4		55	64	63	106
Rank mean =		10.3		13.8	16.0	15.6	26.5

Step 3: Sum of ranks and mean ranks as noted:

Step 4: $$H=\left[\frac{12}{N(N+1)}\sum_{i=1}^{k}\frac{R_i^2}{n_i}\right]-3(N+1)$$

$$N=28, k=5$$

$$H=\frac{12}{28(29)}\left[\frac{124^2}{12}+\frac{55^2}{4}+\frac{64^2}{4}+\frac{63^2}{4}+\frac{106^2}{4}\right]-3(29)$$
$$=0.015[1281+756+1024+992+2809]-3(29)$$
$$=102.9-87=15.9$$

Step 5: $$H'=\frac{15.9}{1-\left[\frac{15^3-15}{28^3-28}\right]}=\frac{7.5}{1-\frac{3\ 360}{21\ 924}}=18.8$$

Step 6: For k – 1 = 5 – 1 = 4 df, $\chi^2_{4,0.05}=9.49$

Since the observed value of 18.8 is greater than the chi-square critical value, there is evidence of significant differences between the well groups. Post hoc comparisons are needed.
Step 7: Calculate the critical difference for compliance well comparison to the background, when the background wells are taken as group 1:

$$C_i=Z_{\alpha\ |\ (k-1)}\left[\frac{N(N+1)}{12}\right]^{1/2}\left[\frac{1}{n_1}+\frac{1}{n_i}\right]^{1/2}$$

for i = 2, …, k. $Z_{\alpha|k-1}$ is the upper $Z_{\alpha|k-1}$ percentile from the standard normal distribution. Since the number of samples at each compliance well is four, the same critical difference can be used for each comparison, namely,

$$C_i=\frac{Z_{0.05}}{4}\left(\frac{28(29)}{12}\right)^{1/2}\left(\frac{1}{12}+\frac{1}{4}\right)^{1/2}\quad \frac{Z_{0.05}}{4}=Z_{0.0125}$$
$$=2.24(8.23)(0.577)$$
$$=10.64$$

Step 8: Form the differences between the average ranks of each compliance well and the background and compare these differences to the critical value of 10.64:

Well 4: $\bar{R}_4 - \bar{R}_1 = 13.8 - 10.3 = 3.5$
Well 5: $\bar{R}_5 - \bar{R}_1 = 16.0 - 10.3 = 5.7$
Well 6: $\bar{R}_6 - \bar{R}_1 = 15.6 - 10.3 = 5.3$
Well 7: $\bar{R}_7 - \bar{R}_1 = 26.5 - 10.3 = 16.2$

Since the average rank difference at well 7 exceeds the critical difference, there is significant evidence of contamination at well 7 but not at wells 4, 5, and 6.

Note: In normal circumstances, it would not be necessary to do the parallel steps in Example 12.5a and b. The presentation here was done solely for the purposes of explanation.

Example 12.6 For the values in Example 12.5a, there is one set of tied ranks, involving two values. Calculate the correction factor:

$$1 = \frac{2^3 - 2}{18^3 - 18} = 1 - \frac{6}{5814} - 0.999$$

The value of H from Example 12.5a from Equation (12.12) is 8.90; corrected for ties, this becomes

$$\frac{8.90}{0.999} = 8.91$$

This represents a very small adjustment. Unless more than a quarter of the values are tied ranks, the effect of the correction factor is small.

When small samples are involved ($n_i < 6$), the χ^2 approximation can no longer be used as a reliable procedure. In these situations, the sampling distribution of h must be used. Critical values of H for use with three small samples are listed in Table A.10. If more than five monitoring locations are involved, the individual companies should be performed using ZA = 0.01 rather than $Z_{\alpha|k-1}$.

12.4 References

Conover, W.J. (1980). *Practical Nonparametric Statistics*, 2e. New York: John Wiley.
Hamby, D.M. (1994). A review of techniques for parameter sensitivity analysis of environmental models. *Environmental Monitoring and Assessment* 32: 135–154.
Hodges, L. and Lehmann, E.L. (1946). The use of previous experience in reaching statistical decisions. *Annals of Mathematical Statistics* 23: 396–407.
Klotz, J. (1963). Small sample power and efficiency for the one sample. *Annals of Mathematical Statistics* 34 (2): 624–632.
Noether, G.E. (1981). Comment. *The American Statistician* 35 (3): 129.

12.5 Problems

12.1 Groundwater concentrations at two locations were located upgradient and downgradient of a landfill as detailed below. Use the Mann–Whitney test to determine if there is a reasonable basis to determine that the landfill is leaking.

Sampling time	Upgradient location (μg/l)	Downgradient location (μg/l)
November 1999	27	29
December 1999	34	45
January 2000	145	132
June 2000	132	145
December 2000	75	54
January 2001	21	22
July 2001	17	56

Well	U	U	D	U	D	U	D	D	D	U	U	D	U	D
Obs.	17	21	22	27	29	34	45	54	56	75	132	132	145	145
Rank	1	2	3	4	5	6	7	8	9	10	11.5	11.5	13.5	13.5

12.2 Ibuprofen is an emerging contaminant now periodically being monitored in the wastewater outfalls in Ontario. The results of the monitoring are listed in the below table.

Month and year	Ibuprofen at location A (μg/l)	Ibuprofen at location B (μg/l)
August 2010	92	112
December 2010	<10	13
August 2011	<10	18
December 2011	15	213
August 2012	21	17
December 2012	<10	17

a) Determine whether the August 2010 data point at location A is likely an outlier.
b) Assuming the ibuprofen data for location A can be described by a lognormal distribution, and assuming the August 2010 data is not an outlier, estimate the mean at location A using the probability plotting paper method.
c) The Mann–Whitney test could be employed to determine if the ibuprofen concentrations are higher at A or B or can be considering to be the same. Why do you think the Mann–Whitney test would likely be an appropriate test for determining whether the quality of Ibuprofen in the wastewater A is different from that in wastewater B? Prepare the data into a table for purposes of the Mann–Whitney test (but do not try to complete the testing procedure).

12.3 Strawberries are imported (hypothetically) from a country named Elbonia. You are concentered with the number of *Escherichia coli* pathogens that have been reported on strawberries from this source. You have requested your staff to undertake and analyze, and they have reported back to you, the results indicated in the table below. You are interested in determining the probability that the number of *E. coli* exceeds 30. Without doing the calculations, itemize the procedures you would rely upon, including a ranking as to which procedures are likely to provide the best estimate.

	Microorganism	Microorganism
E. coli	X	Y
12	22	31
28	15	52
36	<5	75
<5	17	12
38	<5	78
12	12	26
14	15	29
<5	21	15
36	20	71
42	<5	112

12.4 You have measures of three constituents in the air quality within a room, as detailed in the table immediately below.

Constituent A	Constituent B	Constituent C
12	22	31
28	15	22
36	<10	13
12	17	23
38	<10	14
12	12	20
14	15	23
13	21	30
36	20	27
42	<14	14

You are concerned with estimating the probability of constituent B exceeding 25 ug/m^3. List the alternative calculation procedures you might employ to make this determination and assign priorities (1 is best) for your feeling as to appropriateness of alternative procedures listed:

Approach – Priority – Major limitation or concern for the approach in this problem

Appendix A

Risk Assessment: Procedures and Protocols, First Edition. Edward A. McBean.

Companion website: www.wiley.com/go/McBean/RiskAssessProcedure_1e

Table A.1 Areas under the normal distribution curve.

$$F(z)=\int_0^z \frac{1}{\sqrt{2\pi}} e^{\frac{1}{2}z^2} d_x$$

z	0.00	0.01	0.02	0.03	0.04	0.05	0.06	0.07	0.08	0.09
0.0	0.0000	0.0040	0.0080	0.0120	0.0159	0.0199	0.0239	0.0279	0.0319	0.0359
0.1	0.0398	0.0438	0.0478	0.0517	0.0557	0.0596	0.0636	0.0675	0.0714	0.0753
0.2	0.0793	0.0832	0.0871	0.0910	0.0948	0.0987	0.1026	0.1064	0.1103	0.1141
0.3	0.1179	0.1217	0.1255	0.1293	0.1331	0.1368	0.1406	0.1443	0.1480	0.1517
0.4	0.1554	0.1591	0.1628	0.1664	0.1700	0.1736	0.1772	0.1808	0.1844	0.1879
0.5	0.1915	0.1950	0.1985	0.2019	0.2054	0.2088	0.2123	0.2157	0.2190	0.2224
0.6	0.2257	0.2291	0.2324	0.2357	0.2389	0.2422	0.2454	0.2486	0.2518	0.2549
0.7	0.2580	0.2611	0.2642	0.2673	0.2704	0.2734	0.2764	0.2794	0.2823	0.2852
0.8	0.2881	0.2910	0.2939	0.2967	0.2995	0.3023	0.3051	0.3078	0.3106	0.3133
0.9	0.3159	0.3186	0.3212	0.3238	0.3264	0.3289	0.3315	0.3340	0.3365	0.3389
1.0	0.3413	0.3438	0.3461	0.3485	0.3508	0.3531	0.3554	0.3577	0.3599	0.3621
1.1	0.3643	0.3665	0.3686	0.3708	0.3729	0.3749	0.3770	0.3790	0.3810	0.3830
1.2	0.3849	0.3896	0.3888	0.3907	0.3925	0.3944	0.3962	0.3980	0.3997	0.4015
1.3	0.4032	0.4049	0.4066	0.4082	0.4099	0.4115	0.4131	0.4147	0.4162	0.4177
1.4	0.4192	0.4207	0.4222	0.4236	0.4251	0.4265	0.4279	0.4292	0.4306	0.4319
1.5	0.4332	0.4345	0.4357	0.4370	0.4382	0.4394	0.4406	0.4418	0.4430	0.4441
1.6	0.4452	0.4463	0.4474	0.4485	0.4495	0.4505	0.4515	0.4525	0.4535	0.4545
1.7	0.4554	0.4564	0.4573	0.4582	0.4591	0.4599	0.4608	0.4616	0.4625	0.4633
1.8	0.4641	0.4649	0.4656	0.4664	0.4671	0.4678	0.4686	0.4693	0.4699	0.4706
1.9	0.4713	0.4719	0.4726	0.4732	0.4738	0.4744	0.4750	0.4756	0.4762	0.4767
2.0	0.4772	0.4778	0.4783	0.4788	0.4793	0.4798	0.4803	0.4808	0.4812	0.4817

2.1	0.4821	0.4826	0.4830	0.4843	0.4838	0.4842	0.4846	0.4850	0.4854	0.4857
2.2	0.4861	0.4865	0.4868	0.4871	0.4875	0.4878	0.4881	0.4884	0.4887	0.4890
2.3	0.4893	0.4896	0.4898	0.4901	0.4904	0.4906	0.4909	0.4911	0.4913	0.4916
2.4	0.4918	0.4920	0.4922	0.4925	0.4927	0.4929	0.4931	0.4932	0.4934	0.4936
2.5	0.4938	0.4940	0.4941	0.4943	0.4945	0.4946	0.4948	0.4949	0.4951	0.4952
2.6	0.4953	0.4955	0.4956	0.4957	0.4959	0.4960	0.4961	0.4962	0.4963	0.4964
2.7	0.4965	0.4966	0.4967	0.4968	0.4969	0.4970	0.4971	0.4972	0.4973	0.4974
2.8	0.4974	0.4975	0.4976	0.4977	0.4977	0.4978	0.4979	0.4980	0.4980	0.4981
2.9	0.4981	0.4982	0.4983	0.4983	0.4984	0.4984	0.4985	0.4985	0.4986	0.4986
3.0	0.4987	0.4987	0.4987	0.4988	0.4988	0.4989	0.4989	0.4989	0.4990	0.4990
3.1	0.4990	0.4991	0.4991	0.4991	0.4992	0.4992	0.4992	0.4992	0.4993	0.4993
3.2	0.4993	0.4993	0.4994	0.4994	0.4994	0.4994	0.4994	0.4995	0.4995	0.4995
3.3	0.4995	0.4995	0.4996	0.4996	0.4996	0.4996	0.4996	0.4996	0.4996	0.4997
3.4	0.4997	0.4997	0.4997	0.4997	0.4997	0.4997	0.4997	0.4997	0.4998	0.4998
4.0	0.499968									

Source: Adapted from Weatherburn (1957) and Bowker and Lieberman (1959).

Table A.2 Percentiles of student's t-distribution (df = degrees of freedom) (for one-sided test).

F/df	0.60	0.75	0.90	0.95	0.975	0.99	0.995	0.9995
1	0.325	1.000	3.078	6.314	12.706	31.821	63.657	636.619
2	0.289	0.816	1.886	2.920	4.303	6.965	9.925	31.598
3	0.277	0.765	0.633	2.353	3.182	4.541	5.841	12.941
4	0.271	0.741	1.533	2.132	2.776	3.747	4.604	8.610
5	0.267	0.727	1.476	2.015	2.571	3.365	4.032	6.859
6	0.265	0.718	1.440	1.943	2.447	3.143	3.707	5.959
7	0.263	0.711	1.415	1.895	2.365	2.998	3.499	5.405
8	0.262	0.706	2.397	1.860	2.306	2.896	3.355	5.041
9	0.261	0.703	1.383	1.833	2.262	2.821	3.250	4.781
10	0.260	0.700	1.372	1.812	2.228	2.764	3.169	4.587
11	0.260	0.697	1.363	1.796	2.201	2.718	3.106	4.437
12	0.259	0.695	1.356	1.782	2.170	2.681	3.055	4.318
13	0.259	0.694	1.350	1.771	2.160	2.650	3.012	4.221
14	0.258	0.692	1.345	1.761	2.145	2.624	2.977	4.140
15	0.258	0.691	1.341	1.753	2.131	2.602	2.947	4.073
16	0.258	0.690	1.337	1.746	2.120	2.583	2.921	4.015
17	0.257	0.689	1.333	1.740	2.110	2.567	2.898	3.965
18	0.257	0.688	1.330	1.734	2.101	2.552	2.878	3.922
19	0.257	0.688	1.328	1.729	2.093	2.539	2.861	3.883
20	0.257	0.687	1.325	1.725	2.086	2.528	2.845	3.850
21	0.257	0.686	1.323	1.721	2.080	2.518	2.831	3.819
22	0.256	0.686	1.321	1.717	2.074	2.508	2.819	3.792
23	0.256	0.685	1.319	1.714	2.069	2.500	2.807	3.767
24	0.256	0.685	1.318	1.711	2.064	2.492	2.797	3.745
25	0.256	0.684	1.316	1.708	2.060	2.485	2.787	3.725
26	0.256	0.684	1.315	1.706	2.056	2.479	2.779	3.707
27	0.256	0.684	1.314	1.703	2.052	2.473	2.771	3.690
28	0.256	0.683	1.313	1.701	2.048	2.467	2.763	3.674
29	0.256	0.683	1.311	1.699	2.045	2.462	2.756	3.659
30	0.256	0.683	1.310	1.697	2.042	2.457	2.750	3.646
40	0.255	0.681	1.303	1.684	2.021	2.423	2.704	3.551
60	0.254	0.679	1.296	1.671	2.000	2.390	2.660	3.460
120	0.254	0.677	1.289	1.658	1.980	2.358	2.617	3.373
∞	0.253	0.674	1.282	1.645	1.960	2.326	2.576	3.291

Source: From Beyer (1966).

Table A.3 Tolerance factors (K) for one-sided normal tolerance intervals with probability level (confidence factor) $\gamma = 0.95$ and coverage $P = 95\%$.

(a) For sample sizes ≤50

n	K
3	7.655
4	5.145
5	4.202
6	3.707
7	3.399
8	3.188
9	3.031
10	2.911
11	2.815
12	2.736
13	2.670
14	2.614
15	2.566
16	2.523
17	2.486
18	2.543
19	2.423
20	2.396
21	2.371
22	2.350
23	2.329
24	2.309
25	2.292
30	2.220
35	2.166
40	2.126
45	2.092
50	2.065

(b) For sample sizes ≥50: K values calculated from large sample approximation

n	K
75	1.972
100	1.924
125	1.891

(*Continued*)

Table A.3 (Continued)

(b) For sample sizes ≥50: K values calculated from large sample approximation	
150	1.868
175	1.850
200	1.836
225	1.824
250	1.814
275	1.806
300	1.799
325	1.792
350	1.787
375	1.782
400	1.777
425	1.773
450	1.769
475	1.766
500	1.763
525	1.760
550	1.757
575	1.754
600	1.752
625	1.750
650	1.748
675	1.746
700	1.744
725	1.742
750	1.740
775	1.739
800	1.737
825	1.736
850	1.734
875	1.733
900	1.732
925	1.731
950	1.729
975	1.728
1000	1.727

Source: From (a) Lieberman (1958); (b) From US EPA (1989).

Table A.4 Critical values for the t-distribution.

	Level of significance for a one-tailed test							
	0.250	0.100	0.050	0.025	0.010	0.005	0.0025	0.0005
	Level of significance for a two-tailed test							
Degrees of freedom	**0.500**	**0.200**	**0.100**	**0.050**	**0.020**	**0.010**	**0.005**	**0.001**
1	1.000	3.078	6.314	12.706	31.821	63.657	27.321	536.627
2	0.816	1.886	2.920	4.303	6.965	9.925	14.089	31.599
3	0.765	1.638	2.353	3.182	4.541	5.841	7.453	12.924
4	0.741	1.533	2.132	2.776	3.747	4.604	5.598	8.610
5	0.727	1.476	2.015	2.571	3.365	4.032	4.773	6.869
6	0.718	1.440	1.943	2.447	3.143	3.707	4.317	5.959
7	0.711	1.415	1.895	2.365	2.998	3.499	4.029	5.408
8	0.706	1.397	1.860	2.306	2.896	3.355	3.833	5.041
9	0.703	1.383	1.833	2.262	2.821	3.250	3.690	4.781
10	0.700	1.372	1.812	2.228	2.764	3.169	3.581	4.587
11	0.697	1.363	1.796	2.201	2.718	3.106	3.497	4.437
12	0.695	1.356	1.782	2.179	2.681	3.055	3.428	4.318
13	0.694	1.350	1.771	2.160	2.650	3.012	3.372	4.221
14	0.692	1.345	1.761	2.145	2.624	2.977	3.326	4.140
15	0.691	1.341	1.753	2.131	2.602	2.947	3.286	4.073
16	0.690	1.337	1.746	2.120	2.583	2.921	3.252	4.015
17	0.689	1.333	1.740	2.110	2.567	2.898	3.222	3.965
18	0.688	1.330	1.734	2.101	2.552	2.878	3.197	3.922
19	0.688	1.328	1.729	2.093	2.539	2.861	3.174	3.883
20	0.687	1.325	1.725	2.086	2.528	2.845	3.153	3.850
21	0.686	1.323	1.721	2.080	2.518	2.831	3.135	3.819
22	0.686	1.321	1.717	2.074	2.508	2.819	3.119	3.792
23	0.685	1.319	1.714	2.069	2.500	2.807	3.104	3.768
24	0.685	1.318	1.711	2.064	2.492	2.797	3.091	3.745
25	0.684	1.316	1.708	2.060	2.485	2.787	3.078	3.725
26	0.684	1.315	1.706	2.056	2.479	2.779	3.067	3.707
27	0.684	1.314	1.703	2.052	2.473	2.771	3.057	3.690
28	0.683	1.313	1.701	2.048	2.467	2.763	3.047	3.674
29	0.683	1.311	1.699	2.045	2.462	2.756	3.038	3.659
30	0.683	1.310	1.697	2.042	2.457	2.750	3.030	3.646
35	0.682	1.306	1.690	2.030	2.438	2.724	2.996	3.591

(Continued)

Table A.4 (Continued)

	Level of significance for a two-tailed test							
Degrees of freedom	**0.500**	**0.200**	**0.100**	**0.050**	**0.020**	**0.010**	**0.005**	**0.001**
40	0.681	1.303	1.684	2.021	2.423	2.704	2.971	3.551
45	0.680	1.301	1.679	2.014	2.412	2.690	2.952	3.520
50	0.679	1.299	1.676	2.009	2.403	2.678	2.937	3.496
55	0.679	1.297	1.673	2.004	2.396	2.668	2.925	3.476
60	0.679	1.296	1.671	2.000	2.390	2.660	2.915	3.460
65	0.678	1.295	1.669	1.997	2.385	2.654	2.906	3.447
70	0.678	1.294	1.667	1.994	2.381	2.648	2.899	3.435
80	0.678	1.292	1.664	1.990	2.374	2.639	2.887	3.416
90	0.677	1.291	1.662	1.987	2.368	2.632	2.878	3.402
100	0.677	1.290	1.660	1.984	2.364	2.626	2.871	3.390
125	0.676	1.288	1.657	1.979	2.357	2.616	2.858	3.370
150	0.676	1.287	1.655	1.976	2.351	2.609	2.849	3.357
200	0.676	1.286	1.653	1.972	2.345	2.601	2.839	3.340
∞	0.6745	1.2816	1.6448	1.9600	2.3267	2.5758	2.8070	3.2905

Table A.5 Critical values for T_N (one-sided) when standard deviation is calculated from the same sample.

Number of observations n	Upper 0.1% significance level	Upper 0.5% significance level	Upper 1% significance level	Upper 2.5% significance level	Upper 5% significance level	Upper 10% significance level
3	1.155	1.155	1.155	1.155	1.153	1.148
4	1.499	1.496	1.492	1.481	1.463	1.425
5	1.780	1.764	1.749	1.715	1.672	1.602
6	2.011	1.973	1.944	1.387	1.822	1.729
7	2.201	2.139	2.097	2.020	1.938	1.828
8	2.358	2.274	2.221	2.126	2.032	1.909
9	2.492	2.387	2.323	2.215	2.110	1.977
10	2.606	2.482	2.410	2.290	2.176	2.036
11	2.705	2.564	2.485	2.355	2.234	2.088
12	2.791	2.636	2.550	2.412	2.285	2.134
13	2.867	2.699	2.607	2.462	2.331	2.175
14	2.935	2.755	2.659	2.507	2.371	2.213
15	2.997	2.806	2.705	2.549	2.409	2.247
16	3.052	2.852	2.747	2.585	2.443	2.279
17	3.103	2.894	2.785	2.620	2.475	2.309
18	3.149	2.932	2.821	2.651	2.504	2.335
19	3.191	2.968	2.854	2.681	2.532	2.361
20	3.230	3.001	2.884	2.709	2.557	2.385
21	3.266	3.031	2.912	2.733	2.580	2.408
22	3.300	3.060	2.939	2.758	2.603	2.429
23	3.332	3.087	2.963	2.781	2.624	2.448
24	3.362	3.112	2.987	2.802	2.644	2.467
25	3.389	3.135	3.009	2.822	2.663	2.486
26	3.415	3.157	3.029	2.841	2.681	2.502

(*Continued*)

Table A.5 (Continued)

Number of observations n	Upper 0.1% significance level	Upper 0.5% significance level	Upper 1% significance level	Upper 2.5% significance level	Upper 5% significance level	Upper 10% significance level
27	3.440	3.178	3.049	2.859	2.698	2.519
28	3.464	3.199	3.068	2.876	2.714	2.534
29	3.486	3.218	3.085	2.893	2.730	2.549
30	3.507	3.236	3.103	2.908	2.745	2.563
31	3.528	3.253	3.119	2.924	2.759	2.577
32	3.546	3.270	3.135	2.938	2.773	2.591
33	3.565	3.286	3.150	2.952	2.786	2.604
34	3.582	3.301	3.164	2.965	2.799	2.616
35	3.599	3.316	3.178	2.979	2.811	2.628
36	3.616	3.330	3.191	2.991	2.823	2.639
37	3.631	3.343	3.204	3.003	2.835	2.650
38	3.646	3.356	3.216	3.014	2.846	2.661
39	3.660	3.369	3.228	3.025	2.857	2.671
40	3.673	3.381	3.240	3.036	2.866	2.682
41	3.687	3.393	3.251	3.046	2.877	2.692
42	3.700	3.404	3.261	3.057	2.887	2.700
43	3.712	3.415	3.271	3.067	2.896	2.710
44	3.724	3.425	3.282	3.075	2.905	2.719
45	3.736	3.435	3.292	3.085	2.914	2.727
46	3.747	3.445	3.302	3.094	2.923	2.736
47	3.757	3.455	3.310	3.103	2.931	2.744
48	3.768	3.464	3.319	3.111	2.940	2.753
49	3.779	3.474	3.329	3.120	2.948	2.760
50	3.789	3.483	3.336	3.128	2.956	2.768

51	3.798	3.491	3.345	3.136	2.964	2.775
52	3.808	3.500	3.353	3.143	2.971	2.783
53	3.816	3.507	3.361	3.151	2.978	2.790
54	3.825	3.516	3.368	3.158	2.986	2.798
55	3.834	3.524	3.376	3.166	2.992	2.804
56	3.842	3.531	3.383	3.172	3.000	2.811
57	3.851	3.539	3.391	3.180	3.006	2.818
58	3.858	3.546	3.397	3.186	3.013	2.824
59	3.867	3.553	3.405	3.193	3.019	2.831
60	3.874	3.560	3.411	3.199	3.025	2.837
61	3.882	3.566	3.418	3.205	3.032	2.842
62	3.889	3.573	3.424	3.212	3.037	2.849
63	3.896	3.579	3.430	3.218	3.044	2.854
64	3.903	3.586	3.437	3.224	3.049	2.860
65	3.910	3.592	3.442	3.230	3.055	2.866
66	3.917	3.598	3.449	3.235	3.061	2.871
67	3.923	3.605	3.454	3.241	3.066	2.877
68	3.930	3.610	3.460	3.246	3.071	2.883
69	3.936	3.617	3.466	3.252	3.076	2.888
70	3.942	3.622	3.471	3.257	3.082	2.893
71	3.948	3.627	3.476	3.262	3.087	2.897
72	3.954	3.633	3.482	3.267	3.092	2.903
73	3.960	3.638	3.487	3.272	3.098	2.908
74	3.965	3.643	3.492	3.278	3.102	2.912
75	3.971	3.648	3.496	3.282	3.107	2.917
76	3.977	3.654	3.502	3.287	3.111	2.922
77	3.982	3.658	3.507	3.291	3.117	2.927

(*Continued*)

Table A.5 (Continued)

Number of observations n	Upper 0.1% significance level	Upper 0.5% significance level	Upper 1% significance level	Upper 2.5% significance level	Upper 5% significance level	Upper 10% significance level
78	3.987	3.663	3.511	3.297	3.121	2.931
79	3.992	3.669	3.516	3.301	3.125	2.935
80	3.998	3.673	3.521	3.305	3.130	2.940
81	4.002	3.677	3.525	3.309	3.134	2.945
82	4.007	3.682	3.529	3.315	3.139	2.949
83	4.012	3.687	3.534	3.319	3.143	2.953
84	4.017	3.691	3.539	3.323	3.147	2.957
85	4.021	3.695	3.543	3.327	3.151	2.961
86	4.026	3.699	3.547	3.331	3.155	2.966
87	4.031	3.704	3.551	3.335	3.160	2.970
88	4.035	3.708	3.555	3.339	3.163	2.973
89	4.039	3.712	3.559	3.343	3.167	2.977
90	4.044	3.716	3.563	3.347	3.171	2.981
91	4.049	3.720	3.567	3.350	3.174	2.984
92	4.053	3.725	3.570	3.355	3.179	2.989
93	4.057	3.728	3.575	3.358	3.182	2.993
94	4.060	3.732	3.579	3.362	3.186	2.996
95	4.064	3.736	3.582	3.365	3.189	3.000
96	4.069	3.739	3.586	3.369	3.193	3.003
97	4.073	3.744	3.589	3.372	3.196	3.006
98	4.076	3.747	3.593	3.377	3.201	3.011
99	4.080	3.750	3.597	3.380	3.204	3.014
100	4.084	3.754	3.600	3.383	3.207	3.017
101	4.088	3.757	3.603	3.386	3.210	3.021

102	4.092	3.760	3.607	3.390	3.214	3.024
103	4.095	3.765	3.610	3.393	3.217	3.027
104	4.098	3.768	3.614	3.397	3.220	3.030
105	4.102	3.771	3.617	3.400	3.224	3.033
106	4.105	3.774	3.620	3.403	3.227	3.037
107	4.109	3.777	3.623	3.406	3.230	3.040
108	4.112	3.780	3.626	3.409	3.233	3.043
109	4.116	3.784	3.629	3.412	3.236	3.046
110	4.119	3.787	3.632	3.415	3.239	3.049
111	4.122	3.790	3.636	3.418	3.242	3.052
112	4.125	3.793	3.639	3.422	3.245	3.055
113	4.129	3.796	3.642	3.424	3.248	3.058
114	4.132	3.799	3.645	3.427	3.251	3.061
115	4.135	3.802	3.647	3.430	3.254	3.064
116	4.138	3.805	3.650	3.433	3.257	3.067
117	4.141	3.808	3.653	3.435	3.259	3.070
118	4.144	3.811	3.656	3.438	3.262	3.073
119	4.146	3.814	3.659	3.441	3.265	3.075
120	4.150	3.817	3.662	3.444	3.267	3.078
121	4.153	3.819	3.665	3.447	3.270	3.081
122	4.156	3.822	3.667	3.450	3.274	3.083
123	4.159	3.824	3.670	3.452	3.276	3.086
124	4.161	3.827	3.672	3.455	3.279	3.089
125	4.164	3.831	3.675	3.457	3.281	3.092
126	4.166	3.833	3.677	3.460	3.284	3.095
127	4.169	3.836	3.680	3.462	3.286	3.097
128	4.173	3.838	3.683	3.465	3.289	3.100

(*Continued*)

Table A.5 (Continued)

Number of observations n	Upper 0.1% significance level	Upper 0.5% significance level	Upper 1% significance level	Upper 2.5% significance level	Upper 5% significance level	Upper 10% significance level
129	4.175	3.840	3.686	3.467	3.291	3.102
130	4.178	3.843	3.688	3.470	3.294	3.104
131	4.180	3.845	3.690	3.473	3.296	3.107
132	4.183	3.848	3.693	3.475	3.298	3.109
133	4.185	3.850	3.695	3.478	3.302	3.112
134	4.188	3.853	3.697	3.480	3.304	3.114
135	4.190	3.856	3.700	3.482	3.306	3.116
136	4.193	3.858	3.702	3.484	3.309	3.119
137	4.196	3.860	3.704	3.487	3.311	3.122
138	4.198	3.863	3.707	3.489	3.313	3.124
139	4.200	3.865	3.710	3.491	3.315	3.126
140	4.203	3.867	3.712	3.493	3.318	3.129
141	4.205	3.869	3.714	3.497	3.320	3.131
142	4.207	3.871	3.716	3.499	3.322	3.133
143	4.209	3.874	3.719	3.501	3.324	3.135
144	4.212	3.876	3.721	3.503	3.326	3.138
145	4.214	3.879	3.723	3.505	3.328	3.140
146	4.216	3.881	3.725	3.507	3.331	3.142
147	4.219	3.883	3.727	3.509	3.334	3.144

Table A.6 Coefficients (a_{N-L+1}) for Shapiro–Wilk W-test of normality.

i/n	2	3	4	5	6	7	8	9	10
1	0.7071	0.7071	0.6872	0.6646	0.6431	0.6233	0.6052	0.5888	0.5739
2	—	0.0000	0.1677	0.2413	0.2806	0.3031	0.3164	0.3244	0.3291
3	—	—	—	0.0000	0.0875	0.1401	0.1743	0.1976	0.2141
4	—	—	—	—	—	0.0000	0.0561	0.0947	0.1224
5	—	—	—	—	—	—	—	0.0000	0.0399

i/n	11	12	13	14	15	16	17	18	19	20
1	0.5601	0.5475	0.5359	0.5251	0.5150	0.5056	0.4968	0.4886	0.4886	0.4734
2	0.3315	0.3325	0.3325	0.3318	0.3306	0.3290	0.3273	0.3253	0.3232	0.3211
3	0.2260	0.2347	0.2412	0.2460	0.2495	0.2521	0.2540	0.2553	0.2561	0.2565
4	0.1429	0.1586	0.1707	0.1802	0.1878	0.1939	0.1988	0.2027	0.2059	0.2085
5	0.0695	0.0922	0.1099	0.1240	0.1353	0.1447	0.1524	0.1587	0.1641	0.1686
6	0.0000	0.0303	0.0539	0.0727	0.0880	0.1005	0.1109	0.1197	0.1271	0.1334
7	—	—	0.0000	0.0240	0.0433	0.0593	0.0725	0.0837	0.0932	0.1013
8	—	—	—	—	0.0000	0.0196	0.0359	0.0496	0.0612	0.0711
9	—	—	—	—	—	—	0.0000	0.0163	0.0303	0.0422
10	—	—	—	—	—	—	—	—	0.0000	0.0140

i/n	21	22	23	24	25	26	27	28	29	30
1	0.4643	0.4590	0.4542	0.4493	0.4450	0.4407	0.4366	0.4328	0.4291	0.4254
2	0.3185	0.3156	0.3126	0.3098	0.3069	0.3043	0.3018	0.2992	0.2968	0.2944
3	0.2578	0.2571	0.2563	0.2554	0.2543	0.2533	0.2522	0.2510	0.2499	0.2487
4	0.2119	0.2131	0.2139	0.2145	0.2148	0.2151	0.2152	0.2151	0.2150	0.2148

(Continued)

Table A.6 (Continued)

i/n	21	22	23	24	25	26	27	28	29	30
5	0.1736	0.1764	0.1787	0.1807	0.1822	0.1836	0.1848	0.1857	0.1864	0.1870
6	0.1399	0.1443	0.1480	0.1512	0.1539	0.1563	0.1584	0.1601	0.1616	0.1630
7	0.1092	0.1150	0.1201	0.1245	0.1283	0.1316	0.1346	0.1372	0.1395	0.1415
8	0.0804	0.0878	0.0941	0.0997	0.1046	0.1089	0.1128	0.1162	0.1192	0.1219
9	0.0530	0.0618	0.0696	0.0764	0.0823	0.0876	0.0923	0.9650	0.1002	0.1036
10	0.0263	0.0368	0.4590	0.0539	0.0610	0.0672	0.0728	0.0778	0.0822	0.0862
11	0.0000	0.0122	0.0228	0.0321	0.0403	0.0476	0.0540	0.0598	0.0650	0.0697
12	—	—	0.0000	0.0107	0.0200	0.0284	0.0358	0.0424	0.0483	0.0537
13	—	—	—	—	0.0000	0.0094	0.0178	0.0253	0.0320	0.0381
14	—	—	—	—	—	—	0.0000	0.0084	0.0159	0.0227
15	—	—	—	—	—	—	—	—	0.0000	0.0076

i/n	31	32	33	34	35	36	37	38	39	40
1	0.4220	0.4188	0.4156	0.4127	0.4096	0.4068	0.4040	0.4015	0.3989	0.2964
2	0.2921	0.2898	0.2876	0.2854	0.2834	0.2813	0.2794	0.2774	0.2755	0.2737
3	0.2475	0.2463	0.2451	0.2439	0.2427	0.2415	0.2403	0.2391	0.2380	0.2368
4	0.2145	0.2141	0.2137	0.2132	0.2127	0.2121	0.2116	0.2110	0.2104	0.2098
5	0.1874	0.1878	0.1880	0.1882	0.1883	0.1883	0.1883	0.1881	0.1880	0.1878
6	0.1641	0.1651	0.1660	0.1667	0.1673	0.1678	0.1683	0.1686	0.1689	0.1691
7	0.1433	0.1449	0.1463	0.1475	0.1487	0.1496	0.1503	0.1513	0.1520	0.1526
8	0.1243	0.1265	0.1284	0.1301	0.1317	0.1331	0.1344	0.1356	0.1366	0.1376
9	0.1066	0.1093	0.1118	0.1140	0.1160	0.1179	0.1196	0.1211	0.1225	0.1237
10	0.0899	0.0931	0.9610	0.0988	0.1013	0.1036	0.1056	0.1075	0.1092	0.1108
11	0.0739	0.0777	0.0812	0.0844	0.0873	0.0900	0.0924	0.0947	0.0967	0.8960
12	0.0585	0.0629	0.0669	0.0706	0.0739	0.0770	0.0798	0.0824	0.0848	0.0870

i/n	31	32	33	34	35	36	37	38	39	40
13	0.0435	0.0485	0.0530	0.0572	0.0610	0.0645	0.0677	0.0706	0.0733	0.0759
14	0.0289	0.0344	0.0395	0.0441	0.0484	0.0523	0.0559	0.0592	0.0622	0.0651
15	0.0144	0.0206	0.0262	0.0314	0.0361	0.0404	0.0444	0.0481	0.0515	0.0546
16	0.0000	0.0068	0.0131	0.0187	0.0239	0.0287	0.0331	0.0372	0.0409	0.0444
17	—	—	0.0000	0.0062	0.0119	0.0172	0.0220	0.0264	0.0305	0.0343
18	—	—	—	—	0.0000	0.0057	0.0110	0.0158	0.0203	0.0244
19	—	—	—	—	—	—	0.0000	0.0053	0.0101	0.0146
20	—	—	—	—	—	—	—	—	0.0000	0.0049

i/n	41	42	43	44	45	46	47	48	49	50
1	0.3940	0.3917	0.3894	0.3872	0.3850	0.3830	0.3808	0.3789	0.3000	0.3751
2	0.2719	0.2701	0.2684	0.2667	0.2651	0.2635	0.2620	0.2604	0.2589	0.2574
3	0.2357	0.2345	0.2334	0.2323	0.2313	0.2302	0.2291	0.2281	0.2271	0.2260
4	0.2091	0.2085	0.2078	0.2072	0.2065	0.2058	0.2052	0.2045	0.2038	0.2032
5	0.1876	0.1874	0.1871	0.1868	0.1865	0.1862	0.1859	0.1855	0.1851	0.2218
6	0.1693	0.1694	0.1695	0.1695	0.1695	0.1695	0.1695	0.1693	0.1692	0.1691
7	0.1531	0.1535	0.1539	0.1542	0.1545	0.1548	0.1550	0.1551	0.1553	0.1554
8	0.1384	0.1392	0.1398	0.1405	0.1410	0.1415	0.1420	0.1423	0.1427	0.1430
9	0.1249	0.1259	0.1269	0.1278	0.1286	0.1293	0.1300	0.1306	0.1312	0.1317
10	0.1123	0.1136	0.1149	0.1160	0.1170	0.1180	0.1189	0.1197	0.1205	0.1212
11	0.1004	0.1020	0.1035	0.1049	0.1062	0.1073	0.1085	0.1095	0.1105	0.1113
12	0.0891	0.0909	0.0927	0.0943	0.0959	0.0972	0.0986	0.0998	0.1010	0.1020
13	0.0782	0.0804	0.0824	0.0842	0.0860	0.0876	0.0892	0.9060	0.0919	0.0932
14	0.0677	0.0701	0.0724	0.0745	0.0775	0.0785	0.0801	0.0817	0.0832	0.0846

(Continued)

Table A.6 (Continued)

i/n	41	42	43	44	45	46	47	48	49	50
15	0.0575	0.0602	0.0628	0.0651	0.0673	0.0694	0.0713	0.0731	0.0748	0.0764
16	0.0476	0.0506	0.0534	0.0560	0.0584	0.0607	0.0628	0.0648	0.0662	0.0685
17	0.0379	0.0411	0.0442	0.0471	0.0497	0.0522	0.0546	0.0568	0.0588	0.0608
18	0.0283	0.0318	0.0352	0.0383	0.0412	0.0439	0.0465	0.0489	0.0511	0.0532
19	0.0188	0.0227	0.0263	0.0296	0.0328	0.0357	0.0385	0.0411	0.0436	0.0459
20	0.0094	0.0315	0.0175	0.0211	0.0235	0.0277	0.0307	0.0335	0.0361	0.0386
21	0.0000	0.0045	0.0087	0.0126	0.0163	0.0197	0.0229	0.0259	0.0288	0.0314
22	—	—	0.0000	0.0042	0.0081	0.0118	0.0153	0.0185	0.0215	0.0244
23	—	—	—	—	0.0000	0.0039	0.0076	0.0111	0.0143	0.0174
24	—	—	—	—	—	—	0.0000	0.0037	0.0071	0.0104
25	—	—	—	—	—	—	—	—	0.0000	0.0035

Table A.7 Percentage points of the Shapiro–Wilk W test.

n	0.01	0.05
3	0.753	0.767
4	0.687	0.748
5	0.686	0.762
6	0.713	0.788
7	0.730	0.803
8	0.749	0.818
9	0.764	0.829
10	0.781	0.842
11	0.792	0.850
12	0.805	0.859
13	0.814	0.866
14	0.825	0.874
15	0.835	0.881
16	0.844	0.887
17	0.851	0.892
18	0.858	0.897
19	0.863	0.901
20	0.868	0.905
21	0.873	0.908
22	0.878	0.911
23	0.881	0.914
24	0.884	0.916
25	0.888	0.918
26	0.891	0.920
27	0.894	0.923
28	0.896	0.924
29	0.898	0.926
30	0.900	0.927
31	0.902	0.929
32	0.904	0.930
33	0.906	0.931
34	0.908	0.933
35	0.910	0.934
36	0.912	0.935
37	0.914	0.936
38	0.916	0.938
39	0.917	0.939
40	0.919	0.940
41	0.920	0.941

(Continued)

Table A.7 (Continued)

n	0.01	0.05
42	0.922	0.942
43	0.923	0.943
44	0.924	0.944
45	0.926	0.945
46	0.927	0.945
47	0.928	0.946
48	0.929	0.947
49	0.929	0.947
50	0.930	0.947

Table A.8 Percentage points of the Shapiro–Francia W test for n ≥ 35.

n	0.01	0.05
35	0.919	0.943
50	0.935	0.953
51	0.935	0.954
53	0.938	0.957
55	0.940	0.958
57	0.944	0.961
59	0.945	0.962
61	0.947	0.963
63	0.947	0.964
65	0.948	0.965
67	0.950	0.966
69	0.951	0.966
71	0.953	0.967
73	0.956	0.968
75	0.956	0.969
77	0.957	0.969
79	0.957	0.970
81	0.958	0.970
83	0.960	0.971
85	0.961	0.972
87	0.961	0.972
89	0.961	0.972
91	0.962	0.973
93	0.963	0.973
95	0.965	0.974
97	0.965	0.975
99	0.967	0.976

Table A.9 Percentage points of the normal probability plot correlation coefficient.

n	0.01	0.025	0.05
3	0.869	0.872	0.879
4	0.822	0.845	0.868
5	0.822	0.855	0.879
6	0.835	0.868	0.890
7	0.847	0.876	0.899
8	0.859	0.886	0.905
9	0.868	0.893	0.912
10	0.876	0.900	0.917
11	0.883	0.906	0.922
12	0.889	0.912	0.926
13	0.895	0.917	0.931
14	0.901	0.921	0.934
15	0.907	0.925	0.937
16	0.912	0.928	0.940
17	0.912	0.931	0.942
18	0.919	0.934	0.945
19	0.923	0.937	0.947
20	0.925	0.939	0.950
21	0.928	0.942	0.952
22	0.930	0.944	0.954
23	0.933	0.947	0.955
24	0.936	0.949	0.957
25	0.937	0.950	0.958
26	0.939	0.952	0.959
27	0.941	0.953	0.960
28	0.943	0.955	0.962
29	0.945	0.956	0.962
30	0.947	0.957	0.964
31	0.948	0.958	0.965
32	0.949	0.959	0.966
33	0.950	0.960	0.967
34	0.951	0.960	0.967
35	0.952	0.961	0.968
36	0.953	0.962	0.968
37	0.955	0.962	0.969
38	0.956	0.964	0.970
39	0.957	0.965	0.971

(Continued)

Table A.9 (Continued)

n	0.01	0.025	0.05
40	0.958	0.966	0.972
41	0.958	0.967	0.973
42	0.959	0.967	0.973
43	0.959	0.967	0.973
44	0.960	0.968	0.974
45	0.961	0.969	0.974
46	0.962	0.969	0.974
47	0.963	0.970	0.975
48	0.963	0.970	0.975
49	0.964	0.971	0.977
50	0.965	0.972	0.978
55	0.967	0.974	0.980
60	0.970	0.976	0.981
65	0.972	0.977	0.982
70	0.974	0.978	0.983
75	0.975	0.979	0.984
80	0.976	0.980	0.985
85	0.977	0.981	0.985
90	0.978	0.982	0.985
95	0.979	0.983	0.986
100	0.981	0.984	0.987

Table A.10 Critical values of H for the Kruskal–Wallis test.

			Significance level			
n_1	n_2	n_3	0.10	0.05	0.01	0.05
2	1	1				
2	2	1				
2	2	2	4.571			
3	1	1				
3	2	1	4.286			
3	2	2	4.500	4.714	5.357	
3	3	1	4.571	5.143		
3	3	2	4.556	5.361		
3	3	3	4.622	5.600	7.200	7.200
4	1	1				
4	2	1	4.500			
4	2	2	4.056	5.208		
4	3	2	4.511	5.444	6.444	
4	3	3	4.709	5.727	6.746	
4	4	1	4.167	4.967	6.667	
4	4	2	4.555	5.455	7.036	
4	4	3	4.546	5.599	7.144	
4	4	4	4.654	5.692	7.654	
5	1	1				
5	2	1	4.200	5.000		
5	2	2	4.373	5.160	6.533	
5	3	1	4.018	4.960		
5	3	2	4.651	5.251	6.882	
5	3	3	4.533	5.649	7.079	
5	4	1	3.987	4.986	6.955	
5	4	2	4.541	5.286	7.118	
5	4	3	4.549	5.631	7.445	
5	4	4	4.619	5.618	7.760	
5	5	1	4.109	5.127	7.309	
5	5	2	4.508	5.339	7.269	
5	5	3	4.545	5.706	7.543	
5	5	4	4.523	5.643	7.791	
5	5	5	4.560	5.780	7.980	

Table A.11 Values of lambda for Cohen's method.

	Percentage of nondetects						
λ	0.01	0.05	0.10	0.15	0.25	0.40	0.50
0.01	0.0102	0.0530	0.1111	0.1747	0.3205	0.5989	0.8403
0.05	0.0105	0.0547	0.1143	0.1793	0.3279	0.6101	0.8540
0.10	0.0110	0.0566	0.1180	0.1848	0.3366	0.6234	0.8703
0.15	0.0113	0.0584	0.1215	0.1898	0.3448	0.6361	0.8860
0.20	0.0116	0.0600	0.1247	0.1946	0.3525	0.6483	0.9012
0.30	0.0122	0.0630	0.1306	0.2034	0.3670	0.6713	0.9300
0.40	0.0128	0.0657	0.1360	0.2114	0.3803	0.6927	0.9570
0.50	0.0133	0.0681	0.1409	0.2188	0.3928	0.7129	0.9826
0.60	0.0137	0.0704	0.1455	0.2258	0.4045	0.7320	1.0070
0.70	0.0142	0.0726	0.1499	0.2323	0.4156	0.7502	1.0303
0.80	0.0146	0.0747	0.1540	0.2386	0.4261	0.7676	1.0527
0.90	0.0150	0.0766	0.1579	0.2445	0.4362	0.7844	1.0743
1.00	0.0153	0.0785	0.1617	0.2502	0.4459	0.8005	1.0951
1.10	0.0157	0.0803	0.1653	0.2557	0.4553	0.8161	1.1152
1.20	0.0160	0.0820	0.1688	0.2610	0.4643	0.8312	1.1347
1.30	0.0164	0.0836	0.1722	0.2661	0.4730	0.8458	1.1537
1.40	0.0167	0.0853	0.1754	0.2710	0.4815	0.8600	1.1721
1.50	0.0170	0.0868	0.1786	0.2758	0.4897	0.8738	1.1901
1.60	0.0173	0.0883	0.1817	0.2805	0.4977	0.8873	1.2076
1.70	0.0176	0.0898	0.1846	0.2851	0.5055	0.9005	1.2248
1.80	0.0179	0.0913	0.1876	0.2895	0.5132	0.9133	1.2415
1.90	0.0181	0.0927	0.1904	0.2938	0.5206	0.9259	1.2579
2.00	0.0184	0.0940	0.1932	0.2981	0.5279	0.9382	1.2739
2.10	0.0187	0.0954	0.1959	0.3022	0.5350	0.9502	1.2897

	Percentage of nondetects						
2.20	0.0189	0.0967	0.1986	0.3062	0.5420	0.9620	1.3051
2.30	0.0192	0.0980	0.2012	0.3102	0.5488	0.9736	1.3203
2.40	0.0194	0.0992	0.2037	0.3141	0.5555	0.9850	1.3352
2.50	0.0197	0.1005	0.2062	0.3179	0.5621	0.9962	1.3498
2.60	0.0199	0.1017	0.2087	0.3217	0.5686	1.0072	1.3642
2.70	0.0202	0.1029	0.2111	0.3254	0.5750	1.0180	1.3784
2.80	0.0204	0.1040	0.2135	0.3290	0.5812	1.0287	1.3924
2.90	0.0206	0.1052	0.2158	0.3326	0.5874	1.0392	1.4061
3.00	0.0209	0.1063	0.2182	0.3361	0.5935	1.0495	1.4197
3.10	0.0211	0.1074	0.2204	0.3396	0.5995	1.0597	1.4330
3.20	0.0213	0.1085	0.2227	0.3430	0.6054	1.0697	1.4462
3.30	0.0215	0.1096	0.2249	0.3464	0.6112	1.0796	1.4592
3.40	0.0217	0.1107	0.2270	0.3497	0.6169	1.0894	1,4720
3.50	0.0219	0.1118	0.2292	0.3529	0.6226	1.0990	1.4847
3.60	0.0221	0.1128	0.2313	0.3562	0.6282	1.1086	1.4972
3.70	0.0223	0.1138	0.2334	0.3594	0.6337	1.1180	1.5096
3.80	0.0225	0.1148	0.2355	0.3625	0.6391	1.1273	1.5218
3.90	0.0227	0.1158	0.2375	0.3656	0.6445	1.1364	1.5339
4.00	0.0229	0.1168	0.2395	0.3687	0.6498	1.1455	1.5458
4.10	0.0231	0.1178	0.2415	0.3717	0.6551	1.1545	1.5577
4.20	0.0233	0.1188	0.2435	0.3747	0.6603	1.1634	1.5693
4.30	0.0235	0.1197	0.2454	0.3777	0.6654	1.1722	1.5809
4.40	0.0237	0.1207	0.2473	0.3806	0.6705	1.1809	1.5924
4.50	0.0239	0.1216	0.2492	0.3836	0.6755	1.1895	1.6037
4.60	0.0241	0.1225	0.2511	0.3864	0.6805	1.1980	1.6149
4.70	0.0242	0.1235	0.2530	0.3893	0.6855	1.2064	1.6260

(*Continued*)

Table A.11 (Continued)

	Percentage of nondetects						
4.80	0.0244	0.1244	0.2548	0.3921	0.6903	1.2148	1.6370
4.90	0.0246	0.1253	0.2567	0.3949	0.6952	1.2230	1.6479
5.00	0.0248	0.1262	0.2585	0.3977	0.7000	1.2312	1.6587
5.10	0.0249	0.1270	0.2603	0.4004	0.7047	1.2394	1.6694
5.20	0.0251	0.1279	0.2621	0.4031	0.7094	1.2474	1.6800
5.30	0.0253	0.1288	0.2638	0.4058	0.7141	1.2554	1.6905
5.40	0.0255	0.1296	0.2656	0.4085	0.7187	1.2633	1.7010
5.50	0.0256	0.1305	0.2673	0.4111	0.7233	1.2711	1.7113
5.60	0.0258	0.1313	0.2690	0.4137	0.7278	1.2789	1.7215
5.70	0.0260	0.1322	0.2707	0.4163	0.7323	1.2866	1.7317
5.80	0.0261	0.1330	0.2724	0.4189	0.7368	1.2943	1.7418
5.90	0.0263	0.1338	0.2741	0.4215	0.7412	1.3019	1.7518
6.00	0.0264	0.1346	0.2757	0.4240	0.7456	1.3094	1.7617

Table A.12 Percentiles of the chi-square x^2 distribution with df degrees of freedom, x^2_v, p.

p/df	0.75	0.900	0.950	0.975	0.990	0.995	0.999
1	1.323	2.706	3.841	5.024	6.635	7.879	10.83
2	2.773	4.605	5.991	7.378	9.210	10.60	13.82
3	4.108	6.251	7.815	9.348	11.34	12.84	16.27
4	5.385	7.779	9.488	11.14	13.28	14.86	18.47
5	6.626	9.236	11.07	12.83	15.09	16.75	20.52
6	7.841	10.64	12.59	14.45	16.81	18.55	22.46
7	9.037	12.02	14.07	16.01	18.48	20.28	24.32
8	10.22	13.36	15.51	17.53	20.09	21.96	26.12
9	11.39	14.68	16.92	19.02	21.67	23.59	27.88
10	12.55	15.99	18.31	20.48	23.21	25.19	29.59
11	13.70	17.28	19.68	21.92	24.72	26.76	31.26
12	14.85	18.55	21.03	23.34	26.22	28.30	32.91
13	15.98	19.81	22.36	24.74	27.69	29.82	34.53
14	17.12	21.06	23.68	26.12	29.14	31.32	36.12
15	18.25	22.31	25.00	27.49	30.58	32.80	37.70
16	19.37	23.54	26.30	28.85	32.00	34.27	39.25
17	20.49	24.77	27.59	30.19	33.41	35.72	40.79
18	21.60	25.99	28.87	31.53	34.81	37.16	42.31
19	22.72	27.20	30.14	32.85	36.19	38.58	43.82
20	23.83	28.41	31.41	34.17	37.57	40.00	45.32
21	24.93	29.62	32.67	35.48	38.93	41.40	46.80
22	26.04	30.81	33.92	36.78	40.29	42.80	48.27
23	27.14	32.01	35.17	38.08	41.64	44.18	49.73
24	28.24	33.20	36.42	39.36	42.98	45.56	51.18
25	29.34	34.38	37.65	40.65	44.31	46.93	52.62
26	30.43	35.56	38.89	41.92	45.64	48.29	54.05
27	31.53	36.74	40.11	43.19	46.96	49.64	55.48
28	32.62	37.92	41.34	44.46	48.28	50.99	56.89
29	33.71	39.09	42.56	45.72	49.59	52.34	58.30
30	34.80	40.26	43.77	46.98	50.89	53.67	59.70
40	45.62	51.80	55.76	59.34	63.69	66.77	73.40
50	56.33	63.17	67.50	71.42	76.15	79.49	86.66
60	66.98	74.40	79.08	83.30	88.38	91.95	99.61
70	77.58	85.53	90.53	95.02	100.4	104.2	112.3
80	88.13	96.58	101.9	106.6	112.3	116.3	124.8
90	98.13	107.6	113.1	118.1	124.1	128.3	137.2
100	109.1	118.5	124.3	129.6	135.8	140.2	149.4

Source: From Johnson and Leone (1977).

Table A.13 Critical values of D for the Kolmogorov-Smirnov goodness-of-fit test.

	Significance level				
Degrees of Freedom	**0.20**	**0.15**	**0.10**	**0.05**	**0.01**
1	0.900	0.925	0.950	0.975	0.9950
2	0.684	0.726	0.776	0.842	0.929
3	0.565	0.597	0.642	0.708	0.829
4	0.494	0.525	0.564	0.624	0.734
5	0.446	0.474	0.510	0.563	0.669
6	0.410	0.436	0.470	0.521	0.618
7	0.381	0.405	0.438	0.486	0.577
8	0.358	0.381	0.411	0.457	0.543
9	0.339	0.360	0.388	0.432	0.514
10	0.322	0.342	0.368	0.409	0.486
11	0.307	0.326	0.352	0.391	0.468
12	0.295	0.313	0.338	0.375	0.450
13	0.284	0.302	0.325	0.361	0.433
14	0.274	0.292	0.314	0.349	0.418
15	0.266	0.283	0.304	0.338	0.404
16	0.258	0.274	0.295	0.328	0.391
17	0.250	0.266	0.286	0.318	0.380
18	0.244	0.259	0.278	0.309	0.370
19	0.237	0.252	0.272	0.301	0.361
20	0.231	0.246	0.264	0.294	0.352
25	0.21	0.22	0.242	0.264	0.32
30	0.19	0.20	0.22	0.242	0.29
35	0.18	0.19	0.21	0.23	0.27
40	0.17	0.18	0.19	0.21	0.25
50	0.15	0.16	0.17	0.19	0.23
60	0.14	0.15	0.16	0.17	0.21
70	0.13	0.14	0.15	0.16	0.19
80	0.12	0.13	0.14	0.15	0.18
90	0.11	0.12	0.13	0.14	0.17
100	0.11	0.11	0.12	0.14	0.16

Table A.14 K values for Pearson type III distribution.

	Recurrence interval in years						
	1.01	1.11	1.25	5	25	50	100
	Percent chance						
Skew coefficient C_8	99	90	80	20	4	2	1
	Positive skew						
3.0	−0.667	−0.660	−0.636	0.420	2.278	3.152	4.051
2.5	−0.799	−0.771	−0.711	0.518	2.262	3.048	3.845
2.0	−0.990	−0.895	−0.777	0.609	2.219	2.912	3.605
1.5	−1.256	−1.018	−0.825	0.690	2.146	2.743	3.330
1.0	−1.588	−1.128	−0.852	0.758	2.043	2.542	3.022
0.5	−1.955	−1.216	−0.856	0.808	1.910	2.311	2.686
0.0	−2.326	−1.282	−0.842	0.842	1.751	2.054	2.326
−0.1	−2.400	−1.292	−0.836	0.846	1.716	2.000	2.252
−0.5	−2.686	−1.323	−0.808	0.856	1.567	1.777	1.955
−1.0	−3.022	−1.340	−0.758	0.852	1.366	1.492	1.588
−1.5	−3.330	−1.333	−0.690	0.825	1.157	1.217	1.256
−2.0	−3.605	−1.302	−0.609	0.777	0.959	0.980	0.990
−2.5	−3.845	−1.250	−0.518	0.711	0.793	0.798	0.799
−3.0	−4.051	−1.180	−0.420	0.636	0.666	0.666	0.667

Table A.15 Significance levels of Dixon's criteria for testing extreme observations in a single sample.

n	Criterion	Level 5%	Level 1%
3	$\frac{x_n - x_{n-1}}{x_n - x_1}$ or $\frac{x_2 - x_1}{x_n - x_1}$	0.941	0.988
4		0.765	0.889
5		0.642	0.780
6		0.560	0.698
7		0.507	0.637
8	$\frac{x_n - x_{n-1}}{x_n - x_2}$ or $\frac{x_2 - x_1}{x_{n-1} - x_1}$	0.554	0.683
9		0.512	0.635
10		0.477	0.597
11	$\frac{x_n - x_{n-2}}{x_n - x_2}$ or $\frac{x_3 - x_1}{x_{n-1} - x_1}$	0.576	0.679
12		0.546	0.642
13		0.521	0.615
14	$\frac{x_n - x_{n-2}}{x_n - x_3}$ or $\frac{x_3 - x_1}{x_{n-2} - x_1}$	0.546	0.641
15		0.525	0.616
16		0.507	0.595
17		0.490	0.577
18		0.475	0.561
19		0.462	0.547
20		0.450	0.535
21		0.440	0.524
23		0.421	0.505
24		0.413	0.497
25		0.406	0.489

Source: From Dixon (1953).

Table A.16 Significance levels of Dixon's criteria for testing extreme observations.

Sample number	Sampling location										Sampling location									
	A	B	C	D	E	F	G	H	I	J	K	L	M	N	O	P	Q	R	S	T
1	37.2	35.6	21.8	26.7	26.0	35.4	44.6	28.9	37.2	55.9	23.3	33.2	31.1	39.5	47.4	37.3	46.5	41.6	43.2	40.9
2	24.6	38.4	51.6	50.9	34.0	42.4	23.8	35.7	36.3	39.7	43.1	33.6	35.4	25.5	38.2	39.0	22.9	46.9	41.3	31.6
3	36.1	36.8	44.1	45.4	36.6	24.2	18.8	30.7	25.9	31.4	41.9	45.2	56.9	28.2	40.3	34.9	46.3	38.8	31.4	40.3
4	29.2	24.8	33.5	38.5	23.0	32.6	29.8	46.4	31.1	29.7	41.2	40.2	28.4	29.3	36.0	25.8	20.8	34.3	43.3	37.8
5	30.0	36.2	43.4	35.1	25.2	36.9	40.9	29.5	53.3	21.8	38.0	35.4	38.9	37.0	39.2	54.8	25.8	38.9	39.8	40.5
6	34.9	30.0	49.4	27.6	46.8	43.9	11.1	31.6	25.5	44.1	31.4	38.2	26.7	17.1	27.9	31.9	29.5	33.4	23.5	38.7
7	31.1	18.3	24.6	39.1	37.6	33.1	42.5	26.4	20.0	43.6	28.6	24.4	29.6	32.9	25.0	37.3	22.3	43.3	26.7	53.8
8	27.9	34.8	17.2	32.7	33.0	39.8	27.4	43.4	36.1	24.6	43.3	46.3	30.9	16.1	25.2	32.7	18.4	44.4	42.1	30.9
9	50.6	28.3	43.9	25.3	51.8	35.9	34.0	40.1	38.5	45.8	67.5	27.4	38.8	22.9	30.3	26.2	17.5	46.1	45.4	22.1
10	53.3	29.2	37.9	39.1	21.0	30.8	34.2	34.4	34.8	37.7	39.8	43.0	25.7	30.4	34.3	30.6	32.5	39.5	32.9	40.5
11	20.8	32.8	35.7	21.5	41.0	42.6	52.1	36.7	32.8	38.0	41.0	36.1	35.9	29.3	29.0	58.9	31.4	29.0	37.2	41.5
12	25.5	49.7	40.4	34.8	31.9	41.9	34.2	45.6	31.3	41.6	23.5	34.2	30.0	44.3	49.3	28.9	32.5	74.6	38.5	28.2
13	32.7	44.2	36.0	15.8	30.9	29.0	52.1	35.2	55.9	52.8	35.3	32.2	38.5	27.0	26.5	29.7	29.7	34.5	33.2	37.0
14	63.0	42.3	21.9	18.9	34.2	25.3	47.6	48.9	51.6	33.2	26.6	22.6	35.4	30.8	24.3	33.1	42.0	28.0	45.2	34.9
15	45.1	31.3	39.6	21.4	32.6	30.6	32.7	27.4	44.1	34.3	36.7	36.1	36.7	38.3	28.9	17.9	25.8	29.2	38.2	32.4
16	19.0	40.1	32.9	34.0	43.6	38.1	15.0	25.0	49.4	52.5	42.1	40.1	29.2	43.5	41.9	24.1	31.2	27.7	46.3	29.0
17	31.7	44.6	52.8	36.9	30.2	26.9	38.2	33.3	43.9	28.7	39.9	37.9	51.8	25.9	34.0	29.8	54.8	52.2	43.0	40.3
18	36.1	55.9	37.6	53.5	48.2	44.8	49.9	65.5	40.4	47.4	45.4	13.6	35.2	29.5	36.2	45.4	25.6	28.6	34.2	31.5
19	44.2	46.1	31.2	31.6	43.3	51.3	30.6	40.8	39.5	37.1	18.8	44.6	29.0	50.8	30.7	33.5	28.5	34.7	36.1	23.0
20	36.8	39.7	28.6	34.9	38.8	25.3	32.1	31.2	37.6	29.7	32.9	23.2	35.8	21.0	26.8	32.0	23.3	34.6	38.0	33.5
21	29.7	24.6	32.3	45.3	25.1	39.1	38.9	51.6	32.3	34.7	35.4	24.9	35.0	54.9	44.9	29.4	27.8	32.2	28.2	38.9
22	48.6	31.7	37.8	46.9	23.7	37.0	32.8	24.5	50.5	39.0	37.2	40.3	20.1	31.5	18.2	29.2	38.7	35.9	40.3	49.4
23	33.4	36.1	23.1	33.9	34.4	32.1	51.9	30.7	50.8	36.8	41.3	36.7	34.9	34.9	57.7	27.1	26.1	24.7	30.0	42.2
24	33.9	29.7	50.5	32.4	29.4	43.1	55.2	46.8	38.5	20.9	38.5	30.1	38.5	36.9	54.3	40.1	29.7	21.9	31.1	41.2
25	46.5	38.3	43.0	29.1	43.9	39.2	5.4	21.7	5.7	36.0	34.1	46.8	31.7	19.3	46.0	62.3	43.2	32.5	56.9	27.0

References

Beyer, W.H., ed. (1966). *CRC Handbook of Tables for Probability and Statistics*, 2e. Cleveland, OH: Chemical Rubber Company (Sciences, Volume I. New York: John Wiley & Sons).

Bowker, A.H. and Lieberman, G.J. (1959). *Engineering Statistics*. Englewood Cliffs, NJ: Prentice-Hall.

Dixon, W.J. (1953). Processing data for outliers. *Biometrics* 9: 89.

Johnson, N.L. and Leone, F.C. (1977). *Statistics and Experimental Design in Engineering and the Physical Sciences*, 2e, Volume I. New York: John Wiley & Sons.

Lieberman, G.F. (1958). Table one-sided statistical tolerance limits. *Industrial Quality Control* XIV (10): 6 and 7.

US EPA (April 1989). Stabilized Analysis of Groundwater Monitoring Data at RCRA Facilities, Interim Final Guidance. Washington, DC: Office of Solid Waste.

Weatherburn, C.E. (1957). *Mathematical Statistics*. London, U.K.: Cambridge University Press.

Index

Risk Assessment: Procedures and Protocols, First Edition. Edward A. McBean.

Companion website: www.wiley.com/go/McBean/RiskAssessProcedure_1e